Sabina Elizabeth Cox

SOCIAL STUDIES for CANADIANS

By GEORGE A. CORNISH, B.A.
Ontario College of Education, Toronto

and

SELWYN H. DEWDNEY, B.A.
Sir Adam Beck Collegiate Institute, London

VOL. I

TORONTO
THE COPP CLARK COMPANY, LIMITED

FOREWORD

All knowledge is related and its division into different branches is artificial, but serves a useful purpose for the mature mind; however, for the non-logical mind of the child, which crystallizes the knowledge about centres of immediate interest rather than abstract generalizations, it has decided disadvantages. The field of history, geography, civics, and economics may all supply material for the filling in of the facts about a single centre of interest. Therefore the whole tendency for elementary classes is to make logic the servant of interest, and to blend knowledge from history, geography, civics, and economics together into a course on the social studies. This plan is followed in this text.

Facts, suitable for children, are described in detail, sufficient to make them interesting, and in language, simple and graphic enough to make them clear and enjoyable. Yet unfamiliar or partly familiar words are introduced in such numbers that the child's vocabulary is being increased though his memory is not overtaxed.

The text begins with a series of chapters on the manner of life of typical Canadian groups, beginning with that of the most elemental, the trapper, and then treating the fisherman, miner, and lumberman, which show greater complexity. Then in a chapter on life in a Canadian village, the civic influences experienced by children are stated as concretely as possible, no attempt being made to organize them logically into municipal, provincial, and federal factors.

After studying the varying methods of life in a highly developed country like Canada, the book then presents to the children, methods of life, just as different as possible from those of Canada, and selects those forms that will show all varieties of geographical environment. It starts with the simplest forms in the equatorial forest, the Arabian desert, and the Arctic desert, where each family is complete in itself. From these it goes to life in Mediterranean climates, and in the Swiss Alps, which are more complex. It might have gone on to describe lives of increasing complexity in the monsoon countries and in highly developed countries but space forbade.

Before blending the history and geography of the Canadian regions (British Columbia, Prairie Provinces, Ontario, Quebec, Maritime Provinces), a chapter on the growth of Canada is introduced in which is given a running outline of the growth of Canada from a small district around the St. Lawrence to its present area. This includes the gradual winning and consolidation of the country, by means of exploration, wars, settlement, etc., and gives a background of chronology by which historical events under the treatment of the regions can be placed in their proper time relations. The events stressed are those which interest children and will not naturally fall in place under the regions. The high spots in the chief wars are given in detail, but constitutional and political questions are almost entirely ignored.

Then each region of Canada is treated and the history of exploration settlement, and development of farming, fishing, mining, and lumbering related to their present condition. In this way the relation of the past to the present is made clear.

Social studies must be crystallized around topics that appeal to the interest of the pupils, and every endeavour has been made to use interesting topics as centres and to weave the geography and history about these.

The pictures and maps are a unique feature, very different from anything that has ever appeared in a Canadian text before. They have been drawn by a skilled artist, who is also a scientist and teacher of geography. He had the manuscript before him, and every feature in the text is illustrated in the maps and pictures; yet the pictures do not repeat the text, but make it more real, and often supplement the information. The maps are all of a graphical or pictorial character, yet scientifically exact. They appeal to the puzzle instinct in the pupil; purposely many features shown clearly in the maps are not repeated in the text. For example it is not considered necessary to give a list of towns in Ontario with the industries and public buildings of each, which would invite memorization of lists; but these towns with the industries and buildings are all shown graphically in the map, and the very act of puzzling them out impresses these facts in the memory in the proper relationship.

G.A.C.

CONTENTS

SOCIAL STUDIES

CHAPTER I

LIFE OF THE CANADIAN TRAPPER

An industry with a history. Canada owes much to the trapper. He it was who blazed the first trails, and lured the explorers, English and French, into the very heart of the wilderness. The most notable company that ever traded in North America, the Hudson's Bay Company, buffeted the ice-fields of Hudson Strait, and faced the rigours of the frigid winters on the shores of Hudson Bay to trade with these remarkable men. It was to trade with the trapper that posts were founded in most of the key positions in Canada, and because they were so skilfully chosen, they have become the chief cities; *Quebec, Three Rivers, Montreal, Toronto, Fort William, Winnipeg, Edmonton,* and *Victoria* were all fur-trading posts before they were settlements. It was over the trapper's goods that most of the early Canadian wars were fought, and his trade was so magnetic, that fishing, farming, lumbering, and mining were nearly forgotten in the rush for furs.

The golden age is past. Though the trapper's golden age is past in Canada, and he has to take a humble place below men of other industries, the trapper still has far more than half of Canada almost to himself. While the aggressive farmer, as he clears the land and ploughs the fields, has steadily driven the trapper farther and farther back into the woods, the struggle is nearly over, and the parallel of 50° north latitude in east Canada, and 53° in west Canada is going to be the boundary between them. Not that the trapper does not move south of this boundary, for he does. There is scarcely a township in the older parts of Canada in which his traps are not found; indeed the farmer and the farmer's boy have become the chief trappers; nevertheless, north of this boundary line the full-time trapper is nearly "monarch of all he surveys."

The trapper. The trapper of the north is a lonely man (#3). He may be an Indian or an Eskimo, he may be a half-breed, perhaps he is a graduate of a university, who, drifting into the forest wilderness to see what it is like, has found he likes the lonely life in these bleak regions so well that he can not leave. Many of the trappers, however, live in the north only during the winter season

#1

FUR-BEARING ANIMALS OF CANADA

and return to their homes and families in the villages, towns, and cities for the summer. Before he settles down, he spies out the land to find a region well-stocked with most valuable fur-bearing animals (#1); he wants streams for otter and mink; marshes for muskrat and beaver; forests for marten, ermine, and fisher; and a region abounding in rabbits, partridges, ptarmigan, squirrels and lemmings, not only to supply some fresh food for himself, but to attract those flesh-eating fur-bearers, the marten, ermine, mink, lynx, fox, and wolf; though he would probably prefer to be free from wolves (#1) because, though their fur is valuable, they are considered dangerous pests and robbers of traps.

The pest of the forest. The one animal, above all others, that he hates, is the loathsome *wolverine*, well nicknamed the *glutton*. It is a link between the bear and the skunk, and has all the bad qualities of both. Everything it touches is smeared with a filthy smell, worthy of its cousin the skunk, and it touches everything but the trigger on a trap. The wolverine will follow a line of traps, eat every victim caught, set off all other traps, smear them with foulness which repels other animals, eat the bait, and often break the chain by one jerk of its vise-like jaws, and carry off the trap. It is so cunning with its sneaking habits that it is impossible to catch and almost impossible to shoot. Usually when a wolverine moves into a trapper's territory, the trapper has to move out and find a new hunting ground.

The trap-line. Just as towns and cities of south Canada are divided into lots, and the country into farms, each occupied by a family; so the whole of that vast, bleak sweep of country between the Atlantic and the Pacific, and between the fifty-third parallel of latitude and the Arctic Ocean is sub-divided into trapping areas by trap-lines (#2), most of which are followed by trappers. But the trapper does not own his hunting ground nor does he pay rent, indeed, he may move from one to another as often as he likes, because the land belongs to the government, is not surveyed, and is free to be used by every trapper who has a license. But a trap-line covers a broad vague area; it may swing out twenty, even fifty miles in each direction from the trapper's cabin, so that his nearest neighbour is so far away that the trappers may never meet.

The log-cabin. He first selects an opening in the forest alongside a stream, in which to place his cabin (#2). The stream will give him a convenient water supply and an easy means of transportation, in summer by means of his canoe, and in winter over the frozen stream with his dog-sleigh. With his keen-edged axe he cuts logs to build a one-roomed house, and the chinks between the logs are plugged with moss to keep out piercing winds and cutting snow (#3). A log is ripped into planks with a whip-saw, or hewed flat with an

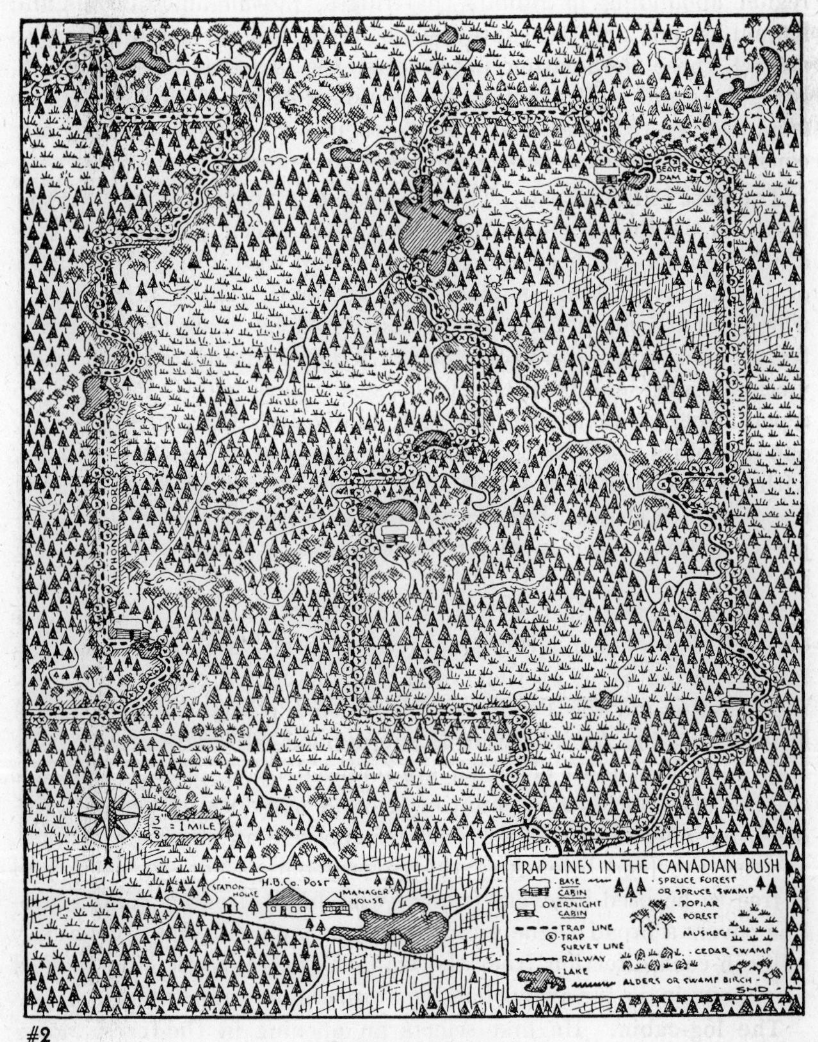

#2

MAP OF TRAP-LINES

Measure the length of Angus McIvor's trap-line. How many traps has he set? How many lakes does he cross? What animals is he likely to obtain in each part of his journey? How do he and Alphonse Dore reach the Hudson's Bay Post? Why is it in a good position? How many kinds of animals can you find in the forest? Find three kinds of deer.

axe, and made into a rough table (#3); boxes serve for chairs; balsam boughs in a bunk in the corner make a bed. Usually he has some kind of stove of sheet-iron (#3), either home-made or bought; and stove-pipes pass out through a hole in the wall. Windows of mica, or netting with the meshes covered by a glassy substance, are just large enough to let one see how dull and dreary the house is, but small enough that the fearful sweep of piercing winter winds

#3

THE TRAPPER AT HOME

Describe all the furniture. Of what is the cabin built? What is the trapper doing?

will not chill the marrow in the bone. Two boards make a door, which swings on creaky hinges. Fire-wood, guns, traps, snow-shoes, paddles, raw skins, and stretching boards, litter the floor; the few clothes and the few pots and pans hang on nails in the wall, whose only other decoration is faded pictures of movie stars cut out of magazines. A shelf, out of reach of the dogs, carries flour, coffee, sugar, rice, beans, salt, and perhaps pepper; on it, always in a covered bottle to guard against wetting, also stand the matches.

The trap-line. Next he has to establish his trap-line (#2). As this is perhaps a hundred miles long, and runs in a loop, which begins and ends at the log-cabin, it may take a week to make the rounds. Therefore he first builds a number of small sleeping cabins along the route (#2), one for each night that he will be out. The path has to be trimmed through the forest and marked by cutting nicks in trees, breaking off branches, or making other trail markers, for the trapper must have cat's eyes, which direct him as easily in the dark as in the light.

The battle of wits. Then begins the expert work, the setting of the traps. The struggle in the forest is keen, bloody, and merciless; the hand,—rather the crushing teeth and tearing claws, of each animal are against those of every other. Each can see farther and more quickly, and can move more quietly and more swiftly than man. Most of them can hear farther than man can see, and smell farther than he can hear. His senses are dull and stupid compared with theirs. The animals are always suspicious, always alert, always ready for a spring. They see man before he sees or hears them, smell him half a mile away if the wind is right, and can smell man's tracks hours after he has passed. The whole life of these dwellers in the wilds is a steady, unceasing training to sneak up on their prey or to escape their enemies. What chance has a dull and sluggish man against these cunning detectives of the forests? Man's keenness of mind more than makes up for his dull senses. He thinks, he remembers, he plans, he makes deadly tools like guns, traps, and snares. Against these the rabbit, the muskrat, and the weasel have no chance, even the sly fox and crafty wolf often stumble, and occasionally the hated wolverine, that essence of all caution and cunning, trips up.

A match for the slinking wolf. The only traps now used are made of steel and have two powerful jaws, which, just as soon as an animal touches the delicately poised pan, snap and hold its foot, head, or neck with a deadly grip between the cruel, toothed jaws. There is no escape, for a chain from the trap is fastened to a tree, stone, or buried log to prevent the victim from dragging the trap away. In order to destroy all smells, whether of iron, rust, or human touch, the trapper buries his traps in earth or manure for several days. He will then smear them, chains and all, with melted tallow, and to keep them free from suspicion, afterwards handles them with gloves (#4). If he is trapping wolves he cuts two logs three feet long and six inches in diameter, attaches a trap to each, and carries them in a piece of canvas to a trail along which wolves' tracks have been seen. The logs are buried, eighteen inches apart, in the earth, and the sod or snow replaced; furrows are cut in the sod for the chains and traps between the buried logs. The chains are covered with earth; the pan of the set trap, carefully placed on the same level as the surface, is covered with a few leaves and

#4

SETTING A TRAP

sprinkled with earth or snow. The bait, a lump of meat or the insides of a sheep, is set, never on, but always above or just beyond the trap. All sod is replaced or snow sprinkled back so that no trace of disturbance is seen even by the greenish grey eye of a crafty wolf. Next a fresh sheepskin, flesh-side down, is dragged up and down the trail to destroy all human smells, and the whole surroundings are left as innocent and inviting as before it was touched, and smelling strongly of tasty sheep. The spot selected for placing the trap depends on knowledge of a thousand facts about the habits of the wolf, every one of which has to be carefully studied. Even then, the wolf may set off the trap, eat the bait, and slink off smiling at how stupid man is.

Trapped or snared. The trapper on his rounds may set a hundred traps (#2), a few under the edge of the steep bank of a stream for mink, and a good number in the paths of muskrat and beaver along the marsh or just under ice. The tiny dots and dashes in the snow show where to hide the traps to catch pure white ermine (#1), whose choice fur decks the robes of kings and emperors. Often he will cut out big chips from the trunks of trees to make suitable shelves for traps to catch the silky marten (#1) as he chases squirrels and birds; and to seize the greedy fisher (#1) as he hunts the marten as well as the squirrel. Traps on the ground will also catch the fisher as he pursues rabbits or even deer; but the trapper will not waste traps on such stupid game as rabbits, but merely puts a loop of limber wire, called a *snare*, along their path. The other end of the wire is attached to a sapling bent over the path and caught by a trigger, which the rabbit loosens as it goes through the noose. The sapling springs up, the rabbit is then caught in the noose, and dangles from the top of the sapling.

Tragedy on the trap-line. Each week the lonely man sets out on his long tramp to examine his traps. The days are short, the snow is deep, the weather far below zero, and he never knows how the day may end. There may be a clear sun at noon, then a wind springs up suddenly, whipping the snow into small spirals. Half an hour afterward fresh snow may begin to fall quickly in large, silent flakes as big as a penny. In a short time a blizzard rages, the wind howls through the trees; it starts to get dark, the snow becomes intense, and the cruel air grows as cold as blue steel. If the poor trapper is unable to find his hut, even his tough flesh and skin become numb with cold and fatigue; he becomes drowsy; he is tempted to lie down and rest just for a minute . . . just for a minute . . . just for a minute . . . When the spring arrives and the snow melts, some constable of the Royal Canadian Mounted Police finds the log cabin empty, searches along the trap-line, and finds the stiff cold body of this victim of the fur trade.

Lifting and setting the traps. The animals are generally dead and frozen when the trapper finds them, for no creature, caught by the foot in an icy steel trap, can long withstand the fierce cold of these regions. He takes the victims from the traps, loads them on the dog-sleigh (#4), and pushes on. If the animal is still alive, he clubs it to death, and at once skins it and feeds the carcass to the dogs, as it saves weight. He is always alert for new tracks in the snow; these are the newspapers of the north, from which his trained eye can read the tragedies of the forest. He knows the tracks of every fur-bearer as well as a child knows its letters; and wherever he finds a promising new trail, he sets a trap. He deceives the senses of the fur-bearers, by a scent which he carries in a bottle. It may be extracted from the scent-glands of beaver or muskrat, usually has oil of decayed fish, and that vilest smelling of all stenches, *asafoetida.* Neither the bait nor the scent is actually put on the hidden trap; the bait is placed near and usually just above, while the scent may be strewed along the trail for many yards, and just a few drops are left above or in front of the bait.

The trophies of the chase. At the end of the week the trapper is back at his cabin with his pelts ready to be cleaned, stretched, and dried. The animals are thawed, the skins removed, scraped free of fat, and put on boards to stretch. If he has a well-populated hunting ground, he may bring in each week a fine assortment of fox, weasel, mink, and marten skins; perhaps he may have had the thrill of the miner who stumbles on a gold nugget, by catching a fine silver fox, for these are the gold nuggets of the forest—and just as scarce.

Game dinners and fish suppers. The trapper usually has abundance of fresh meat. Rabbits are to the hungry people of the north, what bread is to Canadians or rice to Chinamen. A fine moose can often be brought down by a rifle, and a juicy caribou steak is common food. Stew made from ptarmigan, the partridge of the north, breaks the monotony of rabbit pie and venison. If the trapper is near a lake, he may keep two holes open through the ice, a hundred feet apart. A rope extends under the ice from one to the other, and on this is strung a net, which is pulled up through one hole every day and always contains pickerel, delicious whitefish, and the firmest of trout, caught by the gills in its meshes. These are food for man and dogs.

An endless round. All through the icy winter the trapper makes his weekly trips on snowshoes with dogs and sleigh (#4). Perhaps at Christmas he crosses the lake to visit his nearest neighbour, fifty miles away, or may even go down to a Hudson's Bay post (#2), where there is a real celebration, in which Indians, half-breeds, French, English, and Scottish trappers, missionaries, and mounted police all take part and eat Christmas dinner at the Company's expense. But he is soon back at the lonesome hut; there is no person to talk to, he never hears a human voice; no change; no Sunday, no Saturday, for time is nothing in the north; he even fails occasionally to tick off the day of the month from the calendar, pasted on the "grub box," which he always carries on his sleigh; indeed he often gets a few days or even a week behind. But what difference could that make? The work is the same from day to day, the ground is an unbroken spread of monotonous white, the dark trees all look much alike; when darkness blots out the sight of the dead-level day, the heavy stillness of cold night is still more dampening to the spirits; even the stars shine with a hard, cold, dreadful radiance. It is no wonder that the weight of this weary sameness and stillness sometimes presses down the mind, and the trapper goes raving mad. Usually two trappers live together to break the terrible loneliness, and for mutual assistance, if one becomes sick or is injured.

The joys of spring. Spring brings a delightful change. The pleasing trickle of snow-water tells him that the steel bands of fierce winter are being broken by the bombardment of a warm sun, that is now rising higher and higher. He moves down to the marshes and reaps a rapid and rich harvest of muskrat and beaver skins. Then snow falls away before the brilliant sun, ice weakens in the rivers and streams, and refreshing spring rains clear away the last remnants of winter. The swelling current, as it rushes in sheer joy toward the sea, keeps time with the songs of the returning birds. The trees, which yesterday were black and bare, to-day are dense with swelling, silky buds, and in a magically short time will be decked in the greenest of green robes.

Out of the wilderness. The trapper's last trap is gathered in, and all are hung on the wall for the summer. His pelts, or skins, are packed in bales of one hundred pounds and loaded in his canoe. With his gun and dogs, joyfully he starts down stream for the nearest Hudson's Bay post, perhaps three hundred miles away, at the mouth of the river. His load is so heavy that the edge of the canoe is only a few inches out of the water. Often the stream is rapid and treacherous, but his strong arm defies danger, and he enjoys the thrill of swiftly gliding through the boiling waters, away from dull loneliness and toward the companionship of men, and the excitement of trading. He carries his "grub-box," a sleeping-bag, or eiderdown, and perhaps a primus stove. Each night he cooks his meal and sleeps under his canoe. Perhaps a dozen times on the course, furious rapids or roaring waterfalls compel him to carry his whole cargo, including the canoe, through the forest past the danger (#5). Each portage means a number of trips and loads so heavy that they would bring a strong man to his knees, had not long practice made his muscles as tough as the pine roots in the forest.

#5

CROSSING A PORTAGE

Trading at the post. At last he arrives at the cluster of white houses with red roofs. His hard-earned furs are spread out, examined carefully by the practised eye and knowing touch of the Hudson's Bay factor (#6), and fairly valued. Then the trapper selects his equipment for the following year: he must have flour, cornmeal, rice, coffee, salt, etc., for food; new clothes, boots, and perhaps blankets to keep him warm; and cartridges and additional traps to make his living. Tobacco he considers his most soothing companion.

He lingers around the post, tells tales to the other trappers of his adventures along the trap-line; but soon he begins to feel a strange urge; it is that lonely log-cabin three hundred miles up the river with a back yard sloping down to the river and a front yard reaching to the North Pole. Though it is gloomy, cold, lonesome, with bare walls within and disorder without, it is a magnet tugging his heart strings away from the friendliness of the post, the jokes and stories of his fellow trappers, and almost everything that makes

#6

DRIVING A HARD BARGAIN

How are the furs pressed into bales? What is the man wearing the cap doing?

life worth living. The magnet wins; the grim trapper has become the captive of the wilderness, he cannot resist the urge; his canoe is soon headed upstream, and the strong arms, with powerful beats of the paddle, speed him forward against the current. In two weeks he swings around a corner, his eye gleams with a strange light, the paddle hits the seething water with hard swift blows; the canoe grazes the shore; he jumps out; and he is *home.* Another monotonous year has begun.

CHAPTER II

LIFE OF THE CANADIAN FISHERMAN

The greatest fishing fleet on the continent. The quaint little town of Lunenburg, Nova Scotia, has its houses perched row by row on steep, rocky slopes, like seats in the gallery of a theatre. At the foot of the town is a harbour, perfectly protected, and only surpassed on the coast by that of Halifax. Swimming in the harbour, like graceful swans in a pond, is the largest fleet of elegant schooners on the American continent. There is an air of German skill and neatness in the appearance of the humble little houses, and in the long rows of glistening cabbages that deck the trim gardens. It is the second oldest "English" settlement in the province. In 1753, just four years after Halifax was founded, a large group of Germans (from Hanover), and Swiss landed on the shore of this beautiful harbour and cut out homes for themselves on its steep and rocky slopes. For years they suffered from hunger and cold and were tortured and murdered by savage Indians; but there they have been ever since and have become the most skilful sailors and the most daring fishermen on the whole wild coast.

Three modes of fishing. In other villages around the coast fishing is a tame piece of work: they either have small sailing boats, which return each afternoon from the fishing grounds, or at best go out eight or ten miles in motor-boats and return in three or four days. But the daring Lunenburgers are made of sterner stuff. They, in their graceful schooners, push hundreds of miles out to the storm-swept Banks of Newfoundland, often smothered in dense fog, and fish for months on the tossing waters before they return. Let us take a trip on one of these swift ships.

Where sailors are made. These fishermen, as boys, played about the schooners, rowed in the dories, and caught fish in the harbour. On the way home from school, they watched the graceful forms gradually rise above the horizon out of the sea and approach the harbour (#7), with giant white sails like seagulls' wings. As the schooners ploughed through the waves with one edge of the deck swishing through the foaming waters, these boys of the sea longed for the day when they would take their first trip beyond the rim of their world, the harbour, to the Banks of Newfoundland, the sea of their dreams. During the winter they watched their fathers and brothers lay down the long wooden keel and spread out the giant wooden ribs of a new schooner. Then as the timbers were bent and bolted to the ribs, they were thrilled as the graceful form took shape; by spring a new ship, as graceful as a racing yacht, had slid down into the harbour, eager to spread her cloud of canvas to the sea-breeze. The glorious sea is before their eyes from infancy, the roar of it is

#7

WATCHING THE TRIM FISHING SCHOONER

in their ears, and the smell of it is in their nostrils. They take as naturally to the sea as a duck to a pond or a bird to the air. These surroundings make the brave, skilful men that fish on the Banks. Their eyes are keen from scanning the sea; their muscles are tough as iron as they row dories or pull on the ropes; their heads are level as they climb the jerking mast in a storm, or scramble out on the jib-boom in a pitching sea, like a squirrel on a branch tossed by the wind.

Food. In the early spring the whole town bustles as sailors load supplies for a fleet of nearly one hundred ships. The clerks in the stores hustle about night and day; the drayman groans under heavy loads; the ice-man sweats in spite of the goods he handles; the wharves swarm with trucks loaded with bags, boxes, and barrels, and the harbours swarm with dories. Each ship must have supplies for several months. Five bulging barrels of pickled beef and one barrel of fat, salt pork, with perhaps a barrel of pigs' feet for a relish, will give some variety, and with the fresh fish caught every day on the Banks will allow fish or meat for every meal. Eight barrels of flour, a bushel of oatmeal, with cornmeal, and hard-tack will give abundance of bread, rolls, biscuits, and flapjacks to satisfy the boundless appetites of these hard-working, hungry men. But they must have vegetables and fruit. Ten bushels of potatoes, one bushel of onions, a barrel of beans, a bushel of dried peas, a barrel of dried apples, perhaps dried apricots or peaches, and a few boxes of raisins give nourishing first courses and desserts. As we see rice by the barrel, cornstarch by the case, essence of lemon, pepper, salt, mustard, cloves, ginger, sage, and nutmegs, in two and four pound packages, going on board, the mouth waters at visions of fat pies, fluffy puddings, and savoury soups. A barrel of good black molasses and two or three barrels of sugar sweeten everything. Twenty pounds of tea, fifteen pounds of coffee, and several cases of condensed milk give an unfailing supply of hot drinks to which men, chilled to the bone, can help themselves at all times. With all this array of good things to eat and drink, the skilful cook is able to give an endless variety of tasty dishes to these deserving men.

Supplies. Thirty or forty barrels of fresh water are stowed in the hold or on the deck, — wherever a corner can be found. In big bins down below in a special compartment are carefully put fat herring, brown squid, or shining graceful capelin for bait. As the bait has to be kept fresh for at least two weeks, it is packed in alternate layers with chopped ice, and the room in which they are stored is sealed tight, like a refrigerator. Some ships now have an iceless refrigerator for this purpose. A barrel of kerosene, and coal for the cooking stove, nearly complete the supplies.

As fleet as a deer. The fishing schooners are at last gliding out of the harbour with immense sails, white and hard as marble, fully spread, on each of the two masts. Extra sails on the topmasts and jibs in front are all thrown out to the wind, for these ships are built so sturdily that they will stand the strain on the mast of wide sails in strong winds, and their hulls throw off the buffeting of the sea as steel armour turns aside a bullet. They often leave steamers behind. The sailors love the thrill as their little schooner storms along with a broil of white water shearing away from her sharp prow, her mast bending over so that the deck slopes like the roof of a house, the deck rail swishing through the salt waves, her timbers creaking and groaning, and the whole ship bounding from wave to wave like a deer across a field.

Tuning up the gear. But the men are soon busy; as they approach the fishing-grounds they bring out great tubs of fishing gear, called *trawls* or *trawl-lines.* Each is made of a heavy tarred cotton twine, about 1500 feet long called the *ground-line* (#9). Attached every four feet along this is a smaller line, called a *snood,* three feet long, having a strong barbed hook at the free end. These are all examined carefully, for a single flaw might mean a break in the line, the loss of much gear and many fish. Every hook is inspected and sharpened with a hard stone, for success depends on attention to the smallest things. Just before the trawl is set, a lump of bait is put on each hook, and the lines curled up, with the greatest care in tubs, so that they can be let out rapidly without any tangling.

Smelling his way to the fishing grounds. In the meantime the keen eye and keener mind of the experienced captain is guiding the ship through banks of fog, looking for the fishing grounds, which are hid beneath the sea, with no mark above the water to tell their position. They have been sailing for two days, there is no land in view, the sun by day, the stars by night, and the sea by day and night, are blotted out by a wet, sight-defying fog. But the skipper confidently sails along with nothing but a mariner's compass in front of the steering-wheel, a log trailing behind to mark the speed, and an old-fashioned chart on the cabin shelf. The refined instruments of the big ships for finding position, he laughs at; he has been over the route every summer since he was a boy. One would think he almost smelled his way, like a hound. At last he slackens sail, sinks a lead plummet, with some lard on the bottom, finds the depth, examines the mud stuck to the lard, and decides he has a few more miles to go. Another sounding tells him he is over the fishing grounds. The tubs of trawl are all baited, the men are in oilskins.

A dory. Piled on each side of the deck are nests of strange row-boats (#8). There are five in each pile fitted inside one another like berry-baskets. They are called *dories* and are the pride of the fishermen all around the coast, as they are the best, safest, and cheapest small boats in the world. They are flat-bottomed, come to a point in front and behind, but the point at the back is cut off by an upright piece (#8). They have very flaring sides made of over-lapping boards, which run from front to back in a graceful curve. The "flat-bottom" is really curved from front to back like a rocker, and a cross-section will appear like the following diagram ___/. One can easily be built in two days. When you step into a dory, it seems as unsteady as a canoe, but in reality it can

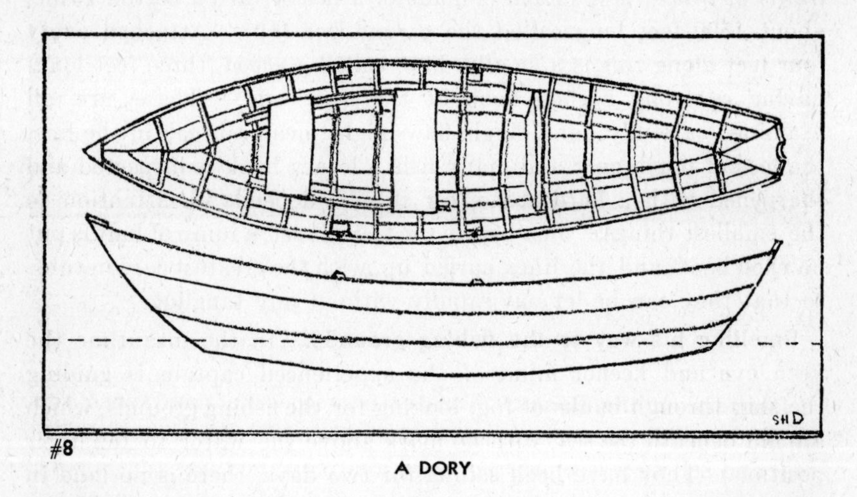

#8

A DORY

hardly be tipped; it glides swiftly through the water; when not too loaded it can ride the roughest seas, like an egg-shell; and when loaded almost to the water's edge with more than a ton of fish (#12), it plunges through a choppy sea without shipping any water.

Setting the trawl. It is now 6 o'clock in the evening, the ten dories are lifted over the edge of the schooner into the sea, one by one; two men in oilskins enter each dory (#9), receive five or six tubs trawl-line, two anchors (*d*), two buoys (*a*), two pairs of oars, a jar of of water, some hard-tack, and a few tools to use in lifting the trawl. The ten dories are rowed away from the schooner for a quarter of a mile in ten different directions. Then each dory begins to set the trawl on the bottom of the ocean. One man rows the dory, the other, with a stick, throws out the line, coil by coil, with great skill, as there must be no tangles, and the snoods and hooks must not catch in the line, or a sharp-toothed fish in seizing the bait might bite the ground-line (*f*) in two. Tub after tub of line goes out until a stout trawl-line, almost two miles long, lies innocently along the bottom; at every four feet branches out a snood with an attractive lump of choice fish or squid, swinging back and forth; but lurking behind each of these alluring baits is a hidden, ugly hook, as sharp as a needle and as treacherous as a trap. At each end of the line is a heavy anchor (*d*), and from this a thin rope rises to the surface, where it is attached to a small striped barrel (*a*) used as a buoy. The buoy at the end of the line nearest the schooner has a white flag, and the other a black ball.

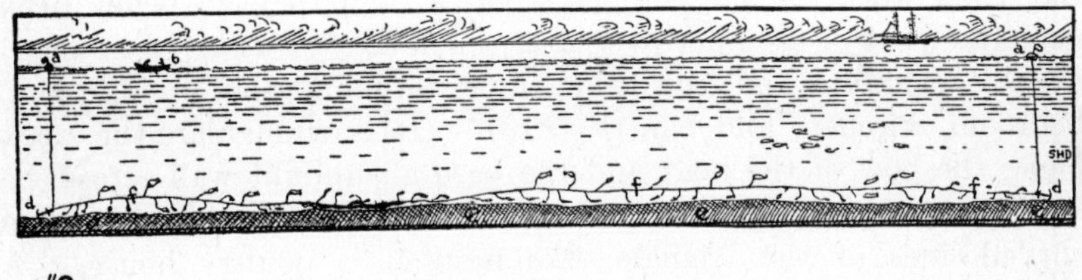

#9

SETTING THE TRAWL-LINE

(a)—Buoys. (b)—Dory. (c)—Fishing Schooner. (d)—Anchor. (e)—Sea-bottom. (f)—Trawl-line.

Dangers on the Banks. To see these cheerful men swing their dories overside on a black winter's night, pull out over a tumbling sea, and set their lines with a torch aflare on the dory gunwale, is a most impressive and awesome sight. The stars may be shining and the air cold, as the men row back. Often there is a dense fog, or in autumn a cutting snowstorm; but these men, by long experience, seem to find their way back as surely as a cat glides through the darkness, or a bird returns from the south in the night. Sometimes a sudden gale rushes in and blows the dory from its track; or a stealthy fog envelopes the fisherman so deeply in its folds, blinding his eyes, and cutting off all marks to guide his course, that he is left afloat to be tossed on the merciless waste. Occasionally a gale drives the schooner from her position, drags her anchors, or snaps the chains, and then the dory is as hopeless as a storm-tossed ship when the lighthouse lights go out. But these brawny Lunenburgers have bodies as tough as leather, and muscles as hard as those in the lion's jaw. If the schooner cannot be found, they do not hesitate to start out in the raging sea for Newfoundland or Nova Scotia, perhaps one or two hundred miles away. They may fight the winds and waves for many days, but their stout hearts never fail. The jug of water and the hard-tack give them scant strength for a few days. Driven by winds or forcing their way with their oars, they are usually picked up by their schooner, land on the coast, or are found by another ship. It is seldom that the schooner creeps mournfully back into the harbour with her flag at half-mast, telling a sad story to the women and children watching the harbour with wistful faces as they look through the small windows of the houses on the hill slope.

The forecastle. When the men return from setting the trawl, the day's work is done, supper is eaten, and they go below deck near the front of the boat to what is called the *forecastle* (pronounced *fōk'-sl*). This dim and crowded chamber is dining-room, living-room, bedroom, kitchen, and pantry (#10). It is triangular, the apex being the bow of the boat and the base an upright wall across the hold of the ship. A double row of sleeping bunks line the two curved sides of the triangle, like nests in a poultry house. A dining-table, part of which can be folded out of the way, occupies most of the space between the opposite bunks. The cooking-stove, cupboards, shelves for dishes and storage bins are all at the base of the triangle and so close together that the cook can reach for nearly everything from his position in front of the stove.

The cook. The cook is a man of character. While the captain guides the ship and controls the fishing, the cook smooths the temper of the men and keeps them happy through their stomachs; he is usually a model in shape, size, and good humour, of what a successful cook should be. He must have plenty of well-prepared tasty food ready for twenty hungry men of large capacity, at least three times a day, and enough pie, cake, etc., left over, that any man may help himself between meals to as much as he wishes; and of course, there must be a big pot of piping-hot coffee, and another of thoroughly boiled tea, ready at hand, twenty-four hours in the day; and a jug of molasses to sweeten these stimulating drinks. This is not so difficult to do on his good broad-topped coal stove when the weather is fair, but it takes a real schooner cook to prepare a meal when the boat is pitching, the stove lurching, and the kettles sliding from one side of the stove to the other with every wave that smashes against the bow. As it takes the cook's two arms and two legs to keep him from being tossed back and forth across the floor, how he manages to spare a hand or two to handle the pots, mix the dough, clean the fish, etc., etc., is his secret. But he masters all these difficulties and many more. It is seldom that the pitching seas are so high as to drive him to cold meat and cold vegetables. The cook also helps the captain on deck when the fishermen are in the dories attending to the lines.

#10

THE FORECASTLE

Tragedies at the bottom of the sea. The fish bite best during the night, and while the tired men sleep soundly in their rows of bunks in the tossing ship, down at the bottom of the sea, where the water is quiet, stretch nearly twenty miles of lines, arranged like spokes out from a hub, which is the schooner. At every three or four feet along these lines is a choice bit of bait, but buried in the bait is a barbed hook as sharp as a whetstone can make it. How fish in the darkness of night at the bottom of a black sea find the bait is their secret, but they do so in great numbers. As they are hopelessly caught one by one, they squirm, drag the line back and forth, and cause many hooks to catch strange creatures settled quietly on the bottom. Here a radiating *starfish*, perfect in shape and tinted with vivid blue and brilliant red, is caught by a hook dragging along the sea floor; again a *sea anemone*, of still more dazzling red, and spreading out like a flower, is ripped from the floor and closes up

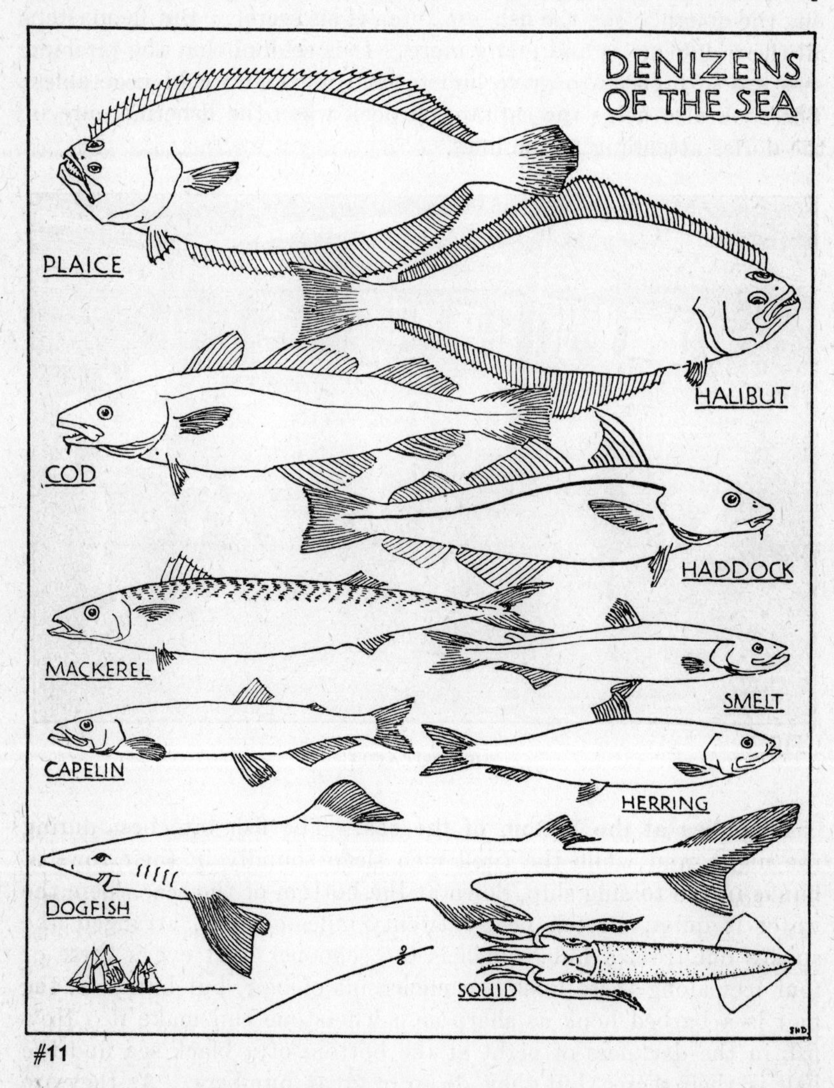

#11

like a big, brilliant strawberry. At another place a starfish, with every arm branched a dozen times and ending in a thousand snake-like twigs, often called a *sea-cancer*, squirms all its branched tentacles about a hook. Soft, slimy, purple animals, the shape and size of fat cucumbers, except that around the mouth is a circle of tentacles, each branched like a tree, are hooked and at once contract to one-quarter their size. Still another hook drags through a group of yellow cylindrical structures which look for all the world like ears of golden bantam corn, but are really clumps of snail's eggs. *Cod* and *haddock* (#11) seem soon to give up the struggle; great flat *halibut* (#11) as big as boys, keep up the fight till they are knocked on the head just before they are taken into the boat. *Dogfish* (#11), from the fisherman's standpoint, are the blue devils of the sea. When they take a hook, it arouses their fear and anger;

#12

HAULING IN THE TRAWL

they swim violently, often tangle the line, or do worse, since, with their saw-like teeth, they bite it in two. If the trawl is bitten in two places, the part between the bites is lost at the bottom of the sea. But the dogfish, which are small sharks, have other nasty habits. A school of them may come along and eat the fish caught on the trawl, so that when the men lift the line, nothing is left attached to the hooks but the mauled skull and backbone of fish after fish. Such are the tragedies of the darkness at the bottom of the sea.

Out in the darkness. The men are at breakfast very early, as there is a long and heavy day's work ahead. It is an impressive sight to watch the ten dories move out into the boundless darkness in ten different directions. Unless the fog is bad they go to the outer buoy, where a white keg with a small black ball above it is the only signal above the white waves. Though the wind may have changed during the night and turned the schooner in such a direction that the trawl that was off the stern may now be off the bow, the fishermen

#13

RETURNING TO THE SCHOONER WITH A HEAVY LOAD

at once head off in the right direction; indeed, the sense of direction of these experienced men is as uncanny as the scent of a bloodhound as he tracks his prey, or the eye of a bird as it migrates a thousand miles to the south.

Lifting the trawls. Each dory finds its own outer buoy just as surely as each bird finds its nest. A wooden pulley is fixed in the bow, the buoy and anchor are hauled on board, and then the exciting but hard work begins. One man stands in the bow as he draws in the two miles of line and knocks off the fish into the dory; the other stands behind him and coils the trawl into the tubs (#12). Hour after hour in fog, in rain, in snow, in cutting wind, this tugging at the line goes on, the men are half-way up to their knees in wet, slimy fish; their hands smart from handling the trawl, soaked with cold salt-water. Two or three times they have to return to the schooner to unload (#13), which they do with a one-pronged pitch-fork as though they were pitching hay. At last the trawls are lifted, and as the ten dories are turned toward the schooner, they look like chicks coming to gather under the mother hen as the darkness of evening comes on.

Curing the fish. The fish have still to be cleaned and salted. One man with a special, sharp knife cuts across the throat and knocks off the head on the edge of the tub as quick as a flash (#14); another slits open the fish, removes the insides, places the liver in a tub, the head and insides are thrown overboard. A third man cuts out the backbone. The washed fish are then taken below and laid in a pile with flesh side up and covered with salt.

The scavengers of the sea. Not a bird is in sight as the trawls are being lifted, but no sooner does the cleaning begin than a hungry army of thousands of gulls, terns, and gannets seem to spring from nowhere (#14) and closely follow the ship. Every fragment of offal thrown over is immediately fought for by this white-feathered army; there is no danger of the sea being polluted, for these garbage gatherers do their work most thoroughly. In former times, if bait was hard to get, the sailors used to throw hooks to drag along the water behind the ship and catch these beautiful birds for bait.

Hard work, long hours. The trawls have all to be baited and set again before the day's work is done. Often when the fish are biting well the fishermen set and lift the trawls twice a day. Then there is little time for rest, often only an hour or so each night, but the men do not grumble, because they are getting a good catch, and their wages depend on the quantity of fish obtained. When bait is exhausted they have to sail to Newfoundland or Nova Scotia to obtain a fresh supply. Frequently much precious time is lost, as bait is often scarce, and they may have to go from village to village seeking a supply.

#14

CLEANING THE FISH

Losses on the Banks. There are many dangers and losses in the fishery. Often part of the gear is lost by the trawl-line breaking, being bitten by dogfish, or being rubbed against jagged rocks at the bottom. Sometimes it gets caught on the rocks and has to be broken. Occasionally in a gale, the anchor chain breaks, the schooner is carried many miles from where the gear is set, and the gear is never found again. This is a terrible loss as a complete new outfit is very expensive.

CHAPTER III

LIFE OF THE CANADIAN MINER

Minerals hard to find. Suppose a piece of gold the size of a pin-head, a piece of silver the size of a pea, and lumps of zinc, nickel, lead,

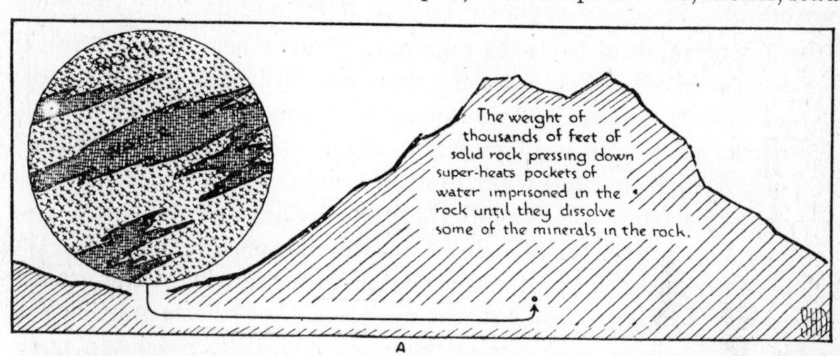

A

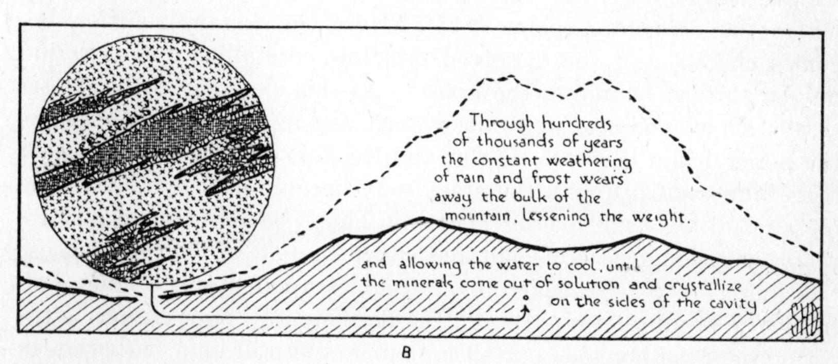

B

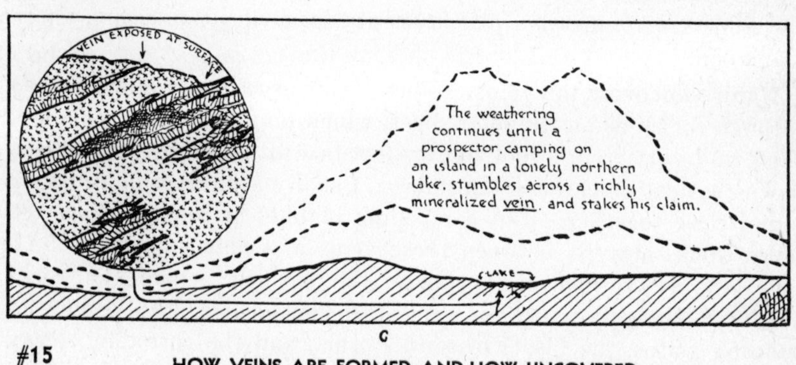

C

#15 HOW VEINS ARE FORMED AND HOW UNCOVERED

copper, and other metals the size of eggs are all ground to fine powder and completely mixed with a large pond of mud. The mud is let stand until it is baked as hard as rock. This would be a fair representation of how thinly these metals were originally scattered through the rocks of the earth. Not one of them could be detected with a powerful microscope, and it would be almost impossible and quite unprofitable to try to separate any of these metals from the mud.

Gathering mineral particles together. Nevertheless these are the proportions in which metallic minerals are still scattered. Almost all rocks contain these metals but in such small and finely divided quantities that they could be no more separated from the rocks than a drop of red ink from the water supply of London. But fortunately for man, kindly nature has not left these good things of the earth scattered so hopelessly. Deep down in the bowels of the earth the rocks are furiously hot; intensely heated water flowing along cracks, or soaking through porous strata dissolves these little particles of metals, as it passes, and carries them along until it reaches a place where the rocks are cooler (#15). Here the watery solution becomes chilled, and the dissolved minerals crystallize out as a film on the surface bounding the crack. As this slow current of water creeps silently along, year after year, and century after century, the mineral film becomes thicker and thicker, and the crack thinner and thinner until finally the crack may become filled with valuable metal, and we have a *mineral vein.* Thus friendly nature gathers these valueless specks of gold and silver, as fine as flour, into these rich fillings of cracks, which man stumbles upon, starts a mine, and suddenly becomes rich. Had it not been for this slow gathering together of these precious particles, man would still be a barbarian in the stone age, for metals are as necessary for civilization as fire, water, and wood.

Nature uncovers the veins. These veins usually run down steeply through rocks, as an irregular sheet, which may be an inch or a foot thick and vary from a foot to fifty feet in width. They may extend down ten feet or perhaps ten miles. Usually a number of veins are found close together. But a repelling blanket of rock, perhaps ten miles thick, may lie between these veins and man's grasp (#15*A*), for they are formed miles below the surface, where the rocks are hot. But the patter of rain, the flow of ice, the erosion of running water, the blasts of sand storms, and the corroding decay of the weather, are all rasping away the surface of the rock just as surely as though it were scraped with a giant's file. Though the tooth of time in this way gnaws away the surface slowly, give it time enough, and it will grate off mile after mile of the hardest rock (#15*B*), until finally that precious vein of glinting gold or shimmering silver, which at first was buried under ten miles of rock, may be exposed (#15*C*).

#16

A LUCKY STRIKE

Veins hard to find. It is these valuable veins that the mineral prospector wears out a whole lifetime searching for. The rock is usually covered by soil, the soil overgrown by vegetation, and the exposed mineral veins so few and far between that it would be much easier to find a needle in a haystack, or a small glass bead lost in the sand of the sea-shore.

Prospector begins his almost hopeless search. The prospector, like the trapper, is a lonely man. Though his quest is almost hopeless, he is not easily baffled; he searches the rough and rocky places far away from roads and railways. Often he moves into the wilderness in a canoe, in which he brings his food, tent, and kit. This consists of an axe, pick, special hammer (#16), a drill, and perhaps a little dynamite. Since no 'find' can be made when the ground is frozen or covered with snow, he begins his long and patient search in the spring. Often it takes weeks and months of dreary tramping over mountains, traversing ugly gorges, paddling against roaring currents, and portaging on foot past murderous waterfalls before a promising field is reached.

A lucky strike. The prospector is almost always a poor, restless spirit, who pours out money like water, as long as it lasts, and who cannot settle down to steady work. The lure of a flash of luck drives him into the wilderness, where for weeks or months he may wander through the lonely forest, the dark gorges, and the swift streams where the sight of a man's face or the sound of a human voice would startle him. He usually follows the streams, as their swift currents wash clean the bare rock. He nervously scans with eager look every coloured streak that resembles a vein. A blow of the hammer breaks off a lump, he wistfully seeks on the clean surface for a flash of colour (#16), but hundreds and thousands of times throws the barren lump away in disgust. He may tramp for months, or even years, and find nothing. Often his whole life is fruitlessly spent in the search. Though the heart grows sick and the body weary, the hope of a rich find that will make him a millionaire in a minute, spurs him on. The chances are almost a million to one against success, but the gambling spirit urges him ever further into the hopeless wilderness. Often a discovery is finally made by pure accident. Perhaps as he slides down a rock, the hobnails of his boots scratch the surface; he turns his head and looks back. What was that colour which caught his wild eye? It was the yellow glint of gold. He can hardly believe his senses. He examines it closely, he uncovers more of it with his hammer; yes, there lies a thread of gold creeping zigzag over the surface of the rock. On his knees he tracks its precious course, hurls the soil and boulders from the rock as though they were snow, and follows its alluring trail, until the covering becomes too thick. He follows it in the opposite direction until it passes under the water of the stream.

Tracing the vein. Time, food, fatigue, are all forgotten in the joy of discovery. He selects a point two feet from the vein, and by ramming the chisel-edge of his hard steel drill against the rock, slowly, very slowly, batters out a small hole toward the vein. He works all day. At last, exhausted but excitedly happy, deep darkness drives him to lie down on the precious rock to sleep. Before dawn he continues furiously to drill the hole. When it is four or five feet deep; he puts in a cartridge of dynamite with a strange paper tube, called a *fuse*, leading a few feet away. He packs the hole above the cartridge with mud, lights the fuse, which contains gunpowder, and runs to a safe distance behind a rock. There is soon a terrific explosion, which echoes through the lonely woods. Rocks fly, and smoke rolls forth. He returns excitedly, removes the broken rock, gives a whoop of joy, and is startled by his own voice; there is the golden vein, wider and richer than ever, at a depth of five feet.

He stakes his claim. That is enough. As he gathers samples from the vein, he looks around stealthily as though his every move was being watched, even though he knows it is twenty miles to the nearest settlement. He places back the rocks, spreads the soil, smoothly sprinkles it with leaves and sticks, and hides every trace of his work; then he cuts four stout stakes with his axe, flattens them, paces off a square, a quarter of a mile wide, around the vein, and drives a stake at each corner. With nervously trembling hand he carves his name, license number, day, and hour, on the north-east stake and probably on all the others. Next he trims the trees along the four lines between the stakes to mark the boundary of his claim. He breathes easier, for the precious thread of gold is his. He notices the position carefully, and even after he has turned his path south, returns several times to make sure that he can locate his fortunate discovery again. Then as speedily as possible he darts off toward The Pas, where he registers his claim. He hands his samples to an engineer, who, after examining them, tells how rich is the gold in the specimens.

The gold rush. The finding of a rich gold vein can no more be kept secret than the arrival of a circus in town; mysterious rumours seem to rise out of the ground, or drop from the clouds. Word soon leaks out among the taverns, cafes, and dives, that the prospector is back; and though he is silent, his looks speak words, and his silence is electric; The Pas is soon in a wild tumult. The shops that sell outfits are packed to the doors, and an excited queue stretches down the street; Winnipeg becomes excited in half an hour, Toronto and Vancouver mining circles a little later, and even certain groups in New York are all agog before the sun is set. In one day The Pas is emptied of a hundred men. Every drunkard becomes suddenly sober and is getting his *grubstake* (see below) at the general store; the bank clerk leaves his wicket, the clerk his counter, the policeman forsakes his beat, the operator resigns from the railway, and even a few young doctors leave their patients, and lawyers their clients. All are rushing to catch the first train north. All types of men leave Winnipeg, Regina, Edmonton, and Fort William for the new gold field, and soon this claim becomes a *Mecca*, and strange pilgrims, hungering for gold, gather in from every quarter, like famishing crows toward the carcass of a horse. In less than a week every claim for miles around is staked; tents whiten all the woods, and soon a mushroom town of tents, log-cabins, galvanized-iron* shacks, and tar-paper huts, springs into throbbing life.

A grubstake. The prospector we have described, like many others, is furnished with a *grubstake*. Some of these men have not enough money to buy a loaf of bread let alone an outfit, which might cost one or two hundred dollars. A wealthy man will pay for the outfit on condition that he shares in the discovery. The outfit is called a *grubstake* because the capitalist stakes, or risks, the grub (food) to get a share in the mine. In nine hundred and ninety-nine cases he gets nothing, but the thousandth may bring him such riches that he could have easily paid for ten thousand grubstakes.

Prospecting, a life job. When the prospector has staked and registered his claim and made the mine secure, his interest is likely to slacken; he soon discovers that the excitement of searching and finding is far more thrilling than the joy of possessing. In nine cases out of ten he sells out for a few hundred or thousand dollars, settles in an expensive hotel in a large city and proceeds to find how fast he can spend his wealth, gets a shock in a few months to find

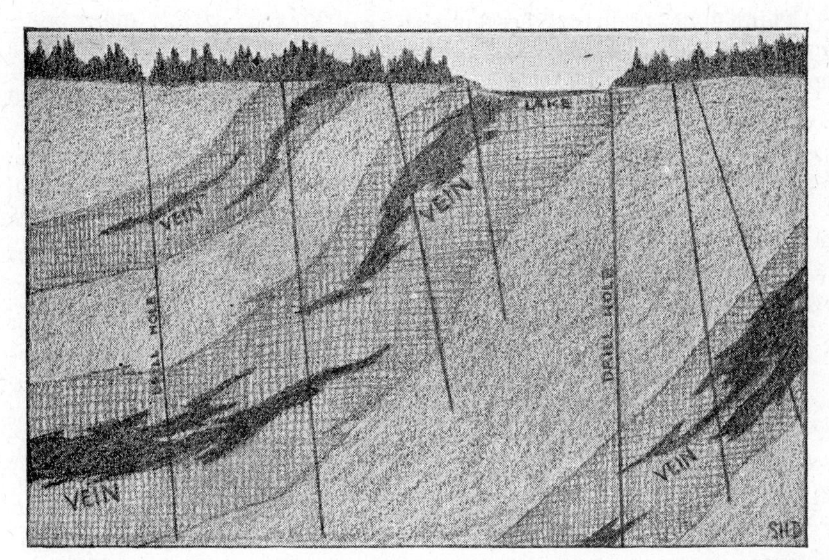

A.— Borings made by a Diamond Drill

B.—Course of Veins as Interpreted from Borings.

#17

EXPLORING A MINE WITH A DIAMOND DRILL

it is all gone, again marches out with a grubstake into the lonely wilderness to search the shores of streams, the faces of rocks, and the steep ramparts of ravines for another "strike" which he probably never finds. Once the alluring search for gold has taken possession of a man, though his life is filled with empty loneliness, the ghost of starvation, the pain of bodily fatigue, the constant torment of biting flies and blood-sucking mosquitoes, and above all the ceaseless anguish of heart-breaking disappointment—in spite of all these, he can never throw off the spell of the search for gold.

Exploring with a diamond drill. Let us return from the prospector to the real miner. First the mine has to be *explored.* Before hundreds of thousands of dollars worth of machinery are dragged in through a rough and roadless forest, a far more thorough study of the number, width, depth, and richness of the veins must be made. For this purpose the miner uses a wonderful little boring machine, called a diamond drill, which is a hollow tube of steel with black diamonds studding the bottom edge. By boring with this instrument (#17), solid cylindrical plugs of rock are cut out and raised to the surface. These plugs when put together give a complete cross section of the rock. By drilling these holes in many directions and to great depths on both sides of the vein, its complete structure, length, width, depth, and richness, can be found, and in all probability new veins be discovered. In a few months the miners are able to decide whether the vein is rich enough to warrant the expense of digging a mine.

Bringing in machinery. Many heavy and expensive machines have to be brought to a region, hundreds of miles from a railway, and without even a trail leading from the nearest settlement. Often this machinery is dragged by truck, or tractor, over the ice of streams and through the snow of the forest. Perhaps much is brought in by aeroplane, and often boats can be used on streams and lakes.

Blasting a shaft. Hundreds of thousands of dollars usually have to be spent before a cent's worth of ore is removed. First a shaft hundreds of feet in depth has to be chiselled out of the solid rock (#18). It is hard enough to dig a deep well through sand and clay, but to cut such a hole through granite is a very different matter. First a number of small vertical holes are made, but these are drilled by powerful machines worked by electricity or compressed air, and not by the hand drill of the prospector (p. 31). In these drill holes dynamite is exploded, the broken rock removed, and then more holes drilled, and the explosion repeated. This slow and expensive process deepens the shaft foot by foot. So skilful are these workers, that although all the rock is cut out by blasting, the walls of the shaft can be kept straight and smooth. It is usually rectangular,

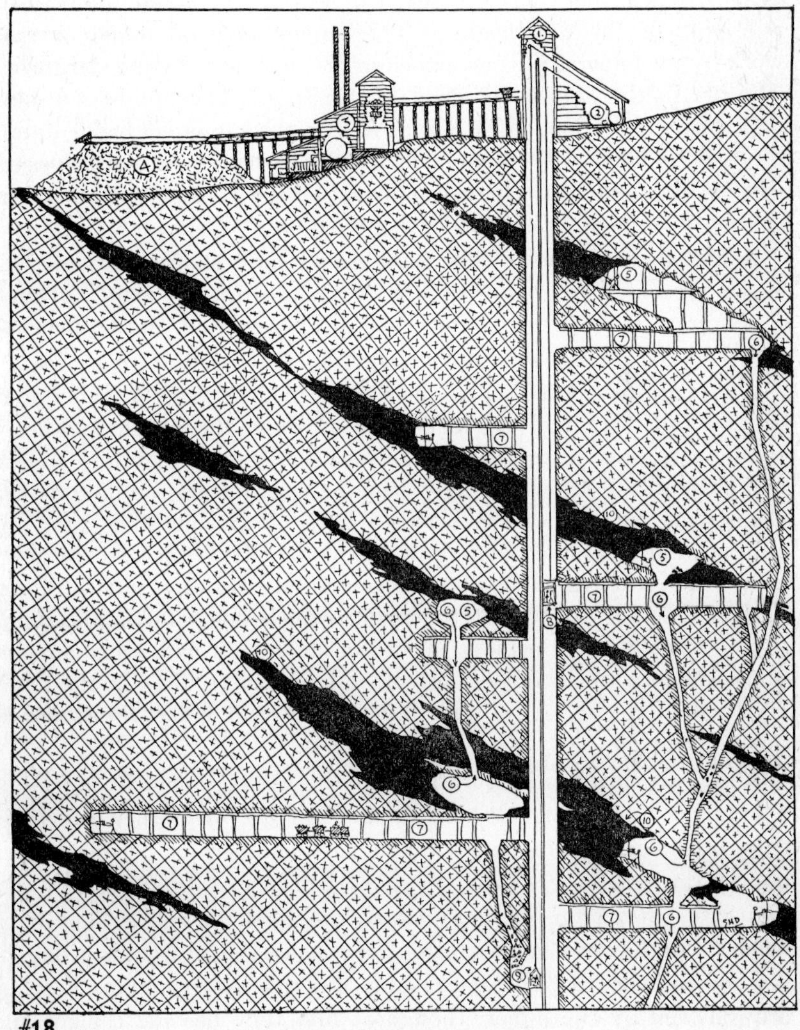

#18

SECTION THROUGH A MINE

1.–Shaft-house. 2.–Hoist. 3.–Crushing and concentrating mill. 4.–Dump. 5.–Stope. 6.–Ore Chute. 7.–Level. 8.–Skip for raising ore. 9.–Cage for miners. 10.–Vein.

lined with lumber, and has *skips*, or elevators, by means of which the men are raised and lowered and the ore taken out of the mine.

The danger of mining. At every hundred feet of depth, horizontal tunnels, called *levels*, are blasted outward from the shaft toward the veins. When the veins are reached, the real work of mining, the removing of the ore, begins. The miners now do much of their work by machinery, but nevertheless the work is hard and dangerous. The ore in the vein is loosened by blasting. The lumps of broken rock have to be loaded into cars (#18); the air is cold and damp; often it is filled with stone dust, which, when breathed, gives incurable diseases, and there is always danger of slides of rock, the caving in of roofs, or explosions of dynamite.

A human ant's nest. The mine grows, new shafts are sunk, old shafts are bored to greater depths, levels are run out in many directions, great holes are left where the ore is removed, until the whole structure becomes more complex than an ant's nest (#18).

A mining town arises. As the vast hidden network of this underground town gradually branches out through the rocks, another is taking form above, to replace the mushroom town of tents, shacks, and log-cabins. Rows of ugly, square, galvanized-iron houses line the streets cut out of the woods. Large brick buildings with tall chimneys arise at the opening of the shaft. Here the ore is ground to powder, the valuable mineral separated, and then put into huge blazing smelters to extract the metal, which pours forth like liquid fire into moulds. Business blocks of fine brick buildings appear; and a modern town, throbbing with life and hope, rises in the midst of the wilderness. A railway and a good road make the improvements complete.

Mines die never to be born again. But mining is not like farming, fishing, or lumbering. Where forests are cut or burned down, in time others will spring up; though vast numbers of fish are taken out of the sea, others grow to take their places; and farmers can reap crop after crop for many years, provided the soil is kept fertilized. But mineral veins do not grow again. Once gold is dug, it is gone forever. In the great mine described, some veins as they are followed by the miner, though at first they become thicker and thicker, still farther on become thin and finally peter out. Other veins are followed to such great depths that it becomes too hot to work, too expensive to raise to the surface, and too dangerous for the miners, as the tremendous pressures cause the rocks in the walls of the levels to explode. Level after level is deserted and allowed to cave in.

#19

A GHOST MINING TOWN

Death of a mining town. Along the streets of the town, houses that in the palmy days brought fabulous rents now have signs in the windows, "To Rent", and no renters appear. The great smelting furnaces and ore crushers are closed one after another. Finally the fatal day comes when the mine is closed; the veins are exhausted; it has made millions, but old age has crept on and now it is dead. The people have gradually moved out, until at last this town of rush and bustle is miserably deserted. The streets that rattled with trucks and tractors, and felt the more delicate touch of twelve-cylinder limousines, now have grass growing up through cracks in the badly patched pavement, and the houses begin to crumble (#19); at last a pitying fire sweeps through this graveyard of a town and lays it in ashes, all but shapeless piles of brick, tangled iron, and stone. Then kindly nature begins to throw a green mantle over this carcass of a town, shrubs and weeds take root, and blot out every trace of what was once rows of houses, stately business blocks, and monstrous mining buildings. In a hundred years it will have again gone back to the forest.

CHAPTER IV

LIFE OF THE CANADIAN LUMBERMAN

A smiling forest becomes a frowning desert. There is nothing more beautiful than a vast forest of giant trees (#20). Their count-

#20

A VAST FOREST OF SMILING TREES

less numbers of shaggy trunks, straight as masts, rise proudly for fifty feet, often without a branch, and spread out above into a glorious covering so dense that only faint light pierces the leafy greenness. One, wandering through the solemn gloom, is reminded of walking amid the columns of an old church with only a flicker of coloured light coming through the stained glass windows. When one thinks that it may have taken hundreds of years for one of these sturdy old trees to grow, it seems a tragedy that man should raid such an army of old veterans, who have successfully battled wind and

#21

A BURNT-OVER FOREST

drought, disease and fire, and that with such unfair weapons as sharp axe and tearing saw, he should bring them down to death in a few minutes, turn the noble forest into a ghastly graveyard of dead and decaying trees, pock-marked with ugly stumps, sticking up like ulcers among the ruins. Then a fire sweeps through the bone-dry *slash* of broken branches, and leaves nothing but blackened skeletons leaning helplessly one on the other (#21). A second fire not only licks up the decaying relics but day by day and week by week eats into the smouldering soil, until at last it becomes a desert, covered by the red, barren ashes of what was formerly a fat and fruitful soil. Then man's awful havoc is complete. The smiling land, proud to bear on its bosom the richest of mantles, is replaced by starving dust trying in vain to shield its nakedness from the burning sun by a scrawny covering of harsh leathery-leaved shrubs and wiry bunch-grass.

An unfailing supply. Man's hunger for wood drives him to deface the beautiful forest. Logs, lumber, and pulp are so necessary that people have a right to use the forest for their benefit; but they should cut only the full-grown trees and protect the others from disease and fire. Then kindly nature would give an unfailing supply, since the growth of the forest is steady and rapid.

Queen Silva and the wooden touch. Little do most people know how hopeless and helpless they would be without the wood of the forest for their well-being and pleasure. Its benefits follow them from birth to death; most people, indeed, are rocked in a wooden cradle and buried in a wooden coffin. Wood serves us every day from the time we get out of bed till we retire for the night (#22). We sleep in wooden beds, and if we are poor, we may sleep on a tick stuffed with excelsior, and have a pillow filled with shavings. As we get out of bed, we step upon a hardwood floor, put on rayon garments made of spruce, stand in front of a walnut bureau, and comb our hair with a celluloid comb (which comes from wood). We warm ourselves at a stove, register, or radiator, which, in nearly all the homes of the world, receives its heat from burning wood, or coal which is fossil wood. We admire the beautiful wall-paper made of wood pulp and gorgeously dyed with colours extracted from coal. We come down the maple stair into the library, all the furniture of which is made of walnut, oak, or mahogany, and every book on whose shelves once formed a piece of the trunk of a small spruce tree. While waiting for breakfast to be cooked over a wood fire, we read the morning paper made from wood pulp. The news for the paper was flashed over telegraph and telephone wires supported on cedar poles, and the paper was rushed by train, whose track rests firmly on cedar ties. If we go to the kitchen we find that the fire was lit by a wooden match. Not only has the fire to be constantly fed by wood, but as the ice is brought in, a few crumbs still clinging to it shows that it too required a bed of sawdust to preserve it from a watery grave. We find ourselves standing on linoleum, which is made largely of sawdust cemented together with oil. Perhaps we try to forget our wooden surroundings by playing a record on the gramophone, but are surprised to learn that the records are made largely of wood. If we attempt to sew,

#22

WOOD SERVES MAN FROM THE CRADLE TO THE GRAVE

Name each article shown in picture and show its dependence on wood.

we put on a wooden or celluloid thimble and use thread wound on a birch-wood spool, and if we knit we find the needles made of wood or celluloid. Even the iron in the stove, the brass in the pots, the lead in the sinks, the silver in the knives, were all obtained from mines by blasting a shaft, and loosening the ore by means of explosives made from woodpulp; and the metal was extracted in furnaces heated with coke, — a product of coal. Even the roofs of the mines have to be propped with wooden poles. It seems as impossible to get away from wood as from air or noise.

The life of the men who go far into the forest, cut these trees, drag them to the rivers, and float them to the saw-mills and pulp mills, is to be described in this chapter.

Deep in the forest. These rough, hardy lumberjacks, as they are called, plunge deep into the northern forests in the autumn. After going to the end of the railroad, they load their bags on wagons and trucks and begin to work their way back through the forest that has been already cut, clearing and making the road as they go. The forest becomes denser, the trees larger, and the streams smaller.

A lumber camp grows up. At last they select a spot near the centre of the timber lands where they are to cut. Here they prepare to put up their buildings facing a small lake (#23). Every man is at once busy, some are cutting down trees, some are clearing the places for the houses, and almost before one can draw one's breath, a portable sawmill has been unloaded, the engine started, and it is buzzing through the logs and shaping them into lumber and scantling. Soon three or four squat buildings begin to take shape (#23), and in a few days the foreman's house, cook-house, bunk-house, and stables are complete, arranged in a row 200 feet back from the water. A stream at the left, fed by a spring, trickles along and empties into the lake. As this is to be the source of water for drinking and cooking, no building is placed close to it. The foreman's office is nearest, then, in order, the cook-house, bunk-house, and away over to the right, the stable for the horses. Behind the bunk-house is the laundry. The squat buildings are all much alike, long, square-cornered, with roofs sloping from a ridge. As they will be used for only three or four seasons they are put together as cheaply as possible.

The houses. The floors and ceilings are of rough boards, and the walls are either of logs or boards. Ugly tar-paper covers the walls and roofs and keeps out rain and cold winds. The larger buildings have three or four windows at each end and one or two windows on each side of the roof. Though they are not plastered or lined within, two big box stoves, one at each end, rammed full of

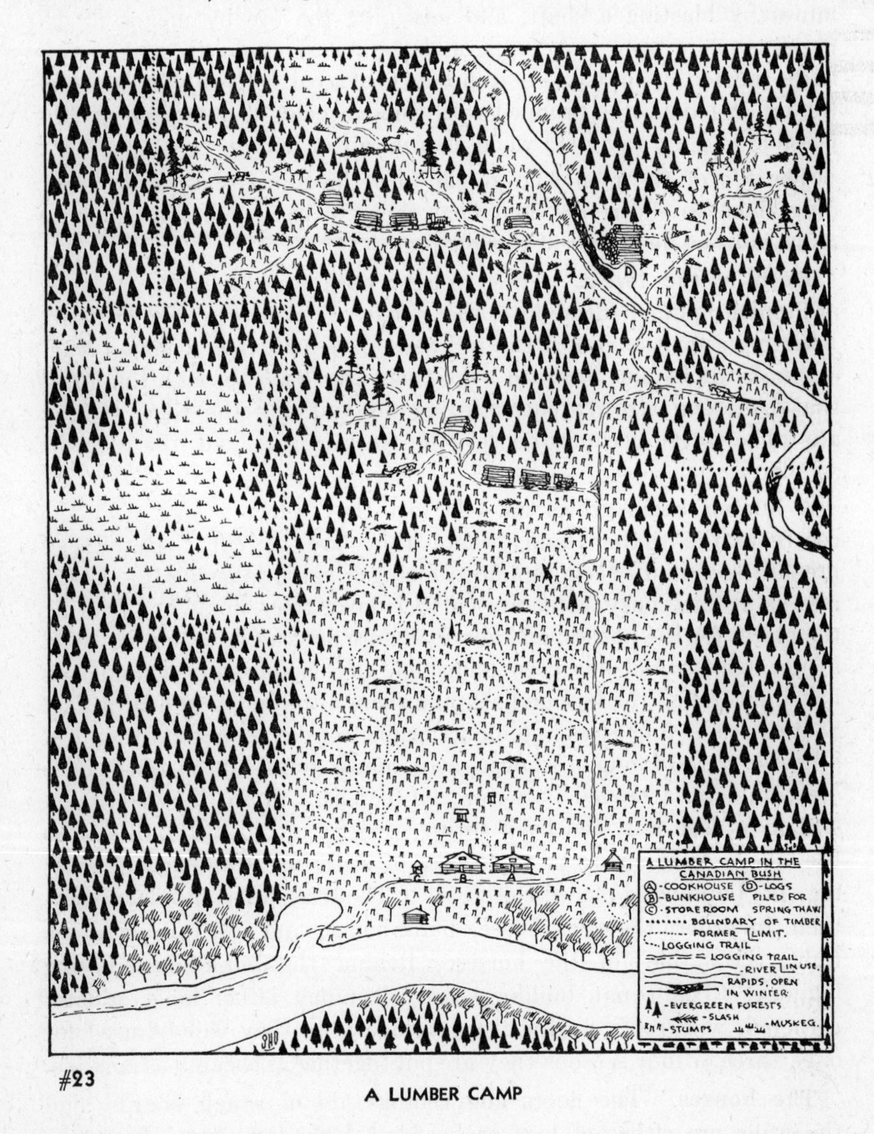

#23

A LUMBER CAMP

wood night and day, make the rooms too hot rather than too cold, even on winter nights when pails of water freeze solid in an hour; the ice on the lake explodes like rifle shots, and furious blizzards howl through the trees. There are ventilators in the roof, and an air vent which brings fresh air up and around the jacketed stoves.

Furniture. The cook-house has two long tables and benches, and a kitchen and storehouse at one end. The bunk-house has double rows of bunks, or beds, all around the outside with narrow aisles between, and each bunk has a mattress, tick, blankets, and sheets. The back part of the bunk-house is separated from the sleeping quarters by a wall. This makes a washroom with a whole row of basins and running water.

Finnish baths. Besides there is often a room forming part of the laundry, which is supplied with Finnish baths. Pail after pail of water is thrown on sizzling hot stones, which loads the room with dense water vapour until it becomes more muggy and humid than the rain forests of the equator and twice as hot. In this steam bath the sweat pours off the languid bathers, and as they go up a succession of steps, it gets hotter and hotter, until every pore in their bodies is streaming out perspiration, which carries along with it the waste matters that make one feel dull and sluggish. Often, as a tonic, the men leave the broiling bath, rush out into the snapping cold air, and roll in the snow under the cold light of the stars. When they have come in and rubbed themselves down, their skins are not only cleansed, but their whole bodies tingle with fresh life and vigour.

Three square meals. The cook is to the camp almost what Santa Claus is to Christmas. His savoury dishes soothe down many a rising temper, and turn seething discontent into peaceful torpor. He is usually a portly, good-natured man, who knows what rough, hard-working men require to eat, and sees that they get it in generous helpings just when they want it. Steaming hot porridge, well-fried pork, eggs, broad flapjacks swimming in corn syrup, stewed fruit with bread and butter, all washed down with an abundance of hot tea, make a breakfast that gives the men a full stomach and a happy start for the day's work. They take lunch in the woods. Either they prepare sandwiches, cheese, pie, and tea, each man selecting as much as he wishes of each; or if the men are not too far away, a 'gorge-car' is wheeled out from the cook-house, and the men gather around and help themselves to pea soup, beef stew, pork and beans, pie, and tea and coffee. After they have all gathered in the evening, the great meal of the day is taken. It starts with a tin plate of pea, rice, or vermicelli soup. Then generous

#24

CHIEF TIMBER TREES

helpings of roast pork or beef, or hot beef stew, well banked up with potatoes, carrots, turnips, beans or cabbage, make a main course that leaves nothing out. But a variety of desserts come next: apple, raisin, and peach pie; great round cakes covered with the most brilliantly-coloured icing; stewed peaches, apricots, and apples; cheese and cookies. These give variety enough to satisfy the taste of every man. But even yet the eating and drinking are not finished, for on a side table are pies and cakes, and on a stove a pot of tea to which the men can help themselves any time of the day or night.

Building dams. Much work has to be done before the snow comes. Gangs of men are sent high up stream to build dams across the branches of the main river so that there will be a reserve of water for the spring. Some are smoothing roads through the woods and down to the river bank, and others may have to clear out any barriers along the main stream that might hold back the logs.

Felling the trees. Even before the snow covers the ground, the cutting of the timber begins. Two men work on each tree (#23). First with their sharp axes they cut a notch in the trunk, which settles the exact direction in which it will fall. Next a sharp crosscut saw, with a man at each end, eats through the giant trunk as though it were cheese, and in a few minutes down it comes with a crash. Then with axes and saw they trim off the branches in a remarkably short time, so that soon the long bare trunk, as straight as a ruler, lies along the ground. They measure it into suitable lengths and cut it across several times into sawlogs. Their work is now done, and they move on to the next tree. If the trees are small and are being cut for pulpwood, one man with a bow-saw may work alone.

Skidding. Next a man with a team of horses drags, or skids, the logs through the woods to the roadside, and by means of a large pulley attached in a tree, builds them up into large piles on platforms, called skidways, to be later loaded on sleighs or trucks and carried to the stream.

A blot on nature. All winter this ghastly slaughter of giants goes on deeper and deeper into the forest; perhaps it is the tall white pines (#24), the princes of the forest, that are being laid low, after one hundred and fifty years of successful struggle and growth to reach the perfectly formed tree. As these lumberjacks gnaw into the trim and stately forest, its beauty and majesty disappear, only the undersized, the diseased, the deformed, the crippled, and the coarser breeds of trees are left standing; in the spring, when the snow melts, the fragrant carpet of dead leaves and mosses is left hidden and disfigured by unsightly stumps, bent and broken trunks, and a disorderly cover of tangled branches and chips, called *slash.* This unsightly mess dies, dries, rots, and chokes the ground so that other trees cannot properly grow to fill the gaps which are left by the trees removed.

The dangerous slash. The greatest danger, however, is that a flash of lightning, a smoldering cigar or cigarette butt, carelessly tossed on the ground, or the embers, scarcely smoking, in a camper's careless fire may gradually kindle this dead slash and let the fire-devil loose. Quickly it licks up this dry mass as though it were kindling-wood soaked in oil. If there is a wind blowing, it bounds forward like a fox chased by hounds. The great flames leap up into the needle-like leaves of the evergreens, filled with gum and resin. The crackling trees, the roaring flames, and the startled cries of terrified animals make a hideous noise. On and on it rushes until there is nothing left to burn, or until a heavy rain drenches the earth.

Paper puts lumber in second place. But the white pine, which was the pride of the forest, is nearly all gone, and the lumberjacks now log the less valuable spruce, hemlock, balsam, and cedar (#24), which make serviceable lumber. The preparing of logs for lumber is a business, however, that is becoming less and less important every year, while the securing of pulpwood logs of spruce, balsam, and poplar is the biggest industry of the forest and is bound to grow larger and larger, while lumbering grows smaller and smaller.

Moving logs from forest to stream. By December the lakes and rivers are frozen, a layer of snow lies deep over everything, and tracks have been pressed down by drags or tractors along the roads. While the cutting, skidding, and piling on the skidway still goes on in the woods, the teamsters are now busy moving the logs from the forest to the river. The first teamsters are up at four in the morning, their horses are fed, and they are off under the light of the stars to the skidway. The logs are loaded on the sleighs, and every teamster prides himself on building a big, even load. They must be skilful men; not only have they to be able to manage horses over narrow, rough roads, around hairpin bends, up and down steep slopes, but they must know how to lift heavy logs to the top of the pile on the sleigh and to arrange them in a firm and balanced load. The sleighs are moved along the rough forest road and then along a permanent road down to the river, where the logs are placed in neat piles on the ice, every log pointing in the direction of the stream. Team after team moves these logs, day after day, week after week, and month after month, as long as there is snow. At last the cutting ceases, the logs are all on the ice, and the stream for miles, perhaps, is one great winding pile of logs.

Tractors replace horses. In this age of progress many changes are taking place in order to rush the work; the horse team is now fighting for its life against the traction-engine and the truck. In suit-

#25

CATERPILLAR TRACTOR

able forests a powerful caterpillar tractor is hitched by chains to several big sawlogs at once and drags them along the forest road to the skidway. Often tractors pull down to the rivers a procession of sleigh loads of logs, all joined together (#25).

The ice breaks up. With spring comes the lumberjack's most exciting times. The rays of the bright sun, getting higher and higher each day, are playing havoc with the deep snow. The musical trickle of snow water is heard in all the forest, bare patches of black appear on the southern slopes, and the ice in the stream becomes uneasy with its heavy load of logs above and its surging current of water beneath. The ice is now wearing thin and soon will break down beneath its terrific load of logs, and sweep forward its confused mass of logs. The dams are closed, and water rises to the brim, making a small lake behind every dam.

The logs go out. At last the exciting day comes when the unequal struggle between the ice and sun is over; the logs begin to settle down, now they are moving restlessly, and at last they rush down stream in the river, which is filled to the top of the banks with spring torrents. The lumbermen, wearing spiked boots, with pike poles in hand, are stationed along the stream. They bound back and forth over the surging sea of logs nimbly as monkeys, and as sure-footedly as squirrels. They loosen one log here, straighten out a knotted jam there, push stranded logs from the shore, and direct all in the course they should follow. These expert men are as much at home on this moving mass of slimy, spinning logs, as though they walked on a paved road.

The dams are opened. Not a moment can be lost. The full flush of water will be past in a few weeks, and the stream will sink to such a low level that logs can no longer be floated, and the season's transport is over. After the first headlong rush of waters is over and the central mass of logs has been swept down, new flushings of the stream are made by opening the dams. Many logs are still piled in quiet bends, adhere to the sides of the stream, or are pushed up and stranded on the banks. The men now roll these into the low water of the sluggish stream, and when all is ready, a signal, or telephone message, is sent to have the dams unlocked. One after another these are opened so that their pent-up waters will all reach at the same time the main stream, now again choked with logs. The surging water rushing down carries all before it and sweeps the logs down stream. The dams are again closed, and this process is repeated again and again, until all the logs are out or the water fails.

Dynamiting a jam. Running logs down stream is often a dangerous business. The logs may become interlocked across a narrow part of the stream and the on-coming rush may pile up a disorderly mass of logs ten feet high. Perhaps the loosening of a single log is like pulling the trigger of a gun, and will free the whole jam, which shoots down almost with explosive force. The one who dares to free that log is a brave man, but these rough fellows know their work so well, and are so alert, that they never hesitate. Where no effort of these skilful men will free the mass, the men strain the whole force of the river to deal the jam one sledge-hammer blow. All dams are closed until every drop of water available is pressing against their gates; a stick of dynamite is then placed deep down inside of this tangled web of logs. The upper dam is opened first, then the others in succession, so timed that the roaring streams unite as they meet and rush downward together. Just as they hit the jam like a battering ram, the dynamite explodes, the logs are belched into the air, and the sweeping waters carry the whole mass down stream.

The logs reach the mill. The logs are kept moving until they reach the sawmill or pulpmill, which is often placed at the mouth of the river, or where the stream is crossed by a railway. Sometimes on large rivers, such as the Saguenay, the mills are far up stream, because large boats can ascend to be loaded. The logs are stored in quiet stretches behind *booms*, made of logs, fastened together, end to end; then they are floated to the mill as they are required, dragged up a sloping way and fed into the hungry jaws of the sawmill or pulpmill to come out as lumber or newsprint.

CHAPTER V

LIFE IN CANADIAN VILLAGES AND TOWNS

There is no place like home. The greatest of all organizations is the *family*. It is the basis of life. All animals have their *homes* or *houses*, in which they rear their young; and so attached are they that they will die defending their young and home. A timid, little bird, which hides at the slightest sound, will fly like a tiger at an intruder's face when he comes near the nest. A bear will usually slink away when she sees a man, but the wise man will beware of disturbing a she-bear with her cubs. Man resembles other animals in this respect, he loves his home. The whole world is dotted with houses, which are the homes of the families of mankind.

Attempts have been made from time to time to do away with the family and home. Children of different families have been gathered together in large groups and fed, trained, and educated by expert nurses and teachers, while their mothers and fathers worked in factories and had little or nothing to do with the care of their children. It has always been a failure. Only a few years ago it was tried in Russia, but to-day that great nation is making feverish efforts to retrace its steps and to restore the family and the home.

Social and unsocial animals. In some groups of animals, each family builds its own home, lives apart from the others, has its own feeding ground, and drives away intruders. The great eagle rears its rough nest high on a lonely crag, many miles from its nearest neighbour. Even our delightful robin will fight any other robin that dares to come into her domain, though she pays no attention to chipping sparrows or yellow warblers even if they build on the same branch. Other birds nest in communities; in the face of a sandbank there may be a hundred holes, each containing a nest of the sand swallow; and a single tree may be loaded down with dozens of great bunches of twigs and branches, each forming a nest of the great blue heron. It is well known by every boy and girl how hundreds of bees live together in the same hive, and thousands of ants in the same mound.

Such animals are said to be *social*, because they live in communities. Now man, like the ant and bee, is a social animal. The community may be a few houses, called a *hamlet*, gathered about a sawmill

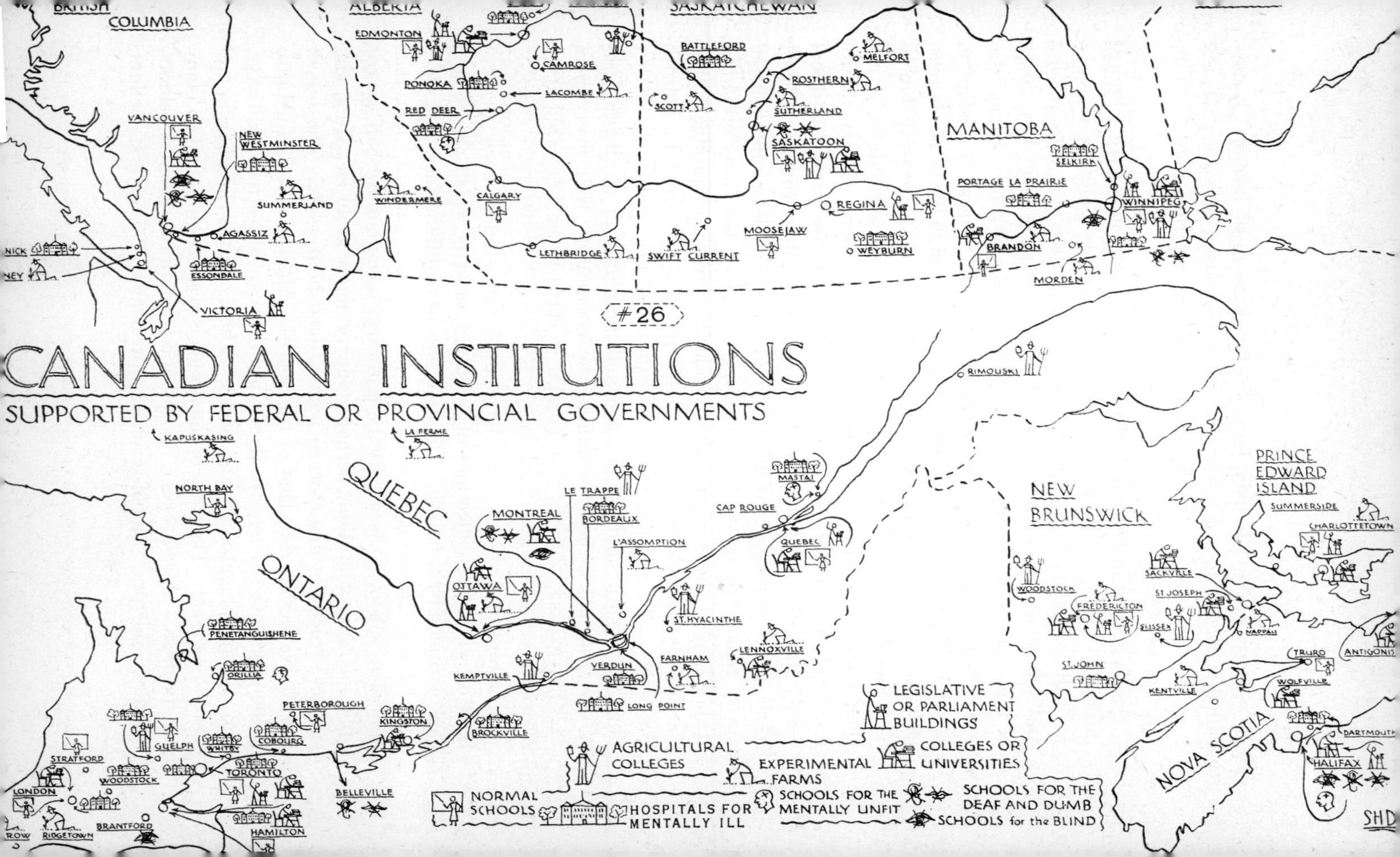

#26
CANADIAN INSTITUTIONS
SUPPORTED BY FEDERAL OR PROVINCIAL GOVERNMENTS
BRITISH COLUMBIA
ALBERTA
SASKATCHEWAN
MANITOBA
QUEBEC
ONTARIO
NEW BRUNSWICK
PRINCE EDWARD ISLAND
NOVA SCOTIA
VANCOUVER
NEW WESTMINSTER
SUMMERLAND
AGASSIZ
ESSONDALE
VICTORIA
EDMONTON
CAMROSE
PONOKA
LACOMBE
RED DEER
WINDERMERE
CALGARY
LETHBRIDGE
BATTLEFORD
MELFORT
ROSTHERN
SCOTT
SUTHERLAND
SASKATOON
REGINA
MOOSEJAW
SWIFT CURRENT
WEYBURN
SELKIRK
PORTAGE LA PRAIRIE
WINNIPEG
BRANDON
MORDEN
KAPUSKASING
LA FERME
NORTH BAY
RIMOUSKI
MASTAI
CAP ROUGE
QUEBEC
LE TRAPPE
BORDEAUX
MONTREAL
L'ASSOMPTION
OTTAWA
ST. HYACINTHE
LENNOXVILLE
FARNHAM
VERDUN
LONG POINT
KEMPTVILLE
BROCKVILLE
KINGSTON
PENETANGUISHENE
ORILLIA
PETERBOROUGH
COBOURG
WHITBY
GUELPH
TORONTO
BELLEVILLE
STRATFORD
WOODSTOCK
LONDON
BRANTFORD
HAMILTON
RIDGETOWN
WOODSTOCK
FREDERICTON
SUSSEX
ST JOHN
SACKVILLE
ST JOSEPH
NAPPAN
KENTVILLE
SUMMERSIDE
CHARLOTTETOWN
TRURO
ANTIGONISH
WOLFVILLE
DARTMOUTH
HALIFAX
LEGISLATIVE OR PARLIAMENT BUILDINGS
AGRICULTURAL COLLEGES
EXPERIMENTAL FARMS
COLLEGES OR UNIVERSITIES
NORMAL SCHOOLS
HOSPITALS FOR MENTALLY ILL
SCHOOLS FOR THE MENTALLY UNFIT
SCHOOLS FOR THE DEAF AND DUMB
SCHOOLS for the BLIND
SHD

or railway station; it may be a scattered group of farmers' homes, called a *township*; a considerable group of houses, called a *village* or *town*; or it may be a dense group of buildings, covering mile after mile in every direction, called a *city*.

Social groups in Canada. Since the great community we are most interested in to-day is the country called the Dominion of Canada, we will first show by a diagram how it is subdivided into smaller and smaller communities (#27). The diagram also gives the names of the capitals, or centres of government, of the larger divisions.

The home the kernel of government. As the success of a country, province, county, and village depends on the kind of life in the family, the constant aim of every good government is to provide homes with such healthy, peaceful, prosperous, and safe surroundings that they will bloom like flowers in a well-kept garden. Every family must first of all have food, clothing, and a house to shelter them from rain, wind, and heat in summer and cold in winter. Most of the great buildings that fill cities and towns, the deep mines blasted out of rocks, the giant trees cut from the forest, the slimy creatures trapped from the sea, and the wonderful variety of products that the farmer extracts from the earth's fertility, are used to nourish the home. To procure these necessities for his family, the father must have work; and the greatest and most difficult of all problems of governments is so to organize the community that every man may have work and yet not become a slave in an organized army of workers.

Structure of a village. A village and a township are each pieces of land on which are collections of buildings, most of which are houses, each occupied by a family. The buildings have different uses. Besides the homes there may be stores, workshops, garages, factories, churches, schools; there may be a library, a community hall, a post office, a railway station, perhaps one or more grain elevators, and smaller structures like storage houses, boat-houses, etc. The use of the land is as varied as that of the houses. Most of it is divided into lots on which the houses are placed, but there are also streets, railways, and perhaps playgrounds and parks. There may also be wharves and a river or a lake. The work of the people is also very different. There are farmers, store-keepers, hotel and restaurant keepers, clerks, lawyers, doctors, school-teachers, preachers, carpenters, tailors, dressmakers, milliners, tinsmiths, blacksmiths, garage mechanics, engineers, draymen, servants, and labourers, and perhaps others,

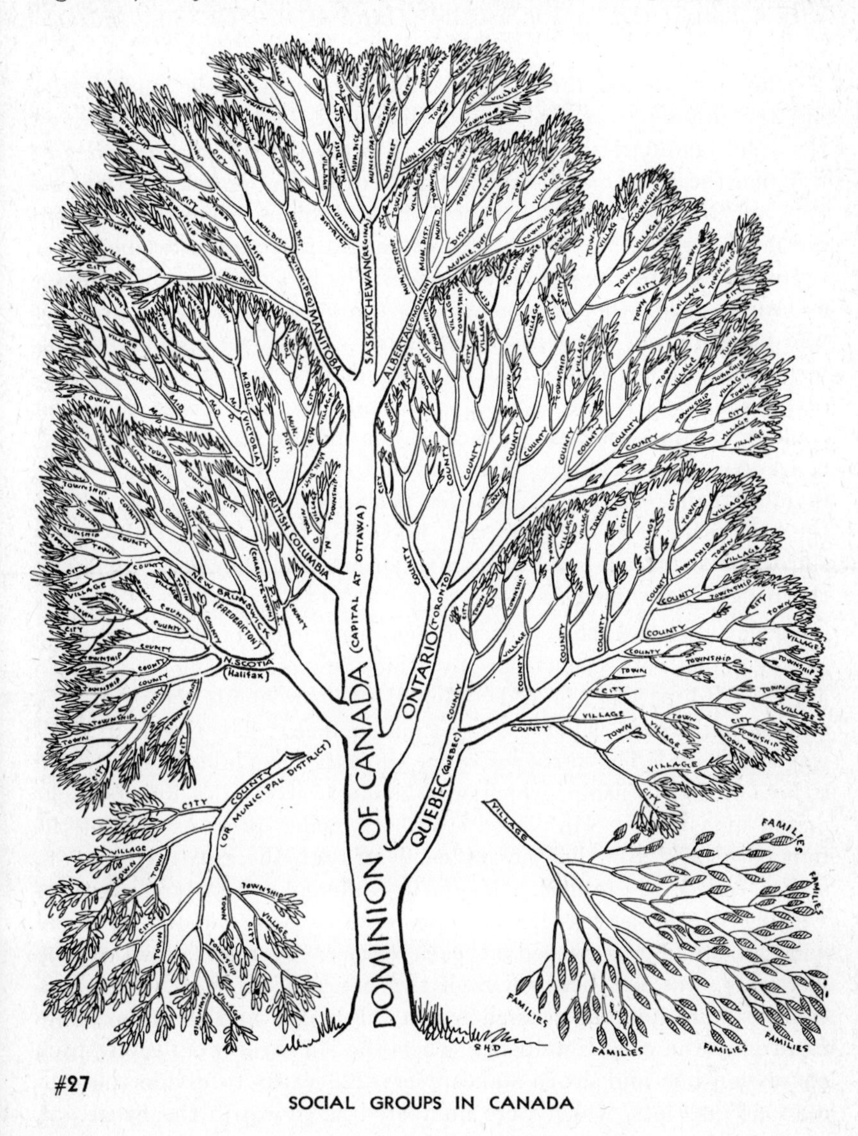

#27

SOCIAL GROUPS IN CANADA

all going about their own work every day. We must now see how all these play their part in the life of the community.

The homes are the flesh and blood of the village; indeed, everything else clings to them. Each home consists of a piece of land with a house; the land is usually a farm in the township and a *lot* in the village.

Who first owned the land? Who first owned this land? How did the different families obtain their plots? How do they know the exact boundaries of their lots? How did the village get its name?

When the country was first conquered by the British, all the land (say of Ontario), was in possession of Indians. The government did not seize a square foot of it, but bought it from the owners, piece by piece, as it was required for settlers, the last purchase being made as late as 1923, when a large area in the north-west of the province was bought from the Ojibway Indians. The government first surveyed it into townships and then divided each *township* into *concessions* and *lots* (#28). These were made of such a size that a lot could be cut into two farms, each of one hundred acres. Townships, concessions, or lots were given or sold to settlers, and they often divided their portion into smaller parts, which they resold. Thus the ownership of every lot in the village and every farm in the township can be traced back through a succession of owners, finally to the King's government.

Surveying a township. As his measuring must be most correct, a surveyor has very exact tools by which he can lay off distances in an exact direction; indeed different measurements between two stakes, miles apart, will not vary by half an inch. The township is always the unit. The surveyor often started from a lake, river, or road to lay out the township, which was about twelve miles long by eight or ten miles wide. He placed stone or wooden posts exactly at the four corners (#28); then he staked it off with the greatest care into strips called *concessions*. These in different townships were of varying widths, seven-eighths, one, or one and one-quarter miles wide; a stake was placed at each corner. A road allowance, 66 feet wide, was left between each two concessions (#28). The concessions were numbered as shown in the figure, and the roads were named by the same numbers. Then the surveyor went along each concession line and drove stakes every 220 yards to divide the concessions into lots, which were numbered as shown in the figure. A road allowance, called a *side-road*, was left between every three, four, or five lots. Thus the lot in black in the figure was known as lot 3, concession V, township of Moore. It had concession road V (not opened) to the north of it and concession road IV to the south of it and a side-road to the west of it. The surveyors from the provincial parliament buildings always surveyed the townships. Their work was not easy. The whole country was covered with dense forest, and for every line they ran, branches and whole trees had to be cut to give a clear view. Often miles of swamp, with bottomless quick-

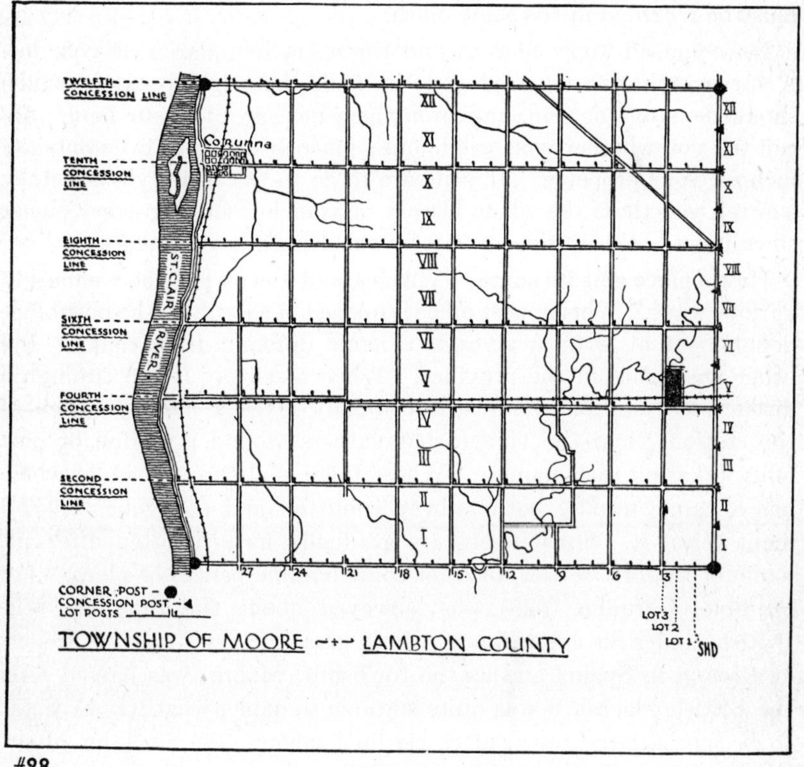

#28

SURVEY OF TOWNSHIP OF MOORE

sands and mosquito-infested marshes, had to be traversed, because a surveyor cannot choose his path but must always follow a straight line.

Surveying a village. A village is surveyed in exactly the same way into streets running at right-angles, and the blocks of land bounded by the streets are divided into lots. Streets are usually named instead of numbered, but the lots are always numbered as in townships. Usually a company or some land agent has a village surveyed. #28 shows a village surveyed at Corunna, but evidently it did not grow, as Corunna is still a very small hamlet. These surveys of townships and villages are always drawn on maps, which are placed in a building called the *registry office*, located in the county town. The maps for Moore township, and Corunna, for instance, are in the registry office at Sarnia. Every change of ownership of land must be recorded in the same office.

Now you all know how to find the exact boundaries of your lot. A surveyor has to be employed; he finds the original stakes, which should be stone or iron, and from them measures to your land. He can tell you whether your neighbour's fence or garage is even one-half inch on your property. If you care to go to the registry office, they can tell you there the whole history of your lot, naming every owner since it was sold or given away by the King's government.

How a place gets its name. Villages and towns get their names in various ways. When a post office is opened in a locality, the post office department at Ottawa gives it a name different from that of any other post office in the province. When a railway is run through a district and stations built, the company gives a name to each of the stations; and the village that gathers around a station or post office takes on their names. Where there is already a hamlet which has a name, usually both railway company and post office department select it. Such names are gradually formed, often after the name of the first settler, or after some notable person or place. For example, Corunna (#28) was surveyed about the time that the British, under Sir John Moore, won a great victory from the French, at *Coruna* in Spain; and as the township (Moore) was named after the British general, it was quite suitable to name, what was expected to become a large town, after his final victory, for Sir John Moore lost his life in the battle. Often when a town or village is first organized, the name, if not very significant, may be changed to a more suitable one. For example the old name for Kenora was Rat Portage, and when it was made into a town they considered it not very elegant and gave it its present name. For similar reasons, Mud Creek in Nova Scotia, was changed to Wolfville, Meyer's Creek to Belleville, Shade's Mills to Galt, and Scott's Plains to Peterborough. You will find it very interesting work to study the origin of the names of all the places in your county.

A man's home is his castle. It is often said that a man's home is his castle. By that is meant that no one has any more right to enter it than he would have had to enter one of the old stone castles, surrounded by a thick wall, lined with soldiers, and bristling with guns. Though there is no stone wall, often not even a fence, nevertheless, nobody has a right to enter your home, to walk through your garden, or even to cut across your lawn. Indeed, it is an offence for which he can be fined. You also have the right to do what you like on your own lot and in your own home, so long as your actions do not prove a nuisance or a detriment to somebody else. You can build a fence or not build a fence, you can paint your home or fence any colour you wish, you can plant a garden, or leave it a lawn, or even let it become choked with tall grass.

Social life brings duties as well as rights. If you lived, like Robinson Crusoe, alone on a desolate island, you could do just as you desired; go anywhere; wear clothes, or go naked; fire off a gun in any direction at any time; keep your house and yard dirty or clean, just as you liked; kill any animal you found; cut down any tree; or destroy anything on the island. But it is not natural for man to live that way. He is a social animal and likes to live with groups of his fellow men. As a result, many things that are not illegal should not be done out of regard for others. While every family has its rights in its home it also has its social duties to its neighbours. Untidy yards, long grass, weedy lawns, ugly fences, unsuitable paint, flaring billboards, loud noises, radios running late at night, barking dogs, prowling cats, and other offences to the eye, the ear, and the nose should be avoided.

Government enters the home. There are, however, three events taking place in the home, which the provincial governments demand every family shall report promptly, and these are the most vital of all occurrences, namely births, marriages, and deaths. In the government offices of the larger provinces is a department with great rows of shelves on which are thousands of books, made by binding millions of reports sent in from all over the province. Each report is of a birth, death, or marriage. The attending doctor should make out the reports on births and deaths and the minister on marriages. These are sent to the clerk of the village or township, who, once a month, forwards them to the provincial parliament buildings, where they are bound into books. These records are of great importance, and all boys and girls should be sure that their births are recorded in this office. Hundreds of birth certificates are sent out every year to boys and girls who wish to take part in organized sports, and in which there is an age limit. Every person is required to prove his age in many other transactions: before a boy or girl can get a driver's license he must present a birth certificate; before a person can collect insurance he must also prove his age. If a death certificate is not sent to the clerk, he dare not issue a burial certificate, which must be presented at the graveyard before the body can be buried; and a marriage certificate often has to be shown to prove that a person is a Canadian, or is entitled to receive property left in a will.

Municipal councils. As soon as a person steps out through his door on to the sidewalk or street his liberty to do what he likes ceases. The sidewalk, street, playgrounds, schools, etc., belong to the whole village or town. It would not be wise to leave the building and repairing of sidewalks and roads to the persons whose lots border on the street. Some would do the work well, some badly, and some not at all. Besides, while a person's house and yard are for his own use alone, the sidewalk and street in front of his lot are for the use of all; he has no more rights on the street in front of his house than any other resident. Therefore the streets and sidewalks are built jointly by those who are to use them though their cost may be charged, in part, against the property they front. Street lighting, waterworks (if there is a system), drains, sewers, schools, playgrounds are for the use of all. Therefore in every village, township, or town there is a *council* to manage those projects which are for common use, though a township is usually divided into smaller sections for the management of schools.

Upkeep of roads. Perhaps a great highway cuts across the village. Along it surges day and night a steady procession of thousands of automobiles and trucks, nine-tenths of which sweep past and do not even check their speed as they rush through the village. This one road has more traffic in a single hour than all the other streets of the village in a week. It would be very unfair to compel the village to pay for the heavy wear and tear of this busy road. Nor do they. It is a provincial highway, properly kept in repair by the provincial government, because it is for use of the whole continent rather than a local road built for the convenience of the people that live on it. The cost of all civic and government work is usually divided fairly according to the use. For improvements of sidewalks and roads on side streets of the village or town, usually the people on the street pay the larger share, while the whole village or town pays the rest; for the upkeep of the main street, bordered by stores, garages, etc., which is used by all the people, the village or town as a whole pays the main cost, while a through highway, throbbing with traffic of tourists and trucks from all over the province, Dominion, and United States, is paid for and controlled by the province.

Government a guide. As you move through the streets of a town or village, enter the stores, schools, library, or "movie," you see few signs of the work of councils and governments; nevertheless they are everywhere. These governments are like good angels hovering over you, invisible, but guarding and guiding you at every step. Their silent men protect you from fraud, disease, and danger wherever you go; they so manage affairs that you get good, clean, wholesome food; they watch the rogues, far too clever for you to cope with, that are trying to cheat you at every turn.

"Pure strawberry jam." Suppose you enter a general, or grocery, store, and look at the articles on the shelves. There are two kinds of strawberry jam in bottles of the same size and shape. You cannot tell which is better. You cannot taste them, because they are sealed; and even if you could you might not be quite sure yet which is better. But one label reads

"Pure Strawberry Jam"

the other,

"Strawberry Jam with added fruit pectin, contains colour."

Now you know which is better. Why did one jam maker leave out the word "pure," and humiliate himself by admitting that he had mixed in a dye to give colour, and had added fruit pectin, which, in ninety-nine cases out of one hundred, is an extract secured by boiling low-grade apples? He only did it because the government compels him to. No man dare put the word "pure" or "genuine" on any can or sealed packet containing food, if it contains a drop of colour, or the slightest trace of any impurity or adulterant. Salt or sugar, which are necessary for preservation, alone are allowed in foods described as "pure." Moreover, skilled chemists in special rooms at Government buildings at Ottawa are examining thousands of these foods every year, and the manufacturers know that they will be fined if they brand their goods falsely. Read the labels on jams, essences, canned vegetables, and fruits, breakfast foods, in fact on all foods that are sealed, and notice that in every case the quantity has to be stated, and the exact quality of the contents. Here is the quiet, invisible government's hand shown on every sealed article, preventing tricky, dishonest manufacturers from cheating simple buyers. They help us where we cannot help ourselves.

Romances written on labels. There are wonderful stories on labels. Look at this one:

"French Dressing
6 fluid ounces
(edible oil, vinegar, sugar, salt, and spices)."

At once the mystery of French dressing disappears, as it contains such ordinary materials. How the manufacturer would have liked the buyer to believe that he had that richest, most aristocratic and expensive of oils, namely olive oil; but he dare not, and has to be satisfied with the tame expression "edible oil."

How particular the government is! Here is an interesting label:

"Fancy Quality, Crushed Pineapple
Net weight 16 ounces
Drained weight 8 ounces"

The government insists that not merely the total weight of such an article shall be stated, which might be largely cheap liquid, but that the weight after it has been drained for a minute and a half shall be plainly printed on the label.

The following two labels tell the same kind of story.

"C———'s Clams
Net weight 16 ounces
Drained weight 6 ounces"

"Mushrooms
Net weight 16 ounces
Drained weight 5 ounces"

Hidden within the can in each case is a deceitful product; it weighs a pound, and one presumes it is clams or mushrooms, but the government forces the manufacturer to reveal that inside the can there are only a few clams or a few mushrooms swimming around in a watery soup.

You may purchase two similar packages of jelly powder, both prepared by the same manufacturer. One is labelled "Lemon" and the other "Pineapple." When opened, they look very much alike, — one as good as the other. But under the "Lemon" is "*pure fruit flavour*" and under the Pineapple is "*Artificial fruit flavour.*" At once you know that the lemon flavour was extracted from lemons, while the other was not obtained from pineapples at all, but made in a chemical plant, probably from tar, distilled from coal.

One more label from the grocery store must do.

"Pash"
New Zealand
Passion Fruit Squash
Shake well and mix with water
or soda-water to taste

and then follows the name of the manufacturer. Here is an attractive looking product from New Zealand, but the label has no weight marked on it. However, the bottle has a second tell-tale label, which was placed on it after it came to Canada. This tells a tale that one would not suspect from looking at the original label or the attractive looking liquid. This was what was said:

"Contents 20 fluid ounces,
contains 1/20 of 1 per cent. of sulphur dioxide."

Evidently New Zealand does not insist that the quantity shall be put on the label. But it is the italics that tell the story. Our government allows such a harmless preservative as sulphur dioxide to be put in foods in small quantities but insists that its presence shall be revealed on the label. But the real romance is how these unknown keen-eyed government officials, from among all the canned and bottled goods that come to Canada from New Zealand, were able to search out in this "Pash" such a small quantity of this gas as one part in two thousand? These silent men must move around like skilled detectives, watching at all our Canadian ports for the arrival of foreign goods. They collect samples that look suspicious and forward them to Ottawa, where the expert scientists perform experiments and test the samples sent to them by the government inspectors. These wise men detect not only one part of sulphur dioxide in two thousand, but one part of impurity in a million. Often they can do it in a minute, but it may take them two months.

The real government. These modest, unknown men have continually to match their skill, learning, and genius against the fraud and trickery of unworthy manufacturers and dealers. As the government experts lay bare their false colouring and cheap but subtle adulterants, the makers seek to find other colours less easy to detect and new adulterants still more subtle to find. It is a never ending battle of wits, and we should always remember that these efficient government men are fighting our battles. It is their great knowledge, long experience, and superior wisdom, that shines forth in the good laws that are passed.

These men also keep their eye on almost everything you buy in the grocery shop to see that the weight is correct, that the components are correctly described on the label, and that even the container is not deceitful. Often the buyer is impressed by the size of the container more than by the weight printed on the outside. Some dealers used to exploit this weakness of mankind for something big by putting false bottoms in berry boxes, and only two-thirds filling tins with baking powder, mustard, etc. But they no longer dare to use such deceitful containers, because they know the government inspector is working day and night searching factories and shops for such false vessels, and has the right to cut the top off any can and put its contents to the most searching test.

They are continually searching baking powders, cereals, butter, cheese, canned goods, cocoa, coffee, candy, figs, flour, syrups, dried fruits, honey, jams, mustard, vinegar, etc., for adulterants and false labelling. In one year they examined nearly 20,000 samples and found many false.

The government in the drug store. If you go into a drug store, the protecting hand of the government inspector is seen on every bottle and on every package. Here are found the most dangerous, poisonous, and powerful of all substances. A few drops of the wrong drugs would snuff out life in a few minutes. One slip by a careless or ignorant clerk in making up a medicine ordered by a doctor might mean the difference between death and recovery. Consequently you will see in a frame on the wall of every drug store a diploma, showing that the man in the store who mixes medicines has had five years training mixing drugs, two of which are in a school of pharmacy preparing himself to do this exacting work. If you buy a bottle of cough medicine, a tonic, a pain-killer, or any other prepared medicine the label must name the amounts of all *strong* drugs and state the size of safe doses for different ages.

Patent medicines. There is a kind of animal in South America called the vampire bat. It hovers stealthily in the air till its victim is asleep, then fans him with its horrid wings, and when sleep is deep, bites open his flesh and sucks his blood. Human vampires used to prepare harmful medicines, advertise them in great flaring headlines in the newspapers, and in lurid letters two feet high on the sides of barns and billboards, as cures for such dangerous diseases as consumption, cancer, goitre, diabetes, and many others. Persons, weakened by these dread diseases, ready to clutch at a straw to save their lives, foolishly believed these lying advertisements, bought these medicines, and became worse instead of better. The loathsome parasites who made these useless medicines became millionaires out of the money snatched from the fingers of their poor dying victims. By persistent and tireless effort on the part of our Government inspectors, it is now almost impossible for manufacturers to offer for sale any harmful drugs under the guise of remedial medicines. No medicine can be advertised to treat, let alone cure any of these stubborn diseases. A patent medicine must be passed by medical experts, and the label must state the amounts of any powerful or dangerous drug contained in it and dare not state that it will cure any of those dreaded diseases. The size of a safe dose must also be clearly shown.

Poisons. Violent poisons, such as arsenic, opium, carbolic acid, etc., cannot be sold except to those who can give a proper explanation how they are to be used, and the buyer has to sign a book for each drug bought. Unannounced, Government inspectors appear in every drug store in the country from time to time to see that the laws are being carried out.

Government stands guard at every store. In every shop it is the same. Every slaughter house is inspected, the health of the animals examined, and every carcass of beef, veal, lamb, or pork, in the butcher shop has the blue stamp of the government's inspectors pronouncing it free from disease, healthy, and wholesome. The government experts even pry into the insides of sausages and hamburger steak to see that they are wholesome and free from cheap adulterants. Every loaf of bread in a bake shop must be either one and one-half or three pounds in weight, there must be a scales on the counter, and any customer may insist on the weight being shown. Even the jewellers' shops are not forgotten. In nothing can buyers be more easily cheated than with articles of silver, gold, or precious stones. If it is good it will be stamped with the quality, as so many karats. The government insists that where such a stamp is used, the name of the maker must also be engraved, and they test the quality of these articles from time to time to see that they are up to the stamping. It is unnecessary to describe the means taken to insure purity and quality in butter, cheese, prepared fish, fruits, flour, vegetables, etc., but these are continually being kept under the keen eyes of efficient inspectors, who move in and out of the villages unknown to all but the sellers of food and drugs.

Frauds in foods. The same care is taken to prevent all kinds of fraud in food. Putting the largest, reddest apples on the top of a box or barrel is a dangerous offence; eggs, butter, and cheese are strictly graded, and the grade must be clearly shown in the shop where they are sold. Feeding sugar to bees, or mixing brown sugar with maple syrup or sugar, condemns the products at once. In fact the words honey, maple syrup, or maple sugar dare not appear on a bottle or parcel that contains the slightest trace of any other product, no matter whether wholesome or injurious.

Inspector of weights and measures. Another of the government's agents appears regularly on the streets of our cities, towns and villages with a mysterious black case in his hand. He visits every store and in most cases he is not much noticed; but a dishonest grocer will turn pale when he sees him, and the butcher, who sells fifteen ounces for a pound, trembles. Two very interesting pieces of metal, enclosed in velvet-lined cases, are almost as carefully packed away in a strong fire-proof vault in government offices at Ottawa as the Ark of the Covenant was in the tabernacle of the children of Israel during their journey through the wilderness. Only once every ten years are these precious pieces of metal taken out of their cases and carefully examined. One is a bronze bar 38 inches long with two little gold plugs imbedded, one near each end. Across each plug is a very thin distinct line. The distance between these two lines when the bar is at 62° F. is defined as exactly a *yard.* The other piece of metal is a cylinder of platinum, more valuable than gold, and its weight is defined as exactly a *pound.* Two other sets of these precious measures, all made by the most skilful workmen in England, are kept in vaults, one in the House of Commons, the other in the Senate. Thus if one should be stolen or destroyed in a fire, the other two would remain. Every five years the two in the Senate and House of Commons are compared, and once in ten years, these two are carefully carried into the precious vault, to be compared with the Dominion standard yard and pound. The inspector carries a brass yardstick and brass pound weight, which have been made exactly the same size as the standard in Ottawa. He also has other sized weights and a gallon measure all compared with the standards, and his yard measure is divided accurately into feet, inches, half-inches, etc. With these and a delicate balance he tests every weight, scales, yard-stick, and measure used by store-keepers, milkmen, pedlars, gasoline sellers, coal merchants and every other person who sells goods by weight or measure. Every measure must have the government stamp, and a dealer who dares to use unstamped weights or measures can be heavily fined. He even tests ten-pound packages of sugar, pound packages of rice, half-pound packages of salts, etc., indeed any package with the weight marked on a label. Occasionally he goes as a customer to buy meat, milk, sugar, or gasoline to see if the dealer gives short measure. As every dealer in Canada is inspected every year, customers receive protection from fraud. Just think of the work! The hundreds of thousands of yard-sticks, the millions of weights, little and big, the countless quart, gallon, and bushel measures, all the way from Vancouver to Halifax, are kept exactly the same size as those precious standards stored carefully in vaults at Ottawa.

If the people buy gas or electricity, the machines, or meters, in the cellar, which measure the quantity, will all be found to have Federal or Provincial government stamps, indicating that they have been tested and found correct.

How a factory is built. Perhaps in the same village or town there is a factory for making boots and shoes, furniture, or some other article. There, amongst whirling wheels, grinding gears, clanking machines, steam-strained boilers, poisonous gases, and electric currents as deadly as a flash of lightning, accidents, explosions, and death lurk around every corner. Let us visit this noisy building, humming with work, to see how our provincial government shields the health, safety, and comfort of the workers. Even before the factory was built, plans were carefully studied and approved by expert architects of the provincial government. This careful examination makes it certain that the walls and floors have sufficient strength, that the walls, and perhaps the ceiling, are well supplied with windows, and that proper ventilators are built to remove foul or poisonous gases. There must be a sufficient number of wide stairways and fire-escapes for all to escape from fire. Doors in the main passage ways must open outwards to prevent a jam in a fire panic.

Comfort and safety in the factory. As one passes through he will see the kindly work of thoughtful government inspectors on every hand. A chair, or seat, must be near the workplace of every female employee in both factory and office, and wherever work can be done while seated, a chair must be supplied to every woman. Every belt, turning gear-wheel, and other dangerous moving part, which might catch up clothing, or in which the hand, foot, or hair might be seized or jammed, must be protected by a screen or fence. All dangerous liquids, such as kerosene, benzine, etc., which might cause explosions or fires, must be kept out of the factory altogether. Elevator shafts must be so guarded that nobody can fall down. Drinking water, washrooms, towels, and soap, in sufficient quantities for comfortable working, must be provided, and a separate room in which women may eat their lunch.

The keen eye of the factory inspector. A provincial inspector quite unexpectedly visits each factory, and insists that it meet all the above conditions. His keen and practised eye also searches the time sheets to check that the number of hours of work of boys and girls, and women has been kept down, that none have started work before 7 a.m. or continued after 6.30 p.m., that all have had an hour for dinner—that no children under fourteen have been employed, that no girls or women have been put at dangerous, unhealthy, or unwholesome work, and that women and girls have received suitable minimum wages. He insists that the factory be kept clean, be well lighted, and have ventilators to sweep away all dust, fumes, and smells, in order that the air may be pure, wholesome, and warm. Since very dangerous and painful diseases are caused by breathing dust, and especially dust from particles of stone, the greatest care must be taken where stone is being chipped, ground, or polished.

A living wage for the injured. When an accident does occur and an employee, perhaps with a family to support, is injured, the government does not forget the poor hospital patient who is able to earn nothing. From a fund to which he and the employer both contribute weekly, he is paid every week a living wage, until he is again ready to work; doctor's and hospital expenses are also included.

Safe travel. No matter where we turn, the help of governments is at our side, but so silent and subtle that we often fail to recognize its presence. Suppose we go for a spin in a motor-car, take a pleasure trip on a steamboat, or make a journey on a railway, our comfort, our safety, and our health are watched over by governments, often as faithfully as are her nestlings by a mother bird hovering above her nest.

The king's highways. The highways are made wide, straight, and smooth, defects are repaired, curves straightened, hills cut through, and hollows filled. They are so wonderfully marked that even a stranger can follow a course from one side of the country to the other by merely reading the signs by the road. Our safety is guarded in many ways. Every driver must carry a license, only obtained after careful tests; brakes, lights, steering, and other features of motor-cars are liable to be tested anywhere on the road by government officers; and the speed of cars is limited in both town and open country. Traffic policemen patrol the roads day and night, ready to help and guide those who seek their assistance, and alert to stop and rebuke all that are careless or reckless.

Water channels made safe. When a trip is taken on a steamboat, the course is more uncertain, as you are subject to changes of weather, swift currents, rocks, shallows, and fogs. But the Dominion government has anticipated all your difficulties and smoothed out all the wrinkles from your path. The *gang-board* by which you enter a boat must have side rails to prevent even the most careless from falling in. The course has hidden rocks, treacherous currents, and many shallows, but in the captain's cabin there is a chart, on which the depths of every part of the course are marked. Every acre of the bottom has been sounded by government surveyors in order to be able to draw such a detailed chart. Then as you pass along, you see the course marked with red and black buoys by day, and lights and lighthouses by night. To fight the sailors' most dangerous enemy, fog-horns, sirens, and bells are placed over treacherous shallows and hidden rocks. Perhaps you may come to an impassable rapid or a defying waterfall. But even these the government engineers have mastered at great expense to give your boat a free passage. You pass through a canal dug around the rapid or waterfall until you find a gate stopping your passage. Then just behind you another gate closes and you are shut in what is called a *lock*. Valves in the bottom of the gates at one end are opened, water pours in from above, your boat begins to rise until finally it has been raised as high as the top of the waterfall. The gates in front of you open, and you pass back into the river above the falls.

Ship captain, a trained expert. It is one of the marvels of the world how a captain can guide his ship in the inky blackness of night, without a star in the sky to beckon him, around hairpin curves, among spray-covered rocks, close to hidden shallows, against fierce gales, through dense fogs, and across baffling currents. But he is only able to perform such skilled work because of hard study, long training, sober habits, and a courageous and alert mind. You will see, in a frame somewhere on the ship, a certificate, which this captain must have obtained through long training and hard work before he can sail a ship. The engineers and the mate must also be trained men with certificates of having passed severe tests as to their fitness. The government is just as strict that captains and engineers shall have qualifying certificates as that teachers or medical doctors shall have proper diplomas.

There are many other devices by which you are protected on shipboard, all carefully planned and inspected on every ship by government inspectors. The boiler is as dangerous as dynamite if not carefully watched. The live steam is pressing against its walls with such tremendous force that if there were a weak seam, a spot nearly rusted through, or some defective rivets, the whole boiler might burst with an explosion that would blow the ship to pieces. Every steamship's boiler is examined by government inspectors while building, and must be tested and passed when finished before it can be used; it is protected from too great pressures by safety devices that act without any attention by man. Not only that, but every year the boiler on every ship must be again tested, and you will find framed, somewhere on the ship, certificates of the inspection of both boiler and engine for the year. These framed certificates in prominent places on a ship, though seldom noticed by passengers, show modestly how governments watch over the lives of passengers and crew.

Life preservers for all. If you further examine the ship on which you are taking a trip, you will find every possible care for your safety against drowning. There are steam pumps and hand pumps with hoses attached ready for action to put out a fire or pump out water caused by a leak. There are fastened on the outer wall of the cabin a number of white structures, the shape of huge doughnuts, with a long coiled rope attached to each. The "doughnut," made of cork, is called a *life-belt* and is to be thrown to anybody that falls overboard. In every stateroom, piled on the deck, and attached to the roof are *life-preservers* made of cork, which, when fastened on like a coat, will keep a person afloat in the water for any length of time. There must be one of these for every passenger, and the boat crew must show passengers how to use them. On the upper deck are large boats, called *life-boats*, which contain oars, and can be swung out and let down into the water if the ship is in danger. The ship is compelled to carry all of these safety devices for fear of a wreck.

Ship inspection. You will find a further certificate fastened on the wall of the steamer, stating that its hull and all equipment have been examined this year and found in proper condition. The life-preservers must be counted and tested, the life-boats examined, the pumps and hose tried out, the hull of the boat tested for tightness and strength, and a hundred other things found right before the certificate is given. How seldom we think of what these unknown government experts have planned and performed for our comfort and safety.

The load-line. Even the loads on ships must be stowed properly; there is on the outside of the hull a white or yellow horizontal line, running through a black circle, the better to be seen. No ship dare carry a load that sinks it below that white mark, called the *load-line.*

Reports on weather. The greatest danger to a ship is a sudden fierce storm, which may drive it against rocks, or batter it to pieces with the terrific beat of the sledge-hammer waves. The Dominion government gives ample warning even of these gales. Hundreds of men, scattered evenly over all parts of Canada and the United States read from very accurate instruments the temperature and pressure of the air, the direction and strength of wind, the movement and character of clouds, and the amount of rainfall or snowfall. Twice a day this army of faithful observers telegraph their information to a stone building in Toronto, called the *Meteorological Office.* So important are these observations that all other messages must be kept off the telegraph lines at these hours, in order that the information may reach the office promptly. At this office, more of these silent, unknown experts speedily mark the data on a map of Canada, and from these numbers scattered irregularly over the map, their trained eyes and minds are able to predict the weather in every part of the country. Records and forecasts are announced through the newspapers and over the radio. If gales are coming, signals are put up at prominent places all along the coast of the Atlantic Ocean, the Pacific Ocean, or the Great Lakes, as required. Every passing ship sees a warning, and if wise, either remains in port or returns to the harbour until the fury of the gale is past.

Education of children. Governments believe that it pays the village, the province, and the state to raise up citizens with alert, happy, contented, educated minds in sound, skilful bodies. We have studied a few of the ways in which they protect health and life, and we now turn to the methods of training the mind. Every village has *an elementary school*, and some have a *continuation* or *high school* as well. These are built and controlled by the village and are considered so important that they have a specially elected body, the *board of trustees*, to manage them. This board hires the teachers, equips the schools, and pays the running expenses. But the provincial department of education plays a big part in making the schools efficient. They train the teachers, prescribe the courses of study, and search out the most suitable text-books. They also send an inspector twice a year to help the teacher and to report to the board of trustees the condition of the schools. These expert educationists never cease searching the whole world for methods and courses that will improve the education of boys and girls. Not only are the ordinary school subjects now taught, but in the larger towns there are *technical, vocational,* and *commercial schools*, in which boys are trained to be carpenters, machinists, auto-mechanics, watchmakers, printers, or to follow almost any other trade, and in which girls are prepared to take positions as book-keepers, stenographers, dressmakers, milliners, and best of all, home-makers.

The public library. But the school is not the only educational institution in our towns and villages. In all probability there is a free library, in which are stored many of the best books. This is an unfailing spring at which every person, old or young, rich or poor, can drink in the most modern knowledge, the noblest ideas, the choicest thoughts of all time, expressed in the most beautiful and attractive language. It brings the whole world, present and past, to the people. By means of this library it is possible for every person steadily to become wiser and more learned.

A marvel of civilization. One other government institution found in every village and town must be described, as it is one of the marvels of civilization. It is the *post-office.* It unites all the people of the earth. You can put a three, or at most a five-cent stamp on a letter and send it to the most out-of-the-way place on earth. It may go to the centre of China, the farthest north settlement in Greenland, a coral island in the centre of the Pacific Ocean, or an Eskimo village at the mouth of the Mackenzie River. It may travel by train, by boat, by dog-train, by aeroplane, by canoe, or on a man's back, but on it goes in its frail covering and is sure to reach its goal in the shortest possible time. Just think of the army of millions of postmasters, dotted over the face of every country in the world, in every small island in the ocean, up in the mountains, far out in the oases of the desert, in little villages choked in jungles and dense forests; just think of all these, speaking hundreds of different languages, but working together to sort out the tangled mass of hundreds of millions of letters, so that each will follow along the right path to reach the person whose name is on the envelope. Is there any more complicated or more efficient machine than the post-office system of the world?

Two dozen truck-loads of bundles of newspapers are dumped into the Toronto, or Montreal, post-office at eleven o'clock at night, by morning the 50,000 papers in the bundles have been separated and are speeding north, south, east, and west, along a thousand different paths, which lead to every village and almost every farm home; by noon the next day the people all over the province—the boarder in the hotel, the hunter in his lodge, the excursionist in his summer home, and the farmers, miles from a village, are reading the daily news. But the post-office does far more than that. You can send parcels, books, even baby chicks, or queen bees, and they will reach their goal. You can pay a sum of money to the postmaster, he will give you a postal order, which you can forward to almost any part of the world and the receiver can cash it in his own money. The Englishman receives it in shillings, the Frenchman in francs, the German in marks, the Jap in yens, or the Indian in rupees. The post-office, since it is connected with the whole world, is controlled by the Dominion Government at Ottawa.

Governments do not forget the dumb and the blind. It is said that in the city of Sparta in ancient Greece, children who were weak, crippled, blind, or sickly were pitilessly put to death as were the old and decrepit. How different are our governments in Canada! We consider such treatment cruel and inhuman, and our provincial government pays special attention to these poor, pitiful creatures, who cannot look after themselves (#26). There may be deaf and dumb or blind children in your village, who can be properly trained only by carefully prepared teachers, and with specially printed books and other equipment. The provincial governments do not forget these handicapped children but have built special schools where they can live and be taught. Many of the dumb soon learn to speak; and while that worst of all drawbacks, blindness, usually cannot be cured, the children with this affliction can be taught to do many things, so that they are often able to earn a living and are thus filled with new hope; all can be taught the pleasure of reading from specially prepared books. During the last year or two, in some of the provinces, special schools are just beginning to be formed, where poor crippled children can be gathered in from all over a province to be trained at different kinds of work, so that they too can make a living.

Mental hospitals. Another great work that has been undertaken by our provincial governments is the care and help being provided for those who become mentally ill. The stigma which once attached to these unfortunate people has given way to sympathetic treatment with the knowledge that mental illness, like physical illness, has been caused by breaking the laws of health, may be cured by skilful treatment, and many of these persons returned to normal life. This sympathetic and friendly attitude toward those who, unfortunately, become victims of mental diseases, has been of untold benefit to numbers of our fellow countrymen.

Training schools for children. We have said that a person's home cannot be interfered with (p. 59). But where parents are brutal or neglect their children, because the little ones cannot help themselves, the firm hand of the government may take the children from such unfit parents and put them in homes where they will be kindly and intelligently treated. If boys or girls become so unmanageable that their parents have no control, the provincial government may lay a firm hand on them too and remove them to training schools, which they have established. Here under firm teachers, with good food, lively games, and wholesome work in gardens, fields, and workshops they are usually so improved that they can soon return to the village and grow up to be useful, law-abiding citizens.

Government, the friend of the poor and the aged. These different governing bodies not only look after the blind, the dumb, the crippled, and the ill-treated, but also after the aged and the poor, who have nobody else to soothe their sorrow, to help their weakness, and to supply their needs. Pensions for the aged are now paid every month, and are the right, not charity, of every needy man or woman over seventy years of age; and comfortable *refuges* in almost every county and city are now provided, where those who have no children to take care of them, may spend their declining years in comfort and without worry. Perhaps in your district there is a poor mother with young children, who has lost her husband by death or desertion, and has no means of support as the children require her at home. The village and the province share in the good work of giving her a monthly allowance, with which she can bring up the children.

The civil service. Sometimes when a member of parliament, or the premier, visits your municipality, there is a great celebration. A procession forms, cheers fill the air, and great honour is shown to such a notable man. And that is proper. But we must never forget that great army of modest, silent men, working in the parliament buildings and government offices, and moving through our villages and towns observing or inspecting. They have worked out all those wonderful laws which protect the ignorant from dishonest salesmen, make employers deal fairly with their workmen, train people in the laws of health, give them education, and look after neglected children, the sick, the deformed, the poor, and the aged. These are called the *civil service*, and though they have no processions while they are alive, nor monuments when they are dead, — they should have our respect and honour; and we as citizens should see that, above all others, only the best are chosen, and that their positions should be secure.

Disobeying laws. Many of the laws to protect people have now been described. These have all been passed by the village council, the provincial parliament, or the Dominion parliament. Those passed by the council are called *by-laws*, those by either of the parliaments are called *statutes.* We have also seen that inspectors are ever on the alert to encourage those who obey and to search out those who disobey these laws. But what happens to the person who disobeys any of these laws? If a grocer uses a pound weight too light, an employer has children under fourteen years of age in his factory, or a canner puts colour in his jam and labels it pure, there are people in the district to deal with them.

The policeman. Probably the village has a *policeman*, or the township *constables*, for both words apply to the same office. They are appointed by the judges. The provincial government has a body of police to enforce its laws, and the Dominion government has the most notable body of police in the whole world, the *Royal Canadian Mounted Police*, who enforce the Dominion laws, especially in out of the way places, such as north Canada, which is not in any province. Indeed, so excellent is this fine body of men, that all the provinces but Ontario and Quebec have dismissed their provincial police and now use the Royal Canadian Mounted Police to enforce their provincial laws.

Arrest law-breakers. The work of the policemen is to prevent the breaking of law and to arrest those law-breakers who persist in the practice. They are the friends and protectors of all those who do right but the foe and dread of all those who break laws. They arrest the thief or robber, the fighter, the reckless driver, and the drunkard, and may put them in jail. But no person can be kept in jail for more than a few hours without being allowed a hearing, and this must take place before a *magistrate* or *judge.*

The judges. A man, no matter what crime he is accused of, is considered innocent until he is proved guilty after a fair trial before an unbiassed judge. Most villages have at least one citizen who can write J.P. after his name, which means that he is a *Justice of the Peace*, and in some villages there is also a *police magistrate*; both of these are appointed by the provincial government. Then in the county town is a still more important man, who is appointed by the Dominion government. He is very learned in law and is called the *county judge.* All of these act as judges whether a man is innocent or guilty of any crime of which he is accused. The magistrate or justice of the peace tries prisoners accused of small criminal offences, and his court meets every week, or sometimes every day. The county judge holds court in the village, perhaps, every month, to try more important cases, and he may have a *jury* of twelve men to help him reach a decision. The prisoner can bring forth any witnesses he desires and have a lawyer to question them and to plead in his favour before the magistrate, judge, or jury, because it is of the greatest importance that no innocent man should ever be punished. The magistrate, judge, or jury, finally pronounces the man guilty or not guilty, and if guilty, the judge or magistrate tells him how much money he is fined or how long he shall stay in jail.

We have now shown the nature of the laws, how they help the people, how people are made to obey them, and how those guilty of breaking them are punished.

Election day. We have not yet explained how the laws are made. The most exciting of all times in the village or town is *election day*. For weeks or months before, the men running for office, called *candidates*, have meetings in the schools or the community hall, make speeches telling what they will do if they are elected, and put up posters advising people to vote for them. Then on election day the men and women mark their *ballots*, usually by putting a cross (X) after the name of the person they prefer. These ballots are put into an iron box, through a hole so small, that no hand could possibly get in to tamper with them. When the election is over, the ballot box is unlocked in the presence of *scrutineers*, and the votes obtained by each candidate counted. Meanwhile, the people outside are very excited. When the vote is announced some of the people cheer, while some others keep quiet and do not look pleased. When the votes from other villages, or other voting places, come in and all are counted and it is known who is elected, his followers light torches, have a procession through the village or town, but next day the district settles down again to its routine life.

The law makers of Canada. Elections for the village or township council are held toward the end of every year. The men elected are called *councillors*, and the head man is the *reeve*, but in a town or city the chairman of the council is called a *mayor*. The elected body is called the *municipal council*. The men elected to the provincial parliament are able proudly to write after their names the capital letters, M.P.P., which stands for *Member of Provincial Parliament*. Sometimes they put after their names M.L.A., which stands for *Member of Legislative Assembly*. Both mean about the same, but the members usually prefer the M.P.P. They are elected for five years, meet each spring for about two months at the capital, receive an indemnity of, in some provinces, as high as two thousand dollars a year, and make the provincial laws. The members of the Dominion parliament are usually elected for about five years, and they have after their name the most desired of all letters M.P., which means *Member of Parliament*. They meet in Ottawa annually, and often have to spend half of each year at the Dominion parliament, for which they receive four thousand dollars a year. As many of these are prominent men, it is a real sacrifice on their part to spend so much time away from their business. They together are called the *House of Commons*, and make the laws for the whole country. But there is a second body at Ottawa, called the *Senate*, whose members, instead of being elected, are appointed by the government for life. A majority of them as well as of the members of the House of Commons have to vote in favour of all laws.

Parliaments as good as the voters that elect them. Now you see how important it is for boys and girls to learn about all the affairs of Canada. When they are of age (21 years) they vote for the members of councils and parliaments, who make our laws; if we want good laws, we must have wise, upright members of council and parliament; and the quality of these men depends entirely on the wisdom and good judgment of the voters.

The cost of government. There is one other very important function of government, which causes more worry to both governments and people than all others combined. It is comforting to have good laws, satisfying to have wise civil servants to protect your health, your life, and your purses, and restful to feel that you have shrewd inspectors to see that the grocer's scales do not weigh against the buyer, and to prevent the employer from ill-treating his workmen. It is a great blessing to have well-equipped schools, skilled teachers, libraries with shelf upon shelf of the choicest books, cool, green parks to play in, and large town halls in which to hold meetings and have entertainments; but all these cost much money, many millions of dollars, and the people must supply the funds.

Property taxes. Every spring, a business-like looking man with a big book passes through the streets of the village or town, enters every house, and asks many peculiar questions which you might refuse to answer anybody else. He is called the *assessor*, and is one of the few men who has a right to enter your house; but you must answer all his questions, which are very personal. He asks the number in the family, the age of children, the value of the house, number of acres of land, and the number of live stock, (including the dog, but not the cat). A little later you receive a tax bill, which is reckoned on the value of your property (#30). This tax is the chief source of money for the council, which they spend in repairing streets and sidewalks, digging drains and sewers, running the schools and the libraries, paying the policemen and the magistrates, and for many other purposes. The council also obtains money from the sale of licenses to pedlars, garages, gasoline stations, auctioneers, etc.

#29

HOW GOVERNMENT MONEY IS SPENT

Revenue of the province. The sign on the filling pump of a gasoline station reads "gasoline 20 cents a gallon, *government tax 6 cents*"; when you go to the moving-picture show, in most of the provinces you receive two tickets, one of which reads "*amusement tax 5 cents*";

if you examine your motor-car license you will find a figure, printed thereon, of $9, $12, or $15, as the fee, and on your driver's license a fee of one dollar. These taxes and fees are all sent to the provincial government and form the largest share of the revenue it

#30

HOW GOVERNMENT MONEY IS COLLECTED

receives to carry on its work. They also receive license fees and inspection fees from many sources, such as hotels, race-tracks, etc. Besides, most of the provinces own vast amounts of wild land, and receive large sums on the minerals produced, timber cut, water-power used, and land sold. Perhaps in your village there is a liquor store. This is operated by the provincial government and is for the sale of alcoholic drinks. The millions of dollars profits from these sales also form an important part of the provincial revenue.

Taxes from the rich. All of the provincial taxes already mentioned are widely spread among the people, but the province has two methods of collecting from the rich alone. When a wealthy resident of your village or town dies, a part of his wealth is passed over to the government in what is called *succession duty*. Sometimes more than a million dollars is obtained on the death of one person. If you live in a village there is probably a railway, a bank, an express office, a telegraph office, and several insurance offices, all of which belong to big companies, called *corporations*, which probably have head offices in Montreal or Toronto. The provincial government levies an annual tax on all these companies, the larger and more wealthy the company, the heavier the tax. This *corporation tax* brings in millions of dollars to most of the provinces.

Dominion Revenues. Excise. On every pack of cards for sale in the village book store there is a little stamp with ten cents marked on it. Stuck to every package of cigarettes, every bottle of liquor, parcel of tobacco, box of matches, box of face powder, lip stick, and other cosmetics, are stamps with *excise* marked on them and perhaps a figure indicating a sum of money or the quantity of goods, such as the number of cigarettes. The manufacturers of all these articles have to buy these stamps and put them on the articles before they sell them to the stores. Manufacturers also have to pay a similar tax on every rubber tire, every automobile, every gallon of beer or hard liquor made. Such a tax, charged on the manufacture of goods, is called *excise* (#30), and the Dominion collects many millions of dollars from this source every year.

Duties on customs. When you buy goods in the United States or other foreign countries and have them mailed or expressed to you, they will not come directly, but will be sent to the nearest town which has what is called a *customs office*. The customs officer will send a card notifying you that it has arrived but that there is a certain amount of *duty* to be paid on it (#30). When you send the duty, he forwards the goods. Customs officers are in all the main ports along the coast of Canada, and at railways, ferries, and main roads crossing from the United States. When passengers cross in boats, trains, or automobiles, a customs officer in a blue uniform stops everybody, searches their valises, trunks, boxes, cars, sometimes even their pockets, for goods which they have bought in the United States, or other country from which they are coming. On all of these articles the duty must be paid before they can enter Canada, though Canadian tourists can now bring back one hundred dollars' worth free. Sometimes foolish people try to hide such goods, which is a dangerous and vicious practice; if they are found cheating the customs, the goods are seized and they may also be heavily fined. From the five or six hundred million dollars' worth of goods imported to Canada the duties collected are twice as much as from excise, and in 1936 amounted to seventy-four million dollars.

Sales taxes.

The Copp Clark Co., Limited,
Publishers, Booksellers, and Manufacturing Stationers,

Sold to James Smith

1 ream Bankhead Bond cap paper	$2.25	
1 " Acme " "	1.10	
Tax 8% on 3.35	.27	$3.62

The above bill was made out by the company that publishes this book, who are also wholesale stationers. You will notice that the person who bought the paper had to pay, besides its regular price, a sales tax of 27 cents, which was 8 per cent of the cost of the paper. Now on almost every article sold by a *wholesale firm* there is a heavy tax (#30). That means that in your village or town on most of the goods bought by the grocer, druggist, drygoods and hardware merchants, as well as all other shopkeepers, they have to pay a heavy tax to the Dominion government, and they have to fix the price to you much higher on that account. When you consider that in every village, town, and city in Canada they are doing the same thing, it is not surprising that sixty or seventy millions of dollars are produced for the Dominion government in that way each year.

Stamp taxes. Perhaps there is a bank in your village. A man stands behind a wicket, and when a customer puts down a piece of paper, called a *cheque* (#30), the teller will count him out the amount of money mentioned on the cheque, but before he does so it must have, according to the amount, a three or six cent stamp attached. If you receive a telegram you will find a stamp attached to it, and if you happen to be travelling on a train at night and hire a berth to sleep in, you will find that a tax has to be paid on your sleeping car ticket. All of these stamps as well as hundreds of millions of postage-stamps are bought from the Dominion government too and the revenue they bring amounts to nearly seventy million dollars in a year.

Income tax. One of the largest sources of revenue is yet to be mentioned. It is the *income tax* to which every person must contribute who has a yearly income of more than $1000 or $2000. These amounts are set for the single and married individual respectively, and the proportion of the income payable as a tax is governed by the total income of the person concerned.

In order that the burden is evenly distributed the government has graded the tax according to the size of the income. It means that our men of wealth must pay a larger proportion of their income as taxes than the average individual.

Every person subject to the income tax is called upon in April of each year to send to the Inspector of Income Tax in his district a declaration of his income for the preceding twelve months.

In this declaration all sources of income must be stated. This may be a salary, interest on money in bank, on mortgages, or on government bonds.

CHAPTER VI

LIFE IN THE EQUATORIAL RAIN FOREST

(*Amazon*)

We approach the mighty river. We boarded an ocean liner at Montreal, sailed out into the Atlantic, and then turned south (#31).

#31

A ROUTE TO THE AMAZON

Day after day the weather grew warmer, and the sun rose higher at noon, until at last there it was directly overhead pouring out its heat like a blast-furnace; then we knew that we were on or near the equator. We turned west, and as the jade-green of the sea gradually took on a yellowish tinge we learned that we were in the mouth of the river Amazon; so wide is the gape of that monstrous mouth that as yet neither jaw could be seen. When at last the mouth of the river narrowed to a throat, (#31) the equatorial forest in all its majesty came into view.

The all-embracing forest. It looks like a limitless ocean of green tossed up in waves of foliage, some of the giant trees stretching their heads in domes above the common level. So desperate is the struggle for space that the tangled network of green bulges out over the water so far that there seems to be no land. As we sail along the river, the bank of greenery shows no break except where a cluster of huts hugs the edge of the forest, and even these villages seem crowded against the stream by the ever expanding leafiness. One can sail up this river 2500 miles in an ocean steamer, and the lofty, silent, confused riot of green life never weakens but rather grows more crowded and more entangled (#32). The tributaries of the Amazon are so many that they look like a thousand ribbons streaming from a main mast, and every ribbon is a mere slit cut through a sea of green. These branches, several of which are longer than the river St. Lawrence, divide and subdivide many times until they finally enter every country in South America except Chile. None of them escape the dense shade of this far-flung forest. Indeed, the terminal threads of the river branches are completely enclosed in an all embracing green archway.

It is the people who have chosen this world of rank vegetation for a home whose simple life we are to study in this chapter.

A tissue of trees. In order to do so it will be necessary not merely to describe the borders of this forest but also to describe its interior. The dense billowy dome of green overhead is supported by the unseen columns formed by the trunks of the trees, which reach up one hundred feet, straight as masts, before they give off their lowest branches. This wild unconquered race of vegetable giants is draped, festooned, matted, and ribboned with rope-like climbers (#35). Some of these are twisted like the strands of a cable, others wind snakelike around the trunks of trees and in time crush out the life as boa constrictors crush their prey. They may form gigantic loops and coils among the larger branches or be bent zigzag to form a series of steps.

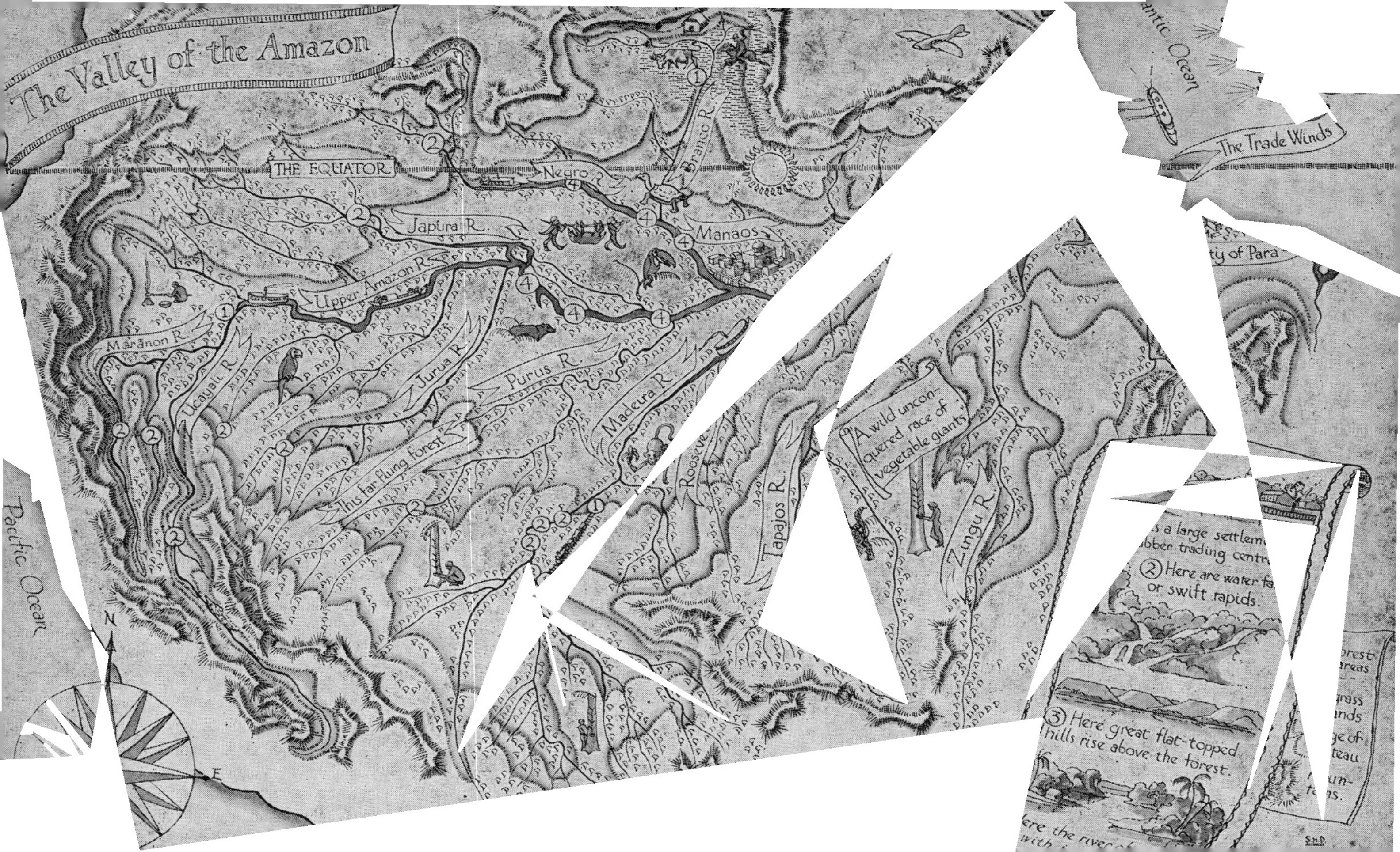

The Valley of the Amazon
THE EQUATOR
The Trade Winds
Negro R.
Branco R.
Manaos
Japura R.
Upper Amazon R.
Marañon R.
Ucayali R.
Jurua R.
Purus R.
Madeira R.
Tapajos R.
Zingu R.
This far-flung forest
A wild unconquered race of vegetable giants
② Here are water falls or swift rapids.
③ Here great flat-topped hills rise above the forest.
Pacific Ocean
N
E

These twiners, called *lianas*, form a warp and weft that weave the whole texture of the forest into a complete barrier. A man can no more penetrate it than he can a hedge of thorns. No nook or corner is unoccupied.

The struggle for life. This dense population of plants desperately struggles below to get rooting space, above to reach the light, and everywhere to get the air. Many give up the struggle to reach the warm, moist soil and are satisfied to nestle on the branches of the trees and let their roots, singly or in tufts, hang down like ropes dangling in the air (#35). On the other hand many a full-grown tree has lifted its roots up in the strife for air and light and now stands on a dozen curving stilts, spread out like the wires of a birdcage.

The floor of the forest. The floor beneath such a dense, leafy roof is as dim as evening. It is either too dark for green plants or just bright enough to be clothed with delicate tracery of clubmoss. If a man could enter this forest he would not be sure whether he was in an entangled mass of the living or the dead. It is both a cemetery and a birthplace. Growth is quick in the warm, wringing-wet soil, the hot, humid air, and the bright sky; but the crowding chokes out the weak, and there is a constant falling of the giants to make a close mesh-work of trunks, branches, and lianas, harder to pass through than the wire entanglements in front of an army. Fortunately this accumulation of the dead rapidly decays, thanks to heat and moisture ably assisted by white ants, which consume the trunk of a tree almost as rapidly as vultures consume the carcass of a dead deer.

Variety in form and colour. The forest is always green. There is no regular shedding of leaves, nor is there any spring opening of flowers. Indeed, flowers are seldom seen. Those of the lianas are delicate little stars hardly large enough to catch the eye, and those of the trees are opened to the sky above and are only visible to the birds and insects that swarm unseen at the roof of the forest. Occasionally a majestic spray of pink and yellow *orchids* of surprising shape catches the eye with its showy flower, or a solitary, crimson *passion-flower*, set like a star in its green mantle, breaks the monotony of the rich, shiny leaf-green. Variety is given by the intimate mixing of trees of different species and various habits. Sometimes on less than a square mile more than one hundred kinds are jumbled together like a patchwork quilt. The palms (of which there are over one hundred species) with their graceful feathery leaves hanging down in great plumes from the top of the stem (#35), are in strange contrast to the wide-leaved trees resembling those of a Canadian forest.

Since many steamers now sail up the main stream of the Amazon, we shall push back into one of the tributaries to learn how the Indians who are untouched by the white people build their houses, clothe their bodies, and obtain their food.

Lavish life. But first of all we must find why the land is choked with such rank plants, why all plants spring forth like mushrooms, why the river and its branches are crowded aquaria, why even parts of the forests that appear deserted are, when closely examined, alive with numberless insects, birds, and mammals, and why the evening air is fanned by the ugly wings of countless millions of disgusting bats, why mosquitoes pour down on the poor people like drops in a fearful storm, and why ants stream through the forest in an army (#37). Not only do the loathsome insects swarm, but glorious butterflies (#35), white, yellow, gorgeous red, flashing green, purple black, azure blue, rose, bordered with metallic lines and spots of silver and gold that flash like mirrors, sail through the dim forests in thousands. (No less than seven hundred kinds of butterflies were found near one town,—more than three times the number in all Canada). Why do these animals grow so large? One butterfly is seven inches across and can be seen one quarter of a mile away. A spider seven inches long (#37) captures little birds instead of flies in its web; bats have wing-spreads as wide as our hens; beetles are big as our turtles, and turtles are as heavy as sheep (#38); caterpillars with coral heads and yellow stripes are as long as lead pencils and twice as thick; the fish are the largest found in fresh water, and the dreaded *anaconda* is a water snake thirty feet long and three feet in circumference (#38).

The path of the equatorial sun. The boys and girls on the Amazon river watch the sun as we do, but it moves across the sky very differently. In our summer it rises toward the north-east (#33), moves upward until noon, but never gets directly overhead, and then sinks below the horizon toward the north-west; it may be shining for sixteen hours. In our winter it rises toward the south-east, moves upward but not nearly so high as in summer (#33 and #34), and after about nine hours, sinks in the south-west. This makes the difference between summer and winter. Every day in the year the Amazonian boys and girls see the sun rise nearly in the east at 6 o'clock a.m. (#34) and set nearly in the west at 6 o'clock p.m. Every day it reaches almost over head at noon, sometimes being slightly to the north, and sometimes being slightly to the south. As the blazing sun at noon shoots rays almost straight down, instead of at an angle

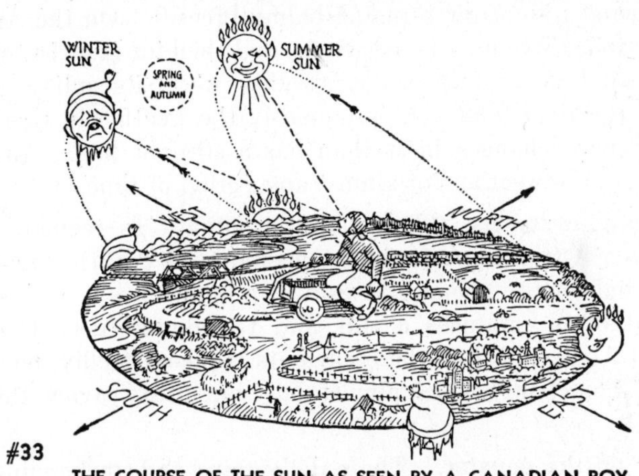

#33
THE COURSE OF THE SUN AS SEEN BY A CANADIAN BOY

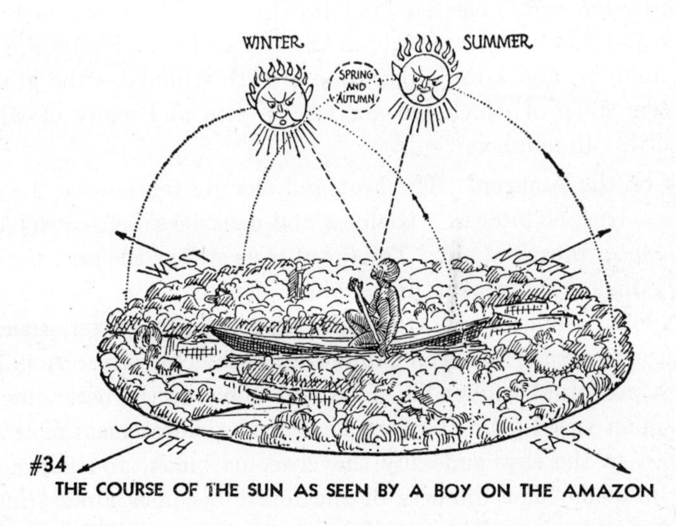

#34
THE COURSE OF THE SUN AS SEEN BY A BOY ON THE AMAZON

as with us, the heat of every day in the year is intense. While our heat goes round the circle of the seasons, that of the children on the Amazon goes on in a straight, monotonous line forever. Our sun, going down slantingly below the horizon, gives us long twilight both morning and evening, but for the boy on the Amazon it goes straight down and almost at once it is dark night.

Steady winds and steady calms. In Canada the winds blow from all directions, the north-west wind to-day bringing cold weather and the south wind to-morrow bringing balmy breezes. On the Amazon for long periods there may be no winds, and then for months together the trade-wind never falters but steadily wafts its sultry, humid breezes up the river (#32). Consequently the weather in this forest has hardly more changes in it than has a straight line. Anybody could forecast the weather for almost any period of time.

Daily thunder-storms. The rainfall is almost as regular as the temperature. Often in Canada on a summer's day the great heat causes the light air to rise as it does over a bonfire; as it ascends it is chilled and often produces clouds and thunder-storms. Over the hot, humid Amazon valley these currents rise continually, cover the sky with leaden clouds, which almost daily burst into savage thunder-storms.

Fast life on the equator. These daily rains and unchanging heat excite the most rapid vegetable growth found anywhere. Plants spring up over night; cleared land breaks out in a green rash the first day and turns into a jungle in a week or two. Plants are the food of animals, and a rank vegetable growth stimulates the growth of animals, some of which are gigantic in size and many of which are countless in numbers.

A day on the Amazon. The heat and electric tension, as the day progresses, become intense. Laziness and uneasiness seize every one. There is not a breath of air. The forest is as still as the grave. The humidity makes movement unbearable. So loaded is the air with water vapour in some districts that salt becomes liquid; sugar, sealed in a tin, turns to syrup; and even gunpowder becomes pasty. A person lives in a perpetual vapour bath. About noon, as the ascending air currents gain speed from the fiery heat, a motionless mass of leaden clouds covers the sky; suddenly they become black, and the sun is entirely hidden. The dimness of the forest becomes almost night. A startling rush of a mighty wind sways the treetops. Then a vivid flash of forked lightning cracks the black dome of the sky; there is a crash of thunder, and down stream the deluging sheets of rain. The storm ceases in an hour or two leaving a beautiful blue sky. Life revives, and almost immediately every vocal animal celebrates with a song. By evening it is deafening. The whirring of cicadas, the shrill scratching of crickets and grasshoppers, the plaintive hooting of tree-frogs, and the screaming of parrots and macaws join the hoarse croaking and drumming of the frogs in the marshes. Add to these the blood curdling cry of the howling monkeys (#35), the hateful buzz of clouds of blood-sucking mosquitoes, and the loathsome hum of armies of bats, and the blending sounds become a monstrous noise, so deafening that shouting is necessary for conversation.

Refreshing night. As quick night descends, these jarring noises fade out; the air becomes delightfully cool and filled with the sweet smell of flower and fruit. This continues till, again at dawn, the sun begins to drive the coolness before it, and animal life once more wakes up with a scream. Such is the succession of to-day, to-morrow, and every day.

Clothing and ornament. In Arctic regions the struggle with cold and want weighs the Eskimo to the ground; on the other hand, the Amazonian tribes are benumbed with nature's abundance. Their wants are few. They wear little or no clothing (#35,#36,#37). In such intense heat the flimsiest clothing is unbearably hot, and with such wetting humidity and daily rainfall, from which they have little protection, clothing is sure to be always wet and a fruitful breeder of colds and consumption. Observers state that natives that look dull and discontented in fine dry weather immediately become cheerful and alert when the rain is pattering down on their naked backs. Even naked man, and especially woman, must have ornaments. Since they cannot satisfy this artistic taste with gay and handsome clothes, they tattoo the body with elaborately coloured patterns (#36), which process causes intense pain and takes years to complete. They also slit their ears, noses, cheeks, chins, shoulders, etc. and insert sticks, which may be ornamental to their admirers but look hideous to civilized man.

Native houses. Their homes (#35) can be put together with little effort. Upright stakes driven in the ground have crosspieces bound together by wooden cords cut from twiners ready at hand. Since two banana leaves are large enough to make a picnic tablecloth for twelve, the labour of thatching the roof with several thicknesses of such gigantic shingles is trifling. Usually the house is quite open at the sides, though in some districts there are mud walls, but these make it depressingly hot. There is no floor. Such a house can be put up in quick order and at no cost.

A house without furniture. Their houses are almost as naked of furniture as their bodies are of clothes. Indeed, houses are chiefly

#35

A HOME ON THE AMAZON

Make a list of the animals in this picture. Find the sleeping hammocks, a woman cooking. A fence enclosing a turtle pond, and a man making a paddle.

to sleep in, since it seems an instinct in mankind, no matter how low, to sleep beneath the shelter of a roof. But in such humid air and rainy weather, mattresses, pillows, blankets, and quilts would be wet all the time, would mildew in a week, and rot away in a month, even if they had sufficient cloth from which to make them, which they have not. From one end of the Amazon to the other and along all its tributaries the chief pieces of furniture are *hammocks*, which are woven loosely from coarse wooden fibres and which are hung between the upright posts of the hut (#35). In these the people sit or lounge by day and sleep at night. Often in small huts, to save space, two or three of these are swung one above the other like bunks in a lumber shanty; and as there are no dishes, cutlery, clothes or other finery, there could be no use for dressers, bureaus, and cupboards. Perhaps there are a few shallow pans made of baked clay for cooking (#35), and scooped-out, pumpkin-like *calabashes* for drinking vessels. If a house and furniture were burned or blown away, a complete new outfit could be put together in a few days without going one hundred feet from the clearing.

Pets. Every house has its pets. Little green *parakeets* run like green mice in and out of the bushes surrounding the house. Gaily coloured and sober-looking parrots, magnificent macaws, and gloriously coloured toucans with cylindrical beaks almost as big as themselves (#32, #35), screech from the rafters or talk back to the people. Monkeys, some as small as squirrels, scamper on the roof and keep the rafters free from vermin (#35). All of these have been trapped in the forest and trained by the women, who have wonderful skill in winning the wildness from these beautiful creatures and training them at many skilful tricks.

Fighting the forest. The most trying of all their work is the fight with the forest. To make a clearing from such a tangled network of twiners and trees would be heart-breaking in a favourable climate, and to do it in these muggy surroundings is almost too much for these languid idlers. The yards and gardens are consequently small and ragged. (#35) Just enough is cleared to grow a small patch of *cassava*, a few *banana* plants, and perhaps a little *cacao*. All of these, when once planted, grow year after year with little attention.

But in this land of mushroom growth, which presses forward for twelve months in the year, it is a constant struggle to keep the hut and clearing from being choked and buried by the jungle growth; usually the forest almost wins, as the garden is nearly smothered with shrubs and weeds, and the wild undergrowth would be looking in at the back door, if there was one. If the native left his hut for a year he would have hard work to find it when he returned. The twiners and small trees would flow into the house, press up the roof, and bury the whole clearing and hut in a dense grave of green.

The food of the forest. The native's most pressing necessity is food. Some he can gather from the forest. Ripe pods, the size of cannon balls, fall down from one of the tallest and handsomest trees, and when open, discharge twenty or thirty *Brazil nuts* (#36), which are very nutritious. Another, the *chicken nut*, much more delicious than the Brazil nut, is also contained in a capsule on a tree. But, unfortunately, the lid of the capsule drops off while the nut is still attached, and the dozen nuts fall separately to the ground, so that they are gathered with difficulty. There are a good many wild fruits quite unknown even by name to Canadians. The most luscious is called the *atta*. Each fruit may weigh ten pounds or more. The scaly rind encloses a rich custardy pulp, frosted with sugar, and its taste is a happy mixture of sweet and sour. The *papaw*, of immense size, is another choice fruit, which grows high on a palm tree. The *cow-tree* (#35), one of the handsomest and tallest in the forest, when tapped, gives out a creamy liquid, which may be used for coffee or custard, although it tastes a little rank.

Food from the garden. However, these fruits are not always near at hand nor easy to gather, and the Amazonian Indian depends chiefly for vegetable food on what he grows in his rough garden. *Plantains* are the chief fruit. These are gigantic green bananas, which, when cooked, far surpass their smaller relation in deliciousness. Just as bread is the staff of life to Canadians, so *farinha* is the chief food of all classes in the Amazon valley (#36). The roots, about the size of parsnips, are dug from the base of the trunk of a small tree, the *manioc* or *cassava*. They are rubbed on a broad, concave, wooden grater, (#36), studded with stones or fishes' teeth to make it rough, and in no time are turned into a mass of pulp. As this pulp contains the most violent poison, it has to be thoroughly washed several times, and the poisonous liquid squeezed from it in a long bag. It is then dried. It looks like a pale, coarse, granular sawdust and is not much better in taste. Nevertheless, as in dry granules or cakes it keeps well and is nutritious, it often prevents death from starvation for months at a time.

Flesh and fish. The streams are swarming aquaria of fishes, turtles, and alligators. The fish are speared (#35), shot with poisoned arrows (#32), or stupefied by the ground pulp of a poisonous plant, which is sprinkled on the water. The flesh of the red-fish, fifteen feet long, when salted and dried, takes the place of beef and pork; and turtle meat, of chicken and duck. Usually every house has a turtle pond (#35). A single turtle of the largest size (#32, #38) is a load for a man. It is tender, tasty, and wholesome but very cloying.

#36

PREPARING FARINHA

The flesh of the breast is minced with farinha or cut into steaks, the stomach is made into sausages, and the entrails are chopped and made into delicious soup. Small turtles are roasted whole in their shells. The natives also hunt *toucans* (#35), monkeys, and other animals of the forest with blow-guns (#37). These are long wooden tubes which contain a small arrow with a point as sharp as a needle, often tipped with poison. This weapon requires a keen eye, a steady hand, and a powerful pair of lungs; but will kill at a distance of sixty yards. The breast of the toucan is a rare treat and dried monkey meat a luxury. But it must not be thought that the abundant fruits of the forest fall into their laps. It is quite the reverse. Both fruits and game have to be searched for carefully and are often not found. As the shiftless and improvident natives never look far ahead, and as food soon spoils in such a climate, it is safe to say that most are undernourished and many near starvation in this land of abundance.

#37

ANIMALS OF THE FOREST

Find a spider attacking a bird, a boa constrictor seizing a large bird, a jaguar, anteater, monkeys, a procession of ants.

The animals of the forest. The wild animals of the forest, with the exception of blood-sucking flies, foraging ants (#37), and biting mosquitoes, give the people less worry than the raids of the unfriendly Indian tribes. The *boa constrictor* (#37), a large snake living in the trees, though often shown in storybooks crushing the life out of men and beasts, in real life is satisfied with birds the size of hens and mammals the size of rabbits. The huge *anaconda* (#38), a water snake, also swallows birds and is not above attacking the poultry roost. Smaller snakes of the rattlesnake breed can deal death blows with their fangs but are not common. Indeed, one may travel a thousand miles up the Amazon and never see a snake. *Alligators* (#32, #38), up to twenty feet long, swarm in every stream; but the natives, men and women, and children, do not hesitate to bathe with these dangerous animals showing their horrid snouts and eyes above the water in the offing. As long as one is on his guard, he is safe,

#38

LIFE ON THE AMAZON

Find an anaconda attacking a bird, men searching for turtle eggs; an alligator, a giant lily, and two kinds of boats.

but occasionally one of these scaly creatures slinks along beneath the water and pulls down a boy or even a man. The *jaguar*, a spotted tiger (#37, #32), is the fiercest animal of the forest. Though often heard by the poor Indians as they sleep in their open huts in the forest, it never attacks and is seldom seen. It seems as much afraid of man as he is of it. The wild animals captured for food or tamed for pets, are friends rather than enemies. The only animal that strikes fear into the hearts of these poor people is one that does not exist. As they sit in their huts in the quietness of the hot day or lie in their hammocks at night, strange and mysterious noises are heard in the forest. It may be the falling of a tree, the rubbing of one branch against another, or the sighing of the wind; all these unknown sounds they attribute to some strange and dangerous creature, whom they always expect to spring upon them, but which never even appears.

Boats on the Amazon. There is only one means of transport and that is by boat and canoe (#32, #35, #38). There are no extensive trails through the forests, and to cut a path would require superhuman effort, in a land where hard work is hateful; even if a road were cut it would close up in this land of luxuriant growth almost as quickly as the wake behind a ship. Many boats are merely scooped-out logs, others are made of four or five boards of proper shape fastened together so badly that water pours through the seams (#35). In these egg-shells, loaded to the water's edge, they paddle with great skill even when every wave would seem to sweep over the top. In slightly larger boats, with a thatched roof over the rear half (#32, #38), they travel hundreds of miles, living on board. But unromantic motor-boats are beginning to penetrate into these twilight parts of the earth, and on the deeper branches of the Amazon, larger steamers now ply regularly (#38).

Trade. These people have hardly any trade. They sell little and buy little. They may gather Brazil nuts in bags, grind and dry more farinha than they require, extract oil from turtles' eggs (#38), gather sarsaparilla roots and pack them in bundles, or tap the rubber plant (#32, #36), and clot the milky juice into a ball over the smoke of a fire, made from palm nuts. Then they carry these commodities down stream, trade unfavourably with rascally traders on the main river, and bring back salt, a little cloth, some ornaments, knives, and perhaps a few drugs. Such is the routine of their little life.

CHAPTER VII

LIFE IN THE ARABIAN DESERT

A naked land. What a difference between the Indian in the Amazon forest and the wanderer through the Arabian desert! Let us take a peep at an encampment (#42). Low black tents stand on a bare patch of red sand, and instead of a circle of lush tree growth there is, as far as the eye can see, a parched land stripped

#39

THE TRIP TO THE ARABIAN DESERT

almost to nakedness of plants and animals. A brassy sun beats down spitefully on sand too hot to touch. Yet a family manages to live around every tent. It is true they look lean and dry, their garments are ragged and dirty, and the tents are as empty as the Amazonian huts; the struggle to live is always grim, and the odds against success are always great. Nevertheless, out of the poverty of such deserts have gone forth conquerors, who have subdued the world, and founders of religions, who though dead thousands of years ago, yet to-day have their devout followers by hundreds of millions of the most enlightened people in the world. Abraham, the father of the Hebrew nation, was a plain man living in tents; Mohammed lived his whole life in the desert; and Jesus, the divine man, never travelled beyond Palestine, which is on the fringe of the Arabian desert. The battle against heat, cold, sand-storms, thirst, hunger, nakedness, and crafty enemies is well worth telling.

A trip to the desert. To find this desert race you must cross Canada to the Atlantic (#39), traverse this great ocean eastward, enter the *Mediterranean* through the *Strait of Gibraltar*, and then sail more than two thousand miles across that vast sea. Before you reach its eastern end, the air becomes chokingly hot, uncomfortable, and smells of the desert; when the boat is passing through the *Suez Canal* into the *Red Sea*, the brown sand of the *Egyptian Desert* shimmers on the right, and the immense mountain blackness of *Arabia* scowls angrily on the left. When you climb over this rampart, anywhere along the western border of Arabia, you are in the midst of the great, silent, lonely, inhuman desert.

A pitiless variety. The surface of the desert is seldom level and is not always sand. Often black lava ridges, or sandstone crags of the most fantastic shapes, stick their lonesome heads above the dreary monotony of the sand (#43, #44). Nowhere in nature is there a more cruel or more horrid desolation than among these bare crags in the pitiless sands. These harsh peaks are sometimes as close together as loaves in a baker's oven. It is from these mountainous shelters that robbers spring forth, and it is among their windings that wild goats feed in safety.

Sand and gravel. But sand covers more than one-third of Arabia. It may be yellow, brown, or red, and finer than sugar or coarser than rice. In many places the winds have blown away the sand so completely that the surface is swept naked and nothing but a disordered sea of harsh gravel and rasping stones is left to play havoc, not only with the bare feet of man, but even with the cow-hide soles of camels.

Sand dunes. The surface of the sand is seldom level. The winds working through the centuries have thrown it into ridges called *dunes.* Very impressive, indeed, are such regions,—a vast ocean of billowing sands, here tilted into sudden frowning heights, and there falling into gentle valleys, merciless to man and beast. The surface of these great billows, exquisitely pure in colour, is carpeted with delicate wavelets. Though in the distance it is a sheet of uniform red or brown, at your feet it sparkles with glints of green and gold.

A living death. No journey is more heart-breaking than over these awful dunes (#42). As the traveller crosses this immense troubled ocean of loose sand he sinks to the ankles, or deeper. In the hollows the traveller is imprisoned in a stifling sand pit, hemmed in by burning walls. When he drags himself up the slope he overlooks what seems a vast sea of fire. There is no rest for the eye or limb, no shelter from torrents of light and heat poured from above, and reflected back from both sides and from below. The effort of toiling through the loose and scorching soil or of lurching on the jolting backs of drooping half-stupified beasts, with the sun blazing down till clothes, baggage, and beast smell of burning, is almost unbearable. Where the face of the desert is level, it is a dreary land of death, so uninteresting that even the face of an enemy is almost a relief amid such utter solitude.

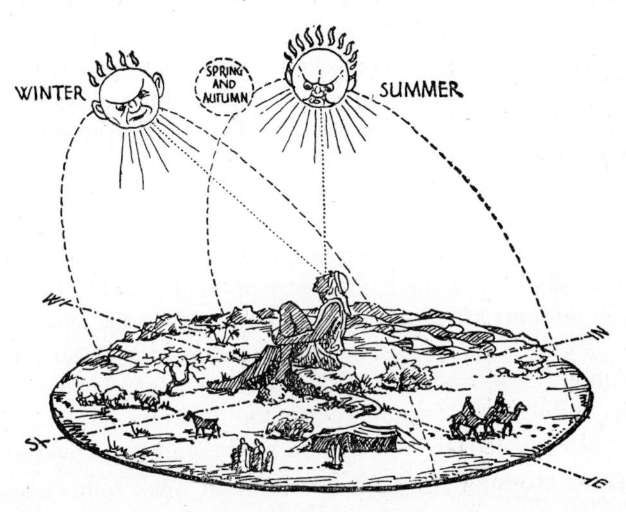

#40
AN ARAB FOLLOWING THE COURSE OF THE SUN

A frowning winter and a smiling spring. The *Tropic of Cancer* cuts the Arabian desert in two. The sun shoots up nearly vertically from the eastern horizon (#40), almost reaches the zenith at noon, and drops nearly straight down to the west. The fiery beams at all seasons meet little resistance from the clear, clean, desert air; even winter days are hot. At night every object shoots back its heat through the unresisting clearness of the air, and in winter it becomes uncomfortably cold; the people shiver through the night in their black tents, which let the heat leak out from the air even more than the scanty bedclothes do from their dried-up bodies. Even the water in their water-skins freezes. Most of the scant rain falls during the cool season. This rain is to the desert Arab as gold to the prospector. By spring it has spotted the desert with lush, green blossoming herbage, which is the life blood of the Arab's cattle (sheep, goats, and camels). Delightful also to the eye is this poor garden of blossom in the desert.

Awful days and serene nights. But the terrifying summer falls on the accursed land like a pall, almost before the spring has well begun. By May the summer has entered with breathless heat and the drooping herbs wither, the wilderness changes colour, and the fleeting spring is past. Each day as the sun's vast flaming eye rises, the suffocating hours crush all activity. In an hour the tent-poles are hot to the touch. By noon this Arabian sky is burning brass above the head and the sand is as glowing coals beneath the feet; the hot sand-blick is in the eye. There is no shade. The dark woven tent gives no shelter, as it allows this rain of sun's rays to leak freely through the roof. One lies helpless in the tent and seems to breathe flames until the lingering day draws down near to the setting of the sun, and the shadows from the western hills at last bring comfortable shade to cover the tents. The day's exhausting heat is soon forgotten. The clear, clean sky allows the heat to stream back quickly into the thin air; in half an hour after sunset it is blue night, the clear starlight with its bright girdle of the milky way lights up the sweet pure desert air; the moon rises suddenly like a mighty blazing beacon, a new breathing coolness comes, and the serene Arabian night brings tranquillity to frayed nerves and rest to weary and worn bodies.

Refreshing storms. Day follows day and week follows week with little change. Very rarely a dark, frowning cloud gives hope of a refreshing rain, and perhaps once during the summer a violent downpour is triumphantly ushered in by cracking thunder and flaming chains of lightning. Then is seen a rare and glorious sight, the dry billowy desert is filled with silvery patches of sweet rain-water. Every pool is gulped by man and beast; water-skins are emptied of the foul well-water and refilled from the hollows of the sun-cleaned sands.

The terrible simooms. Occasionally the dreaded *simoom* piles agony on pain (#41). It is noon with an unclouded sky over a scorched desert. A rumbling like a thousand horsemen is heard in the south; abrupt and burning gusts from this direction bite us in the back. The horizon darkens to deep violet and draws in close like a curtain. We are in a whirling cloud of dust. We pull our clothes over our heads, rush for the tent, and fall flat until the storm is past. Every camel lies without, as though dead, his long neck stretched away from the blast. A still heat, like that of a red-hot iron slowly passing over, is felt. The wind moans and roars. The fine sand passes through the tent as though it were a sieve. It irritates our ears, nose, and especially our eyes. It goes through our garments like air and rasps against the skin. It covers every object in the tent, enters every box and mixes with the food. It covers the tent two inches thick. The camels are almost buried but do not seem to be disturbed. In half an hour it is gone almost as suddenly as it

#41

CAUGHT IN A SIMOOM

appeared. If the Arabs are caught on the march, it means death to the weak, and serious injury to the strong, among both men and beasts; and the injured are left behind upon the loathsome trail. Fortunately in Arabia the simoom is a rare visitor.

The black tents of the Arab. Let us look more closely at the home life of these poor, enduring people—if a tent can be called a home. This black dwelling is made of a cloth (#43) every square inch of which is spun and woven by the women from the wool sheared from the sheep and the hair clipped from the goats and camels. The goat's hair gives it the dark colour. The poor despised women spin the yarn even as they trudge along beside the camel or ride on its

THE ARABIAN DESERT.
#42
PORT
CITY
OASIS
FERTILE LAND
MOUNTAIN
SEMI-DESERT
WADI
RIVER
RAILWAY
CARAVAN ROUTE
Mediterranean Sea
CYPRUS
Antioch
Nile Delta
Beirut
Port Said
Haifa
Jaffa
Damascus
Jerusalem
Euphrates R.
Tigris R.
Baghdad
Kerbela
Babylon
Jauf
Sandstone crags of the most fantastic shape
A disordered sea of harsh gravel
Red Sea
Hail
Bosra
the great, silent, lonely, inhuman desert."
Medina
Bereida
Jidda
TROPIC OF CANCER
Mecca
Riyadh
El Katif
El Hofhuf
Persian Gulf
MID-SUMMER SUN
A vast ocean of billowing sands!
the loose and scorching soil
Red Sea
Day after day, over wide areas of nakedness!
Aden

back, and they weave it during their resting periods in the camp (#43). The fabric is so well woven that a tent lasts for a generation. The tent is composed of a roof raised on poles and fastened by ropes made of twisted fibres of palm leaves, and a skirt around three of the four sides. A partition separates the rectangular space into two unequal parts; a smaller for the women and children, and a larger one for the men. The men's quarters being wide open in front, inviting all who wish to enter. Can such a space be called a home, for it is almost empty? There are no chairs, no tables; there is not even a bench

#43

AN ARABIAN ENCAMPMENT

How is the tent-cloth woven? Who milks the sheep? Describe the churning. What are the men doing? Describe the saddle of the camel. Describe how the tent is erected? How many rooms are in it? What plants are seen on the desert? Describe its surface. Describe the clothing of the Arab men and women.

to sit on. The camel saddles, tossed about on the sand, serve to support the elbows as the men sprawl on the ground, or perhaps the less poverty-stricken may have a coarse mat. In the cool evenings a fire may be built in the men's compartment, where the Arab's one luxury, coffee, is sipped (#43).

Uncouth furniture. What a drab and dreary prison is the women's room, the *harem*, shut in by four black walls, through whose gloom only a faint twilight filters from the meshes of the roof. It contains the ugly sacks in which the tent is packed; some barley and rice in plain bags; a few lumps of rock salt; some wool and yarn; a little calico bought in an oasis; a few roots, the extract from which is used to tan goat skins; and a box containing the women's cheap finery and some medicines. This box holds a poor comb and dull mirror, dear to the heart of even the poor grimy Arab women; nose rings, sometimes of silver; a few other ornaments; and perhaps a "Sunday" dress. A few of the husband's small things are also put here as he has no pockets. How precious is this dear box to the little vanity of these simple women-folk! Unsightly masses of stale, clotted dates, in thick camel skins, and skin sacks filled with the tainted well-water, make up the rest of the uncouth furniture of this woman's living-room. Perhaps a big brazen pot, never washed since the day it was made, may deck the apartment, and the metal coffee pot and cups are the signs of quality of every family.

Baking bread. Their food is as uncertain as the water in the desert wells. The staple is unleavened bread. First a fire is kindled without matches. A bar of steel is struck against a piece of flint, picked up in the desert. The sparks start a dry rag smouldering, which is cherished by dry camel dung and perhaps straw, then little sticks coax it to break into a flame. A few handfuls of flour moistened with dirty water are kneaded to a thin and slimy dough by the unwashed fists of one of the men, a few mouldy dates for a toothsome tit-bit may be added to this appetizing mass, and it is flattened out to a cake, an inch thick and six inches in diameter. This is allowed to sag on the glowing coals; after a suitable time — a few minutes — it is turned over; then it is put in the hot sand beneath the fire, and the glowing coals heaped on top. After a fit term of burial the loaf is resurrected, a little of the sand is brushed off though most of it is tightly cemented to the cindery crusts. At last this loaf, half-kneaded, half-cooked, well-sanded, and burnt all round, is broken among the hungry family and washed down, while piping hot, with a few draughts of dingy, luke-warm water. When the supply of flour is gone, as it often is, they sometimes replace it by minute reddish seeds, extracted from a round pod the size of a pea. Large quantities of this free gift of nature are gathered in July and stored for use. It is ground to flour, like barley, and cooked as cakes.

Salads. In such a parched land any green thing is eagerly seized as a vegetable; a small leek is much relished, and little tubers, which grow underground, taste like potatoes when baked.

The staff of life. But the source of health and strength to these wandering tribes is the milk of the camels, ewes, and goats. There is no more pleasant sound than that of the milk spouting from the udder into the bowl at milking time. The children gather round the vessel, brought in foaming, and dip out the sweet froth with their fingers. The camel is milked once a day, and as it gives milk eleven months during the year, it is priceless to the half-starved Arabs of this languishing country. Since the camel's milk cannot easily be churned into butter, it is drunk whole, or evaporated to a powder and carried long distances to tide the Arabs over periods when other sources of food fail. The milk of ewes and goats, which is only obtained for a month in spring when the winter rains have stimulated the growth of the sweet, juicy desert herbs, is churned to butter (#43), and only the buttermilk is used as food. The churn is a skin bag, which the women rock patiently on their knees, or from a tripod of poles, until the butter appears. But this butter is too valuable to be eaten. It is boiled in the brass kettle, and the clarified product is packed in skins to be traded in the oasis for barley, calico, coffee, tobacco, and dates.

A cloying fruit. Dates are not so important to the wanderers over the desert as to the dwellers in the oasis. Dried dates are a poor food, heating, inwardly irritating under a sultry sky, and so soon sickening that the people search the desert pastures for the grateful sour sorrel to counteract its cloying sweetness.

Luxurious foods. It is seldom that their famished stomachs are sated with savoury meat. The goats, sheep, and camels are their chief wealth and must be exchanged with traders for the silver that buys their clothes, guns, coffee, dates, tobacco, and the other few necessities, which they cannot make themselves. However, when a camel sickens, no vultures in the sky need watch, for the famished Arabs eye it wistfully. Its throat is cut, and the skin flayed off; the flesh is hacked to pieces, divided equally among the group, and each portion is soon in part stewing in the pot, and in part hung in the sun in thin strips to dry. Occasionally a giant lizard comes out of its burrow (#42), is pounced on, killed, and broiled whole on glowing coals. It is the event of a lifetime when a nest of ostrich eggs (#43) is found buried in the sand. Each egg is equal to twenty or thirty hens' eggs. An omelet of such an egg, cooked with butter and flour, may make the mouth water for many a day.

The cup that makes merry. The outstanding luxury of these humble desert tribes is the drinking of coffee, which has long been a product of Arabia and is brought up from the south-west coast by boat and camel. The head of the tribe is as serious in making coffee as in judging a dispute between two of his people (#43). He first sets the water on the fire to warm, then roasts the green coffee beans in a small ladle to a nice rose brown, grinds them in a hollowed stone, and drops them into the boiling water, to which fragrant spices, such as saffron, are added. The news of the preparation of such a royal drink soon spreads through the camp, and his tent becomes a little Mecca to which all the men are welcome among these hospitable people. The little metal cups (four sips each), are filled, and the fragrant and stimulating coffee is relished above every other beverage by these Mohammedan people, who drink no alcoholic liquors. In the poverty of the summer, when the niggard supplies of coffee are all gone, there is not much merry making in the tents at night.

Unchanging habits. The habits of these Arabs have been as unchanged for thousands of years as their eternal deserts. If Abraham, the father of Isaac, could come back to the Arabian desert he would find few surprises.

Women servants to the men. People, cradled in poverty, reared in semi-starvation and uncertainty, with long hungry marches on the backs of jerking camels, are tough but not often robust; and it is seldom that one sees even a young woman with the rose blood in her cheeks. The men look thin, dry, and muscular. The women seem old and are wrinkled even while they are still young. It is no wonder. They are slaves to their masters. The women take down and pack the heavy tents in the morning and erect them at night; they harness and saddle the camels and hold them ready to mount, while their masters sit around the morning fire till everything is ready for the march. The women gather the wood for the fire in the evening, and the men sit around its cheerful and fragrant flame; men would be disgraced, even in the sight of the women, to be seen doing any of these mean services. The men herd and milk the flocks, defend them against robbers, and attack their enemies. In some respects they are highly honourable. Clothing, found hung on a bush, or a skin of dates hid under the roots of a desert shrub, they would not touch; it is against the morals of the desert. If some of the men reach a meagre well before others of their group, though almost dead with thirst, they wait till all arrive in order to share the water equally.

The diary of the desert. Of course none receive any education but what they can read from the sands of the desert. But in its way these sands are a public diary, a newspaper in which are recorded the passing events. The footprints on the sand are the alphabet, and in reading this strange script the desert Arab is a skilful scout. He can tell every camel in the caravan by its tracks and knows whether it is sick or well, whether light or heavily loaded, and whether it has passed three hours or three days before. He is able to say that three days ago a flock of gazelles rushed across the waste, that shortly afterward they were followed by a wolf, and that less than three hours ago a giant lizard crept across the trail and entered its burrow.

The despised sex. The wives keep much in retirement, do not mingle with the men, are cuffed and beaten, and often cast out by their husbands. Even at twenty years of age, golden youth may be faded to autumn in their childish faces, and merry laughter have become wooden. It has been truly said that in the desert, woman is the only female creature not more valued than the male. A girl baby is considered an unprofitable mouth added to the hungry eaters of a slender supply.

Costumes. The men wear very dirty shirts reaching to the ankles, black cotton handkerchiefs over their heads, fastened on by a twist of camel's hair. They may wear a striped cloak, and a leather girdle much the worse for wear. The women wear long, loose, blue garments reaching nearly to the ground, often with embroidered sleeves. They have bracelets and nose-rings.

Where there is not enough water to quench the thirst, there is often little, or none, for cleansing the body or laundering the clothes. Both are often dirty, though the Arab cleanses his body at times by rubbing it with sand.

Beds and bed-clothes. They have no beds and little bed clothing, though the nights, even of hot summer, are cold. They lie on a mantle or tent-cloth spread on the ground, and take off their single garment and spread it over their tender bodies, the better to keep out the cold.

The play of children. The poor little children are born into a world of hunger and sorrow. The baby girl is unwanted; the baby boy is the joy of the mother and the pride of the father. Their only education is what is learned under the tent roof and in the desert. The Arab wife is a tender, but often too mild teacher. The children have no toys but stones, and they play few games. While their fathers herd the camels, goats, and sheep, the children pretend the desert stones to be alive and order them as their parents do the cattle. In the spring, to their great delight, they are allowed to tend the gentle lambs and frisky kids.

Health in the desert. Though these thirsty people are necessarily careless about cleanliness and let filth lie around, the purifying sun, the clean, clear air, and the dry sand choke disease, and they have little sickness. The constant glare of the burning sun and the brazen sky, with the irritation of fine dust from the desert sand, cause widespread eye disease, and occasionally loathsome smallpox or the still more dreaded cholera may almost wipe out a tribe.

Racing with starvation. The wandering Arab's wealth is his cattle stock, which consists of camels, sheep, and goats. Their food and drink guide his every action. These herders must live together in bands, under a leader, or *sheik*, for mutual protection against roving troops of cattle thieves. Fifty to one hundred families, with several hundred camels, form an encampment. Since such herds quickly crop clean the lean desert herbage and drain the miserable wells in a few days, they must be ever on the march to seek fresh pastures and full wells. That is why they live in tents. Such wandering tribes are called *nomads*, and these Arab nomads are named *bedouins*. Often during the dry, hateful summer they have to make long forced marches to reach sufficient pasture. They are never more than a few days from starvation for man and beast. During such a march they are found day after day urging their camels to their utmost speed, fifteen hours together under a nearly vertical sun; then after two or three hours for rest and sleep, the hunger urge spurs them up again, while the night is still dark, to push the jaded beasts still forward. Day after day may wear like a delirious dream as they move over wide areas of nakedness, scanning the bare horizon for a sign of dried herbage.

On the march. When the women have packed the tent and saddled the camels, the camels kneel to receive their loads. The mothers, children, lambs, and kids, the aged, sick, and bed-rid folk ride on the camels (#44), and even after many hours of jostling on these ambling beasts, they show no great signs of weariness. The herdsmen, with sheep, goats, and milking camels, drive forward first, then the women and children on their camels, next the armed men and their sheik often on a desert steed. Frequently the men walk all day through the heavy sands and even gather herbage for their camels as they go. The panorama of a limitless desert with the nomads and their stately camels moving noiselessly along in the blazing sunlight is a noble sight (#44), which becomes brilliant when lit by the sunset turning the sky in rapid succession to opal, rose-pink, orange, and grey.

#44

A CARAVAN ON THE MARCH

Find a dog chasing a hare, a flock of ostriches, an eagle, a man hunting gazelles, a lizard. Which is the sheik? Find a long-horned antelope on a rock.

Uncertain wells. The wells are usually along the bottom of long ravines, called *wadis* (#42), which may extend for hundreds of miles, and the wells are often mere hollows where water oozes up slowly through the sand. Sometimes they are deeper and walled with stone. The water is always warm, usually fouled by the camels coming too close, and frequently the wells are dried up altogether, as they are fed by the uncertain winter rains.

Animals by the way. On these journeys, in spite of the barrenness, the nomads see many wild animals. Numerous small *snakes* and dried up *lizards* (#44) bask in the desert sun and start at every step.

Some snakes are so poisonous that a bite will kill a camel in an hour. A bristling *porcupine* is occasionally knocked with a club near its hole in rocky deserts and, broiled on the coals, quills and all, is considered a great delicacy. Little *jumping mice* swarm almost everywhere. The howls of *wolves* may be heard in the tents at night, and the herdsmen have to be alert that their sheep, goats, and even camels are safe; ugly *hyenas* stand by, glaring and gaping to devour any sick animals, as soon as their breath is gone, and the *fox* and *jackal* are always ready to assist in the feast. Often on the march one of the dogs will catch a desert *hare* (#44).

The commonest of the larger game and the only one usually seen, except by the skilled hunter, is the graceful *gazelle* (#43 and #44), which stops in flocks to gaze at the caravan and then bounds away. The nomad seldom hits one, as he is a poor shot, missing nine times out of ten. The *wild goat* of the rocks, and the magnificent *antelope* with tapered horns, as long as a walking stick (#44), are only obtained by the skilled hunter. As has already been said, the ostrich is now a very rare sight (#43, #44).

The pleasing twitter of little birds is seldom heard in this treeless region. Crows, hawks, eagles (#44), and falcons hover overhead and the last is often taken young from the nest, tamed, and trained to hunt.

The plague of locusts. Fortunately these poor bedouin are not tortured by mosquitoes and biting flies, though scorpions lurk under stones (#42), and can deal a poisonous bite. However, they sometimes are visited with worse insect plagues, which strip clean the lean desert. Great bird-like *locusts,* fluttering upon their glassy, feeble wings are carried by the south wind of summer and fall about the camp. Sometimes clouds of this vile brood, wreathing and flickering as motes in the sunbeam, fly over for days, thick as rain, and ranging from near the earth to great heights. They alight as birds, letting down their long shanks to the ground. They eat every blade of herbage (#42), they even, in their hunger, bite the bare feet and shins of the Arabs. The women and children rush out and gather great heaps, some of which are toasted at the watch-fires and eaten; but great quantities are singed by the women in shallow pits with a weak fire until cured; they are then salted and packed in leather sacks for future food.

Plants of the desert. The plants of the desert are everywhere scant; but nevertheless, in even the driest parts, some spiny shrubs and harsh herbs are able to pierce deeply enough in the sand to suck up water (#41, #43, #44); from these every animal draws its life. In wadis may even be found thorny, stunted trees, the most valuable of which are the *acacia*, and the *frankincense* (#42) in the south. The twigs of the former weep tears, which harden to make *gum arabic*; from the latter, when it is cut, drops of that esteemed fragrant gum, *frankincense*, are obtained to gladden the sense of smell of worshippers in many parts of the world.

The ship of the desert. The camel is to the desert, what the steamboat is to the ocean, and the railway to the land. If the camels die on a desert journey, their masters die also. They are peculiarly adapted to the dry sand; their long necks give the eye a wide range, and they can reach far for shrubs as they jerk along. Their feet, with two broad-cushioned spreading toes (#44), are admirable for soft sand and gravelly soil. The heavy overhanging lid shelters their eyes from the blinding glare of the bright sky; the long lashes filter out the dust, and the lips are protected by a fence of strong bristles The slit-like nostrils can be closed by valves to shut out the dreaded dust of the simoom. They are so strong in the back that they can carry a load of one thousand pounds for many hours day after day, with little food and less water. At a single drink they can consume seventy gallons of water, part of which is stored in eight hundred honeycomb sacks in the stomach and eked out as required. In the spring, when the fodder is juicy and covered with dew, the camel can go months without a drink, unless it makes long journeys every day; but during the dry summer, when even the pasture plants are as dry as paper, it must have water every third or fourth day. This hardy animal not only stores water, but stores food in its hump. A camel lays up flesh and grease in the hump during the spring for the languor of the desert summer. Indeed, this ugly lump is a barometer; when it goes down, the health of the camel goes down and its life is in danger. Camels find food where even a goat goes hungry. In their leathery mouths coarse, prickly thistles, briers and thorns are more savoury than soft green fodder.

A bad temper. The camel is not an attractive animal. He is ugly, heavy, awkward, cold-blooded, sullen, and stupid, having little feeling, no affection, and an evil eye. A traveller always makes a friend of his horse but never of his camel. The speed depends on the breed. The pack camel will travel two or three miles an hour, but the aristocratic *dromedary* will go five miles with ease. Indeed, the dromedary is the race-horse of his species, thin, elegant, fine-haired, light of step, easy of pace, and much more enduring of thirst than the woolly, thick-built, heavy-footed, ungainly, and jolting camel. A caravan of camels is a freight train of the desert, but a caravan of dromedaries is a limited express.

The oasis. Not all the people of the desert are black-tented nomads. Occasionally in their wanderings, when rice, wheat, and barley are nearly gone, and the last mouldy date has been eaten, these wandering Arabs turn toward an *oasis*, loaded with their butter to be sold and all the camels, sheep, and goats they can spare to be traded. What a delightful sight it is at last to see in the distance a long green streak in the dull brown carpet of the desert. The smell of water, or vegetation, has quickened the pace of the camel even before the human eye detects the distant greenness. At last the oasis can be seen in all its contrasting glory. Below in the hollow are blotches of colour as though an artist had spilled out all his greens. Then the details crystallize out. Every place down to the farthest winding of the ravine is studded with tufts of palm groves and clustering fruit trees in dark green patches (#45). Small round turrets and flat-topped mud-brick houses are half buried in greenery (#45), the whole plunged in a flood of vertical light and heat. It is truly a lovely scene, and seems yet more so to the eyes, weary and strained for weeks, or months, with the dreariness of the desolate desert.

A smiling garden. An oasis is really a garden village or a row of villages, often having a population of fifty thousand. Each little garden home is enclosed by a brick wall. The grove of stately date-palms is the capital wealth of every enclosure. But fruiting plum trees, apricots, peaches, gnarled olive and fig trees grow everywhere, and climbing clusters of luscious grapes line the garden walls. Between the trees grow tall dignified corn, and a leafy riot of great pumpkins, sweet smelling melons, dull cucumbers, and vetches, splashed with flaming tomatoes; not one square foot of precious soil dare be wasted. A maze of little streams of water passes from tree to tree and furrow to furrow, and each date-palm is at the centre of a bowl-shaped hollow filled with the muddy, tepid water (#45).

The heart of the oasis. To find the secret of this laughing happy oasis in the midst of a dead sullen desert one must go to the source of the water. There it is near the centre. A dark square well, forty feet wide and so deep that it is lost in the darkness. Night and day camels, or oxen, pull ropes passing over creaking rollers down into the dark depths (#45). At each pull a great horn-shaped skin-bucket of water splashes forth a little roaring cataract of water

#45

AN OASIS

What shape are the houses? Of what are they built? How is the water drawn out of the well? Describe the date-palm. How is the fruit picked? How is it packed for market?

into a channel leading from the well. Forty camels pull up the irrigating streams night and day from the four sides. This well is the great throbbing heart of the oasis; these radiating channels are the surging arteries through which the life-giving stream is carried to every garden, to every tree, and to every blade of grass. In some oases bubbling springs, from far below, carry the precious water to the surface. It flows in a stream through the oasis, and is drawn off by laterals to every garden, until the last drop has been exhausted.

Luscious dates weaken the body. Date groves are the chief wealth of the owner of land in Arabia, and they are the bread of the oasis. In August the crowns of thousands of palm trees are richly loaded with purple-ripe, yellow, and brown clusters of this wonderful fruit. The hard dates have melted to ripeness, have softly swelled under the sun, and are covered with honey moisture. They are mellow and full of sappy sweetness. If left till over-ripe, they are not so delicious. Climbers in the trees cut off the giant bunches and let them down with a cord, where they are picked, sorted, and packed in baskets for the bedouin or in boxes for the foreign market (#45). Those who only know the dried scaly dates bought from shop-windows, can hardly imagine how sweet and mellow are the just-ripe fruit when plucked from the tree in an Arabian oasis. There its richness does not quickly bring satiety. Nevertheless it is not a healthful food, and the dweller in the oasis holds all dates in a kind of loathing; well they might for nature has set a sorry mark upon all the date-eating village folk. It makes them blighted, and sallow, with a hollow cast of visage which once seen can never be forgotten. In the streets of Damascus these miserable, feeble folk can be at once sorted out from the more vigorous and tougher bedouin, nourished on camels' milk.

In a few days the dates are all picked, the uncomfortable bedouin have got their supplies and are eager to leave the crowded oasis and march back into the seemingly empty desert, which is their home.

CHAPTER VIII

LIFE IN THE ARCTIC DESERT

All aboard for the Arctic Ocean. Our train speeds over the cold, flat, snow-covered prairies until we reach the capital city of Edmonton, Alberta (#46). There we transfer to a rough frontier railway (#46), which carries us far north, through the dark forest to McMurray on the Athabaska River. Though we are two hundred

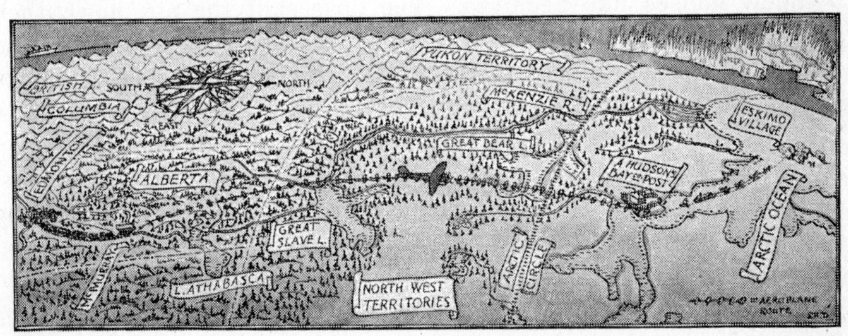

#46

AEROPLANE TRIP TO AN ESKIMO SETTLEMENT

miles north of Edmonton and eight hundred miles farther north than Toronto, we have just started on our trip to the far Arctic. We now leave the railway and enter an airplane headed for the forbidding north (#46). As it rises, the whole billowy surface of dark evergreen forests is spotted white with countless snow-covered lakes and winding river valleys, whose frozen streams are buried deeply in the drifted snow. On we glide through the intense cold, hour after hour. But a change is taking place. The white of the landscape is gaining on the dark green. The billowy trees are getting smaller; open land, covered with snow, seems to be crowding out the forest, until at last it forms only a sombre fringe along the white river channels (#46). At last the cold and wind have shrivelled the forest to nothing, and we see below us a continuous carpet of pure white, except where blizzards have swept the rocks and hill-tops clean. After a trip of eight hundred miles the airplane slackens speed, circles about, and lands at a Hudson's Bay post (#46), a small collection of wooden buildings, painted white, with red roofs. But we have not yet reached the people of the twilight for whom we are searching. We dress in a double layer of furs and make ourselves comfortable on a long, narrow sleigh, drawn by a team of eight burly dogs, called *huskies* (#46). On we skim northward over the snow and ice; but soon we notice a change, the bare patches of rock have disappeared, no stiff weeds stick through the snow, there are no hills, but level whiteness as far as one can see, except for an occasional ridge. Reports, like the firing of guns, come from below, we cross a crack through which water is oozing; it tastes salty. We are riding over the Arctic Ocean!

Snow houses like beehives. The dogs sniff the air, howl, then bound forward, like a shot from a gun, toward a ridge of ice with a group of white beehives in its shade. Before we reach the ridge, imagine our surprise to see bouncing up out of the snow, first dogs, which snarl and snap at our team, then men in furs, smiling women with babies on their backs, and boys and girls like big teddy-bears. Although the air is biting cold and the raw blizzard cutting the face, they are all wild with pleasure to see the white men and invite us into their houses.

The Eskimo. These are the Eskimo, the most northerly people in the world. Here, during the winter, they live on the ice, miles from shore, and with no living thing in sight. What do they eat? How do they get clothes? And above all, how can they keep warm in a region with ten months of winter? In this chapter we shall try to answer these questions.

The sun lights up the drama. These people are slaves of the sun. It controls every action, fixes their food, orders their house building, guides their tailors and dressmakers, and even regulates their play. Let us show how the sun-man lights the stage for this yearly drama of Eskimo life, filled with the laughing comedy of good nature and tinged with the dark tragedy of murderous blizzards, gnawing hunger, and sometimes death from starvation.

The sun in March. It is the early morning of March 21st. The whole east is touched with a rosy tint, though the sun will not rise for two hours. Promptly at 6 a.m. the ruddy upper disk of the sun shows right on the eastern horizon (#47). But instead of ascending, it seems to glide south along the horizon's edge like an aeroplane trying to rise. At last it shows as a fiery ball and moves slowly south, still hugging the horizon. At noon it is directly south but so low that its little heat has no effect on the bleak, wintry air. It sets in the west at

6 p.m. but moves along so little below the rim of the horizon platform that the brilliant twilight lengthens the day to eighteen hours; the

#47

THE COURSE OF THE SUN IN ESKIMOLAND

scant six hours of darkness are only a pale gray that does not reach velvet blackness even at midnight.

The midnight sun. Each successive day the sun follows a parallel course, only at a higher level, the daylight lengthening, the dark-

#48

A CARIBOU LIT BY THE MIDNIGHT SUN

ness shrinking. By May it is so high that even at midnight the sun's upper disk passes across the north horizon like the light of a distant ship (#47, #48). Night has disappeared. The sun continues its course, ever higher and higher, across the sky until June 21st (#47); but even then it scarcely rises as high as in southern Canada during the winter.

Stern spring becomes laughing summer. The continuous sunlight, even though the rays are so slanting, begins to drive back fierce winter. Snow melts away from southern slopes, spring gales have nearly exhausted themselves, and the sun sails hot across the southern sky; the twittering of a few little birds is heard, and the white *ptarmigan* cocks, the partridges of the Arctic, are drumming love to their mates. Before May is past the ice has melted at the edges of the lakes, the snow is slushy, the musical trickle of snow-water is heard everywhere, and the ducks, geese, and shore-birds are sailing northward overhead to find a nesting place on the Arctic islands. In June the twenty-four hours of sunlight rushes spring into summer. The snow is all gone except in sheltered shade, the unlocked waters of lakes and rivers race in foaming torrents to the sea, and great blue gaps are opened in the Arctic ice by the winds. The fine weather wakes nature out of her sleep at a single stroke. Plant growth is almost magical. The bare buds on the stunted willows are changed to leafy branches almost in a day, and the brown heathy surface of the soil becomes patterned carpets of many-coloured flowers overnight. Sheltered river valleys become green meadows, and even the sallow *reindeer moss* which covers most of the land, takes on a greener complexion. Bees hum over the flowers and moths and butterflies appear. But the warmth of the water quickens to life those scourges of the north, the *black fly* and the *mosquito*. The latter spread like a quivering cloud and vex both man and beast.

Winter springs out of summer. Before August is over, the summer has exhausted itself. Already at midnight the sun drops below the horizon in the north, and scums of ice form on the water. During late September there is a real night, the sun's course sinks lower and lower (#47), and its heat becomes less and less. The lakes and rivers freeze over, the gaps in the Arctic ice close in, the flowers and grass become buried in snow, and the frightened shore-birds and water-birds fly screaming on their long journey to the south.

Dark, stormy winter. During October the path of the sun across the southern sky grows shorter and sinks lower, and the hours of night grow longer and darker. It is now a land of long twilight soon to be swallowed up in night. At last the flush of sunrise and sunset blend in the southern sky. Before November is well on its way the thin disk of the sun just slides across the southern horizon at noon and then it is gone for more than two dark months (#47). The cold increases daily and, though light and darkness wrestle for a while in blood-red sunsets, the cold shadowless twilight and the gray dusk of night soon settle down. Some years the winters are quiet, but usually they are plagued by heart-breaking blizzards, often lasting for four or five days. For insane fury the south-west wind is worst and gathers new force with every hour; it whips icy showers over the naked surface and picks the snow up bodily and

#49

NORTHERN LIGHTS

What is the shape of northern lights? How many fish has the Eskimo caught? Find the huskies. Describe the winter dress of an Eskimo.

hurls it in blinding horizontal sheets across the land. One dare not go out, the wind is like cold steel, and the ice crystals cut like a sandblast. At this period, because it is almost too dark to hunt, often the lamp that warms the house has to be turned low, and the shrinking food has to be doled out in semi-starvation quantities.

Northern lights. Even these cold, dark, heartless winters have their attractions. The icy stars flash and twinkle with a double brilliance in the steel-dark sky; the milky way is a glorious flood of luminous spray across the sky; but most arresting of all is the dazzling display of the northern lights (#49). In these latitudes it is no vapoury luminous phosphorescence; it flares and quavers, shooting to right and left, then this glorious display waves like lighted curtains or shoots like luminous snow; now a giant arc tinged with rose-red spreads right across the sky; in a moment it is changed to a dazzling, waving drapery with flashing streamers changing from rose and red, to gold, green and white. Such splendid displays often light up this black sky.

Contrasted life on land and in the sea. There is no greater contrast in all nature than the abundant growth in the Arctic Ocean and the nakedness of the adjoining land, even in the midst of summer. Most of the land is almost bare; indeed, the whole of northern Canada was formerly called the *Barren Lands.* But it is not so naked as that name implies and is now called the *Arctic Plain.* The hollows are, more or less, covered with a pale-green sickly growth, called *reindeer moss.* In sheltered places, as has been said, there are patches of flowers in the short summer, and along the river valleys there are even short-lived meadows. Much of the surface is bare rock. It is all treeless. But as one looks down into the icy ocean bordering this northern desert, he sees a watery oasis. Luxurious masses of seaweeds in rich greens, strong browns, and brilliant reds, cover every rock, clothe the bottom, and almost fill the tidal pools. Here is the secret to the teeming aquatic life in the Arctic. The swarms of birds, the seals, polar bears, and most other creatures, including man, snatch their food from this oasis of the north.

Stabbing seals. Let us return to our snow hut on the ice in March just after sunlight has returned. There is not a break in the icy covering, and yet each morning the Eskimos go out on the ice to obtain their food. Under that thick covering live *seals,* which keep numbers of small breathing-holes open (#50). These are well hidden by a covering of snow. The keen-scented huskies (#50), however, soon smell out the breathing places, and an Eskimo selects a station at one and inserts a little stick, which bobs up when hit by the nose of a seal. There he stands, or sits on a block of ice, with his eye on the marker and his hand on the harpoon. With a fierce wind in his back and a temperature of 40°-50° F. below zero biting in his face, he remains hour after hour tensely waiting. At last he hears gurgling bubbles of air expelled from the seal's lungs, and then he sees the tell-tale stick, or bone, bobbing. A fierce stab buries the barbed harpoon in the poor creature's skull, and the Eskimo tugs on the line attached to the harpoon head the same as though pulling in a big fish. Often a dozen men may not get more than one or two seals in a day. But their whole life depends on the catch. The seal's flesh is the chief food for humans and dogs, its fat, or blubber, supplies fuel for heating and cooking, its skin makes clothing, tent-covers, and sides for boats, its sinews make sewing thread, and its teeth and bones are turned into ornaments, harpoon heads, buttons, needles, and many other household necessities. An Eskimo woman in a few

#50

BREATHING HOLES FOR SEALS

How does the Eskimo find the breathing holes? How does he catch the seal?

minutes flays off the skin, strips the blubber, hacks the flesh into pieces, often without getting a drop of blood on her clothing or on the floor of the hut.

Killing seals on the ice. A little later in the season the seals come out through openings which appear in the ice and bask and sleep in the serene sunshine. Then the wily Eskimo imitates his victim by dressing in sealskin, crawling forward on his stomach, and answering grunt with grunt, until close enough to cast his fatal harpoon. The Eskimos living nearest to the trading posts, who have rifles, use a bullet instead of a harpoon; if they do not fire a dead shot, however, the wounded seal may slip into the water and sink.

The kayak. Still later in the spring, when the sun and wind have opened great rifts in the ice-cover of the ocean, the Eskimos in

THE ARCTIC "DESERT" IN SUMMER
#51
S.H.D.
NORTH POLE
PEARY REACHES POLE-1909
TYPES OF HOUSES AND SLEIGHS IN WINTER
1. SNOW HOUSE
2. STONE AND EARTH HOUSE
3
4
EDGE OF ICE CAP
EDGE OF ICE CAP
ELLESMERE I.
GREENLAND
THE GLACIERS OF GREENLAND, ICEBERGS IN THE MAKING
BAFFIN BAY
BAFFIN I.
BOOTHIA P.
VICTORIA I.
ALASKA
DAWSON
YUKON TERRITORY
WHITEHORSE
MACKENZIE R.
GREAT BEAR L.
COPPERMINE R.
ARCTIC CIRCLE
THELON R.
GREAT SLAVE L.
NORTHWEST TERRITORIES
SOUTHAMPTON
HUDSON STRAIT
UNGAVA BAY
HUDSON BAY
CHURCHILL
HERE FRANKLIN'S TWO SHIPS WERE FROZEN IN, 1846
-LEAVING THEIR SHIPS HE AND HIS MEN WENT SOUTH.
-HERE 40 SURVIVORS OF THE TRIP STARVED TO DEATH.
-MISSION
-TOWN
-TRADING POST
-R.C.M.P. STATION
-SKIN TENTS
-NORTHERN LIMIT OF TREES
-MACKENZIE ESKIMO
COPPER ESKIMO
-LABRADOR ESKIMO
-GREENLAND ESKIMO
-CARIBOU ESKIMO
BAFFIN ESKIMO
-POLAR ESKIMO
NORTH MAGNETIC POLE
RUINS OF ANCIENT VIKING FARMS
x=x=x=x=x=x= -BOUNDARIES

many regions hunt seals in the open water from *kayaks*, undoubtedly the most remarkable invention of this skilled race. The kayak is a

#52

HARPOONING A WALRUS

How is the walrus held after being harpooned? How do the other men help? What animal is behind the mass of ice?

small skin boat so light that it can be easily carried on the head. The paddler fits himself into a round hole in the centre and binds the bottom of his waterproof coat so tightly over the opening that no water can leak into the boat (#52). In this frail egg-shell he chases and harpoons seals in the open water, even in rough weather. If the boat upsets with him fixed in it, he can right it with ease.

Struggle with the walrus. In Greenland and Hudson Bay, great *walrus*, in the open water, are also harpooned from the ice (#52). When the hunter gets one of these monsters, perhaps weighing a thousand pounds, and joins battle with him at the end of a line, he has to act quickly. One end of the handle on the free end of the line is propped in a nick in the ice and a loop over the other end anchors the struggling victim until it can be killed with darts. A walrus is a rich treat; it has immense quantities of blubber, and its flesh is choice, especially when it has been kept to ripen; indeed, the Eskimos value such decayed flesh as much as we do very old mouldy cheese. Another dainty titbit is the clams extracted from the stomach of the walrus, which even white men find delicious. The tough walrus hide makes the best soles for boots, and its two immense tusks make masses of good ivory.

Caches of blubber. The blubber collected in the spring and early summer from seals and walrus is stored on the shore in big skin bags for winter use. The bags are placed high on perched rocks, or else piles of stones (#53) must be placed over each *cache*, as it is called, in order to protect it from foxes; if there are any *wolverines* (see page 3), lurking about, stones as heavy as two men can lift are none too big to keep out this persistent robber.

Polar bear. Occasionally the sprawling tracks of the *polar bear* excite to a high pitch both men and dogs, who chase this dangerous foe (#52). He may be killed with a rifle or a hailstorm of sharp darts. One of these brave men will not hesitate to bring a full-sized bear to bay with a strong knife tied on the end of a long handle. Its hide is the warmest of all furs, its blubber is supposed to have great virtues and bear steak, raw or cooked, is a holiday dish.

Fishing. Salmon, trout, whitefish and other less important fish swarm in many lakes and are fished with bone hooks through holes in the ice during spring (#49), and are often trapped in nets or speared during the summer (#53). The fish are dried in the sun on high racks to keep them away from dogs and they tide both men and dogs over many a hungry period when game is scarce. Often, after a long hard day's fishing, the women work until midnight cleaning the fish. They put the fillets on the racks to dry, cast the backbones and fins to the ravenous dogs, and themselves gnaw at the raw fish-heads.

The caribou. All the animals mentioned get their food from the sea, but there is one very important species, the *caribou* (#48, #51), which crops the beggarly reindeer moss growing in the hollows among the rocks. This is the same animal that has been tamed in northern Europe, as the *reindeer*, and has become the camel of the cold desert of the north. The Laplanders eat its flesh, drink its milk, use its skin for clothing and tents, and hitch it to their sleighs. Among the Eskimo the caribou has never been tamed and is only hunted for its flesh and skin. As the warm sun of April and May

#53

SPEARING FISH

Why is the sleigh raised on piles of stones?

uncovers their feeding grounds, these splendid animals migrate northward for the summer to the Arctic coast and the Arctic islands along regular routes, which the Eskimo know only too well. Formerly the men lay hidden behind a screen of stones and poured volleys of arrows into the dense herds; when rifle bullets replaced arrows, the attacks became so deadly that there was great danger of the animal's complete destruction, until the government undertook their protection. These animals are also hunted throughout the summer in their northern feeding grounds, and again during their autumn southern migration. The flesh is dried in summer, and in autumn carcasses are often sunk in fishless ponds, where the covering of ice keeps away hungry foxes and wolves and greedy wolverines. The skin, when tanned by Eskimo women, makes the best garment for the north. It is surprisingly light, as soft as a glove, and can be tanned by these skilful women in most attractive browns, yellows, and even pure white.

Food of Caribou Eskimos. The Eskimos who live near the coast eat all of the animals just described, but the Caribou Eskimos (#51), who live on the Arctic Plains far from the coast, eat only caribou, and fish obtained from lakes. They use for fuel the stunted shrubs growing throughout this district and the driftwood along the rivers; but since wood fires would melt snow-houses, these poor creatures, often nearly starving, have to huddle together in their fireless snow huts throughout the sunless winter with the temperature many degrees below zero.

Permanent houses. The Eskimos, like other people, build their houses of whatever materials are cheap and handy. As the swift current of the Mackenzie River, bordered by forests for a thousand miles, sweeps vast numbers of tree trunks out into the ocean, whence they are carried eastward by the wind, the Eskimos on both sides of the delta secure these logs and build log huts (#51), which they occupy summer and winter. In Greenland where flat stones are abundant, and there are patches of turf, permanent sod houses with flat stone roofs covered with earth, are found along the west coast (#51). The filth which collects in such houses during the winter is prevented from festering into unbearable foulness during the heat of summer by the people deserting them at this season, taking out the windows, and sometimes breaking down the roof in order that they may be purified by light and air; they live during this season in skin or canvas tents (#53).

Building a snow-house. During the winter, in northern Labrador, northern Greenland, and northern Canada, except near the mouth of the Mackenzie River, the best available building material is snow (#49, #51). It is clean, light, abundant everywhere, easily worked, and keeps out the cold. A house can be built anywhere in an hour or two. The builder with a long snow-knife made of horn, cuts large blocks from the snow, made very firm by the impact of the blizzard. Working from within, he arranges in a circle these blocks, the first being very low and the height gradually rising wedge-shaped. Then he goes right around, building the wall spirally with blocks. When about four feet high the spiral is gradually built inward to form a semi-globular dome. Last of all, the keystone block is placed on the centre of the dome. All holes are plugged with snow. As the blocks have been taken from the interior, the house is partly beneath the level of the snow. An opening at one side, large enough to crawl through on hands and knees, is continued as a long snow tunnel, which the dogs are allowed to enter on cold nights.

#54

INSIDE THE SNOW-HOUSE
(One wall is partly cut away)

Describe the stove. Tell what it warms. What are the people sitting on?

Simple house furnishings. Opposite the door is a snow platform (#54), occupying more than half the floor space and covered with cosy rugs of caribou, musk-ox, or even polar bear. This is a bed for all at night, and the living and working room during the day. To the right of the door is the lamp, which also serves as a stove. It is a semicircle of soft stone, hollowed out and partly filled with crushed blubber. Along the straight edge is a wick of moss or cotton-grass, which gives a clear, smokeless flame, with much heat. Above the lamp is a stone or metal pot, and above this a drying rack for damp clothing, so that no heat may be wasted. The rest of the contents is a disorderly collection of frozen fish, hides, vessels, bags of meat, etc., scattered over the floor.

A wonderful house. The house is so tight that this small lamp keeps it comfortable. It contains all the climatic belts. The feet are always in the Arctic, the middle of the body bathed in temperate weather, while the upper part of the body may extend a good way into the tropics. The thawing and freezing of the roof may in time cause a coating of ice and sparkling icicles to form, and a hole may even thaw through, which is easily patched. Indeed, it is a small job to re-roof the whole hut. Undoubtedly this snow hut, or *igloo*, is one of the most useful Eskimo inventions. When men travel during the winter, they need have no worry about sleeping places, since they can build a new hut every night; when they are threatened by a blizzard on a journey, they throw up an igloo and defy the death-dealing wind, for days if necessary. When spring comes on and snow huts are threatened, these people throw up huts of various shapes and sizes made of caribou skins (#53).

Dog sleighs. Transportation during summer is by skin boats (#51) along the coast, and by foot on the land. In the latter case, the tent, furniture, and all other equipment is tediously carried on the backs of men, women and children, as well as dogs. But it is in the winter that the Eskimo dogs are most useful. These long-haired, burly brutes (#50, #52, #53) are almost as wild as the wolves from which they have probably sprung. All meats, skins, boots, small animals, even babies, have to be placed beyond the range of their hungry maws. They will even eat their harness. The Eskimo sleigh, which they draw, is composed of two long wooden runners joined by cross-pieces (#49, #53). On the bottom of each runner is frozen a smooth layer of peaty clay, over which every morning is sprayed from the mouth, warm water, which at once freezes to a thin coating of ice. This smooth finish is almost without friction on the frigid snow, where metal would bind as though dragged over cement. Each dog can draw several hundred pounds on such a sleigh. As in Greenland, Labrador, and in the eastern part of the Arctic Plains, one long trace goes from each dog to the sleigh, the six or eight dogs are spread out like a fan (#51). Along the Mackenzie River they attach the dogs in pairs to a middle trace (#51), as the Indians do farther south. With such teams, Eskimos can travel considerable distances, day after day. But do not think that the people ride; usually the men and women are also harnessed to the sleigh to assist the dogs (#55).

A cheerful people. The Eskimos are not short, fat people with a yellow complexion, as is commonly believed. They are of medium size, equal in height to the French or Spaniards, and while stout and powerfully built, are not fat. Their bodies are long, their legs short, and their hands and feet small and well formed. The coarse, black hair does not grow long on either the head or face. The parts of the skin exposed to the clear, continuous sunlight of the summer are soon burnt brown, but in winter the true colour, yellow-brown

#55 TRAVELLING IN SPRING

or light chocolate, appears, or would appear if water were not too scarce to wash off the dirt and grime. They are the most cheerful people in the world, though they may be living in a region most harsh and unfriendly and lacking in everything necessary to live. Even the Caribou Eskimos of the northern plains in April, with their food all gone and death by starvation staring them in the face, keep their spirits bright and cheerful in their cold and comfortless snow-houses.

Clad in furs from head to foot. The clothes of the men and of the women are very much alike and the most suitable in the world for cold weather (#49, #50, #52, #53, #55). They both wear loose, inner, fur garments covered by an ornamental outer fur smock with a hood on the back, which can be pulled up over the head. The woman's hood is so large that the baby can be carried in it. In the summer the outer smock is not worn. Caribou skin is ideal for garments, as it is light, tough, flexible as cloth, and very warm. Sealskin is much harsher and not so warm.

At night all completely undress and crawl into fur sleeping bags made of two layers, the inner with the fur in and the outer with the fur out.

Flesh eaters. Their food has been already described. Since fuel, either blubber or drift wood, is generally scarce, fish, seals, and caribou are often eaten raw, but sometimes are boiled in the pot that always hangs over the lamp (#54). So common was this habit that the Indians to the south gave these people their name Eskimo, which means an eater of raw flesh. About the only vegetable food they eat is the ball of partly digested reindeer moss cut from the stomach of the caribou, which they greatly relish. Like all meat eaters they drink large quantities of water but find enough salt in the flesh eaten.

The Arctic fox and the lemming. Living among the grass roots all over northern Canada is a rat-like little animal, called the *lemming.* Indeed, there are two species, one of which turns from brown to gray in the winter, the other retains its brown coat throughout the year. During the winter they tunnel under great banks of snow for their food. But even here they are not safe. A very dainty fox, which will attack any bird or small animal that it can catch, preys chiefly on lemmings. The abundance of these foxes is in proportion to the abundance of their prey which they search out under the snow. This fox has two colour forms, the blue and the white, which may be present in the same litter; in northern Canada there are many more white than blue foxes, but in Greenland they are more nearly equal in numbers.

The Eskimo goes modern. The Eskimo is rapidly changing his simple habits. The extravagant tastes during the rush of good times after the Great War created such a demand for the pure, clean, soft fur of both these foxes that the fur companies pushed their trading posts among the most remote of the Eskimo tribes and offered such fancy prices that these people are forsaking the seal, the fish, and the caribou. Instead they are setting lines of traps everywhere to catch the Arctic fox. The trading-post exchanges the goods found in city stores for fox skins. The primitive native, who twenty years ago would trade his wife for a handful of rusty nails, is now driving hard bargains over the counter for his fox skins. As a result the high-powered rifle has entirely replaced the bow and arrow, motor boats are replacing kayaks and skin boats, tea and badly cooked flapjacks are being munched in igloos instead of fish and seal meat, aluminium pots and sheet-iron gasoline stoves are driving out the home made utensils, and the sound of the phonograph disturbs the quiet air of the Eskimo encampment. Some of the men even have typewriters and the women sewing-machines. Whether the simple minds of these gentle people can accept this sudden onslaught of white man's goods without injury only time can tell.

Reindeer on the move. The Canadian government is trying to make the Eskimos east of the Mackenzie River into herdsmen of tame reindeer. This is urgently needed because the scarcity of caribou in recent years has led to acute famines. A herd of three thousand of these reindeer was bought by the government from Alaska, and trained herdsmen have driven these animals 1000 miles across the north of Alaska to their new pastures east of the Mackenzie River. This most dangerous and difficult task required four years of patient work and has now been completed. Eskimos who have spent their whole lives killing wild animals for food and clothing, are now learning to tend the tame reindeer, drive them in sleighs, prepare their milk, and lead them from pasture to pasture.

CHAPTER IX

LIFE IN MEDITERRANEAN CLIMATE

(*Palestine*)

So small yet so great. Palestine, the country sacred to three great religions, is so little yet so great. It is much smaller than one thinks. In a motor car you can cross it in an hour and go throughout its length from "Dan to Beer-sheba" in about three hours. From prominent positions one can see far beyond all its boundaries, — the green Mediterranean on the west, the fierce mountains of *Moab* on the east, the snow-capped *Lebanon* on the north, and the burning wilderness of *Sinai* on the south,— all outside Palestine.

A narrow plain and a stony ridge. The country consists of four strips running from north to south (#56). Along the Mediterranean coast is a plain, a few miles wide, broken only by the spur of *Mount Carmel* thrusting itself into the sea. Next comes the central hill range that runs like a spinal cord from one end to the other, except for a break made by the beautiful *plain of Esdraelon,* dotted by prosperous Jewish villages of red-roofed brick and stucco houses, surrounded by green plantations, silvery irrigation channels, and rich black soil. This ridge lets itself down by irregular terraces to the coastal plain. Never were there terraces so stony; even the soil of the vineyards, gardens, and grain fields seems to have more pebbles than earth.

A gash in the rocks. The third strip, *the Jordan Valley,* is the most remarkable plain found anywhere in the world. It is a deep rift between the central hill range and the stern steep edge of the escarpment to the east. In the bottom of this deep chasm is a winding green ribbon of river jungle with the mirror of the Jordan showing here and there. Snow-capped *Mount Hermon,* like a white cloud, plugs up the northern end of the rent, then comes the sad silent *Sea of Galilee* brooding over the destruction of a necklace of beautiful cities, which once sparkled on its shores. To the south, the Jordan expands into the deep blue waters of the *Dead Sea,* which occupies the bottom of the deepest and almost the hottest pit on earth. Although the Jordan and other streams pour over six million tons of water into this scalding kettle daily, it has no outlet, except the water

lifted by a scorching sun from a brazen sky. It is well named, since the bitterness of the intensely salt water repels every living thing.

Trans-Jordan. Beyond this deep gash in the surface of the earth rise steeply the heights of the fourth region of Palestine, which is really *Trans-Jordan,* whose impressive tablelands are clothed with greenery in the north (#56), but beyond the Dead Sea are bare; yet in its nakedness this region becomes weirdly beautiful toward sunset, when the level light turns its stern gray to exquisite purples and a tender lilac that at length deepens to violet. We are to study how the people live in this strange country.

The old and the new. We land at the busy port of *Jaffa* (#56), where we see crates of oranges, sacks of wheat and sesame, and bales of sheep and goat skins, being loaded on to ships. This city is older than Noah and looks the part, with its narrow tangled streets, crumbling walls, tottering buildings, and festering dirt, together with a certain attractive quaintness. As we whirl through this ancient maze, suddenly we enter a new world; the streets are straight and paved, the rows of stucco and brick buildings with red roofs, display in plate-glass windows almost all that can be bought in a departmental store in Montreal or Toronto. Elegantly furnished cinema theatres, modernly equipped schools, and dignified domed synagogues mingle with the comfortable homes, supplied with electricity, running-water, and all the improvements found in Canadian cities. Even the people have changed; languid Arabs, dressed in Oriental robes dawdle along the stuffy streets of Jaffa; hustling Jews, dressed like London business men, hustle through the avenues of this modern city, which is called *Tel Aviv,* and which has sprung out of the earth almost full grown. When Lord Allenby in 1918 conquered Palestine, Tel Aviv did not exist; to-day it is a hustling industrial city, containing 60,000 people.

Orange groves of the plain. But we hasten through these stifling cities into the free open plain, which is only a few miles wide. Everywhere on its surface we see rows of trees with bright oranges (#56) flashing among the light green leaves, and among the trunks lazy channels of water, which has been raised from wells (#57) to bring fertility to the soil and life to the vegetation.

Ancient villages. Now we are climbing the terrace of the western slope of the central range of hills. Scattered white villages, old as time, cling to the oblique cliffs, or cover convex domes of rock. Here we find people who do not change. They are creatures of the soil and climate and live much as they did two thousand years ago when the divine man, Jesus, dwelt among them, or three thousand years ago when the children of Israel entered the land for the first time.

Building material. Palestine never had anything but a sparse covering of trees, most of which were destroyed in the ancient past; and the greedy appetite of the bedouins' goats for sprouting twigs

#57

RAISING WATER FROM WELLS FOR IRRIGATION

The camel, moving round, rotates the big horizontal wheel with a circle of stakes projecting upward. These cause the upper axle to rotate. As this is connected with the wheel to the left over the wall, it also rotates. But it has an endless chain of buckets which are being constantly lifted from the well and dumped into the water channel.

have kept the hills bare and the valleys empty ever since. The only possible building material is the white limestone, whose layered structure makes it easy to work, and whose abundance everywhere invites its use.

House and furniture. A square, or rectangular room, often built on a slope, and so excavated that the hill itself forms the back wall, is the whole house (#58). The roof is usually domed (#59), and as it also is built of stone, and is supported by the walls, they have to be a yard thick (#59) to prevent the roof from thrusting outward the top of the wall. Though the roof is domed on the inside it is often built up on the outside to be a flat platform of packed earth.

#58

INSIDE THE HOUSE

Sometimes after the rainy season this becomes covered with grass. As the back edge of the roof is often on a level with the slope, it can be reached without a ladder, and is used for many purposes. In the summer it makes a good living-room in the day and evening and a sleeping porch at night. Occasionally it has a railing around, and the sheep and goats are tethered there for the night. A stone platform, raised about six feet, occupies all the floor of the house except a small part near the door (#58), and steps lead down from this floor to the entrance. This higher portion is kitchen, bedroom, and living-room for the whole family. Clay bins along one side con-

#59

WOMAN SPINNING AS SHE WALKS

tain the wheat, barley, rice, millet, lentils, dried figs, and other foods. Large earthenware jars on another side contain olive-oil, pickled olives, perhaps honey, and one near the door contains that most precious liquid, water. An arched recess in the wall, hid by a curtain, receives each morning the bedding, consisting of a mattress stuffed with wool, cotton, or rags, a pillow of straw, and wadded quilts. These are spread on the floor at night. On the floor are also wooden kneading bowls, cooking vessels, and a stone grinding mill (#58). Shelves, made by removing stones from the thick walls, contain various articles, and pegs driven in between the stones support wicker baskets, garments, and a disorderly array of whatever other meagre articles the family possesses.

The space below the raised floor, with an entrance from the outside, is separated by arches into a number of stalls, in which sheep, goats, asses, cattle, and camels are fastened at night (#58). It was in such a stable that Jesus Christ was born. Sometimes the larger animals are kept at night in the recess just inside the door. We might wonder, with animals on the roof, in the basement, and sometimes in the ante room, whether it is a stable or a house. When times improve or a son gets married a new room is added, either to one side or on top of the first, and in the course of years or centuries the house may become a number of rooms arranged around a rectangular court (#59). Many of these very old houses, built when times were so unsettled that robbers slipped in through any opening, are either without windows, or with small barred windows so high and so few that the rooms are in twilight even at the middle of a sunny day (#59).

Dry summers, rainy winters. These simple, humble village folks are nearly all farmers; but farming in this climate is vastly different from ours. As the sun shines nearly vertically at noon during the summer, the days are so hot that olives, figs, oranges, and almonds flourish, provided water is supplied. The dry heat of the parched land, during this season is relieved every afternoon and evening by cooling breezes from the Mediterranean Sea, but as the north-west winds get warmer as they advance, the summers are unfortunately rainless and the land is painfully dry and bare. The hated simoom, already described (p. 102), with its fine sand-blast occasionally rushes in from the *Syrian Desert* to the east, spreads over the hungry land like a pestilence, and brings irritation to man and wilting blight to crops. Winter is like an oasis in the desert; the sun is far enough to the south for Jerusalem occasionally to be covered with snow for a few hours, and water may on rare occasions freeze at night. The moist wind, now coming in from the Mediterranean Sea, because it is south-west and hence blowing away from the equator, deluges the country with winter rains, which are most intense in the late autumn and early spring. Palestine has consequently two seasons, a rainy winter and a dry summer and, both are productive, because the winters are warm enough to grow such crops as wheat, barley, millet, and lentils.

A renewed life. Through hundreds of years of neglect, under Turkish rule, forests have been hacked down, precious soil has been washed away, irrigation ditches have been broken, and the land which once "blossomed like the rose," has been so stripped and starved that its bare bones stick out, and almost everywhere it is only the carcass of its former self. But, under twenty years of the healing rule of Great Britain, the flesh of fertility is beginning to fill out the leanness, and the life-blood of irrigation water is again beginning to flow over

#60

SHEPHERDS

Notice the terraced fields. What two colours are the sheep? What are the two boys doing?

the dry places, and already this sacred country has begun to "flow with milk and honey."

A luscious spring and an arid summer. The winter rains always clothe the nakedness of the land in the spring with a rich garment of colour across the fields, a flush of green over the rocks, and bright pink ribbons of glorious oleanders in the *wadis* (p. 111). But the broiling summer's sun soon licks up the remains of the moisture that has not filtered through the porous limestone, and for nine months of the year all is gray or dull brown, the grass is withered away or scorched, and this stingy pasture must satisfy the bony cattle, sheep, and goats.

Terrace cultivation. Though the farmers live in villages, their farms consist of every little island of soil found among the sea of rocks. Each family has a number of these poor fields, often scattered far and wide. They usually have no fences. Much of the farming is done on terraces on the slopes of the limestone hills. A stone fence a few feet high is built along the hill, and the earth above is flattened to form a shelf, level with the top of this fence. The whole side of the hill may be ribboned with a succession of these terraces (#60). A cistern, often a pit in the rock above the highest terrace, collects the rainfall or may be fed by trickling springs. During the long dry season, from May till October, a stream is led from the cistern, along the highest level, down to and along the second terrace, and so on until the last trickle of the lessening stream is exhausted at the end of the lowest terrace.

Two crops. Two crops can be grown annually, one of wheat, barley, lentils, or millet during the cool, wet winter, when no irrigation is needed; the other of olives, figs, grapes, almonds, oranges, melons, and all our common vegetables during the summer, when water from cisterns, springs, wells, or streams must be spread over the parched land (#57).

Growing cereals. The drifting of the first clouds of autumn from over the green sea is a warning that the rains are coming and the grain must be sown; this process is always done by hand, the same as when the sower of New Testament times went forth to sow, and more of it now falls upon stony ground than in those days. Indeed, one wonders how anything can grow in such a rubble of loose stones. It is only after the grain is sown that the fields are ploughed, which process does not turn up the soil, but loosens the stones and soil to the air, causes the seed to drop down into the crevices to be covered, and permits more eager absorption of rain-water. Their homely plough, a wooden wedge coated with iron and stuck on the end of a wooden handle, is excellent for stirring up the stones, but is slow to operate. Oxen, donkeys, camels, or even women drag it up and down the fields. The abundant moisture causes a rapid growth even in the niggard soil; and the crop, none too heavy, is ready for the reaper by the end of April, when the rains are over and the sun is bright. The straw is all cut with a sickle, a handful at a time, and carried on animals' backs to the village threshing-floor. As these living bundles go along the road, one can hardly make out under each mass a very small donkey.

Threshing. The straw is piled high on a level stone or hard earth platform, and oxen, donkeys, indeed all animals available, tied together tramp with their iron-shod feet back and forth for many hours until the straw is ground to fine litter and the loosened grain has settled to the bottom. Sometimes a board or toboggan-like structure is drawn over the straw (#61). The afternoon sea-breeze blows away the chaff and the litter as the threshed straw is tossed into the

#61

THRESHING WHEAT

How are the sheaves carried to the threshing floor? How is it threshed? What is the man doing?

air. The wheat or barley, after much sifting, sorting, and hand-picking, is placed in the house bins for winter use.

Greenery on the terraces. The summer crops on the terraces form a beautiful contrast to the dull brown and gray of the dry uncultivated tracts of rock. Olive trees (#62), fig trees, and vineyards are on the outer half of the terrace, the vines hanging over the edge; melons, cucumbers, and other vegetables spread over the inner half of the terrace. In the early summer, the picture, as one looks from below, of cascade after cascade of brilliant green vines, relieved by the darker green of the olives and figs, the warm red-brown of the soil, and the gray of the stone walls peeping through here and there, is very beautiful.

Olives and figs. The olive tree bears fruit for hundreds of years and, under the hammering it gets yearly to knock off the fruit and the ruthless pruning chiefly for firewood, it is no wonder it looks gnarled, distorted, and time-worn (#62). The olives are gathered from the ground in baskets and either pickled in brine or crushed and pressed under a roller revolving over a stone floor to extract the

#62

EXTRACTING OLIVE OIL

A roller is rolled over the olives, the expressed oil runs through an opening in the side.

oil. Pickled olives are eaten by the handful, and the oil replaces butter and lard for cooking. The figs are pulled and generally dried for winter food.

The lonely shepherd boy. The farmers are usually herdsmen also, and have sheep, goats, cattle, and sometimes camels. As fields are often not fenced, a *shepherd* (#60), often a boy, has to watch his flocks continually. Frequently he takes them many miles from the village to find suitable pasture in that miserly land and may be away for weeks at a time. So tender is he to his flock that they follow him to the pasture, and so many the dangers, that they need his constant care day and night. By day they may stray away, become entangled in thorns, or break a leg in a crevice; and by night they may be seized by stray wolves, lurking hyenas, and bedouin robbers. The shepherd may be alone in the solitary wilderness for days or weeks. He lessens his loneliness by playing on a homemade wooden flute (#60) and by spinning wool into thread, while he guards his flocks. He carries a short light rod and a long heavy hooked staff, from which a pull on the neck of a wandering sheep turns her in the right direction. In a bag, made out of the whole skin of a kid, he carries a flint, steel, and tinder to make a fire, his lunch of olives and bread, a knife, and a sling from which he can throw pebbles with great skill (#60).

The return of the Jews. The steady stream of Jewish settlers, who have entered since Palestine came under British rule and who have brought with them from central Europe more modern methods, possess farms mostly along the level strip bordering the Mediterranean Sea and in the fertile plain of Esdraelon. They have increased the production of oranges many times (#56), and as Jaffa oranges have a flavour equal to the best, they are finding a ready market. Their well kept vineyards are producing increasing quantities of grapes (#56), which are crushed to wine, or dried to raisins. All their farms are fenced, and they use machinery similar to that used on small farms in Canada.

Village life in Palestine is almost as unchangeable as the face of their limestone rocks, and as simple as when Jesus shared their humble life in Nazareth and on the shores of Galilee.

The grinding of the mill may be heard. Their food is largely the fruits of their own soil. Wheaten flour is the staff of life. It is ground in a hand mill as ancient as Abraham, and as wide-spread over south Asia as cotton clothes. It consists of a lower stone with a saucer-shaped hollow in which sits the upper hollow cylindrical stone (#58). The wheat, rice, barley, millet, or other cereal is fed in at the top of the cylinder, which is rotated back and forth around its axis by one or two handles. The flour or meal is forced out around the edge of the bottom of the cylinder. So slow is the grinding process, so numerous are the members of a family, and so large a part does wheat flour play in the daily diet, that the mother spends four or five hours every day at this dull, hard work, and the sound may be heard late into the night. No sound is heard so frequently and so steadily by half the population of the world as the rasping rubbing of this weary mill.

Making bread. The flour is kneaded into dough, shaped into bread or cakes, and taken daily to the village oven to be baked. The women gossip while they wait their turn, and not a minute's use of the oven is wasted, for fuel is hard to get in these arid lands. Bunches of weeds, twigs of olive or fig, and dried dung are the chief food for the flames of these ovens. All this fuel may have been carried long distances and stored with great care. This bread is either as large around as a hat brim and about as thin, or the shape and size of a thick scone.

The daily meals. Bread, with a few olives or figs for breakfast; the same, with perhaps the addition of curds from sheep's milk at noon; and a hot meal in the evenings with cooked onions, peppers, and perhaps tomatoes, is their daily food. Lentils, like small peas, cooked with oil and onions and seasoned with pepper and salt, is a much relished dish. They eat little or no meat except when a sheep or fowl becomes ill or an ox or camel falls sick beyond recovery; then the poor peasant hurriedly invites in all his friends, and they feast on the flesh of these animals, snatched from a natural death by the butcher's knife. At a meal, all sit around on the floor, with the bread and bowl of food in the midst. The loaf of bread is broken apart by hand, and each takes his share of vegetables in his hand from the common bowl. After the meal the hands are washed.

Water is precious. Water, one of the most precious commodities, is carried in earthenware jars on the womens' heads from cisterns, wells, or springs, often miles away. To save the carrying of water, the soiled clothes are brought to the spring or cistern to be laundered. Ashes or sandy clay are used instead of soap, which they cannot afford.

The clothing of both men and women is very similar to that worn by the Arabs of the desert (p. 109), which has been already described. In the chilly winters they may wear sheepskin coats.

Village trading. These peasants have a small income. The women churn milk in skins as they do in the desert (p. 107). Butter, sheep, goats, and cattle, olives, olive oil, some lentils, oranges, vegetables, and perhaps figs are sold in the larger towns. The spun goat's hair and sheep's wool are woven into cloth by the village weaver. A good part of the money obtained for these commodities is required to pay high interest on money borrowed and often as rent for their farms. With the rest they buy clothing, matches, tobacco, coffee, medicine, and a few other necessities.

Change in the unchanging east. But change is seen on every hand even in this, till now, changeless land. Good roads, motor cars, and railways are rapidly altering this ancient part of the unchanging east (#56). Motor cars are putting awkward camels and tiny asses into the discard, even tractors are beginning to replace yoked oxen and asses in front of the plough. Reapers are doing the work of a hundred sickles, and aeroplanes now fly with the eagle.

CHAPTER X

LIFE IN THE MOUNTAINS OF SWITZERLAND

Along the border of Lake Geneva. We have just left the beautiful city of Geneva, which is as old and historically interesting as Julius Caesar (#63). The Lake of Geneva is set as a sky-blue crescent among the green hills sloping gradually toward the north, but rising more suddenly and sternly in the south, until they form a brilliant setting of rugged snow-capped peaks, glistening in the sun. If the sky is clear, the eye can just catch a glimpse of those mighty giants, Mont Blanc and Rosa (#63), the first the highest point in Europe, the second, a little to the east, the highest peak in Switzerland. A more careful study of the northern slope reveals richly clad fields, containing lines of grape-vines and vividly beautiful gardens; prosperous, well kept, smiling villages are dotted along its surface. As our train speeds eastward along the north edge of the crescent lake, a quickly-changing panorama of ancient villages, lively towns, splendid cities, and solid old castles delights the eye.

The valley of the river Rhone. Now we have left the lake and are ascending the valley of the river Rhone. The waters rush because the grade is steep; the train puffs forward frantically, grinds against the high grades, swings around hair-pin curves, plunges through dark tunnels, and trembles along the edge of frightful precipices as it moves up the river. The valley, two miles wide, has fierce mountain masses frowning down on it from both north and south. Seething, muddy water, in foaming rapids and splashing waterfalls, plunges down gaping lateral gorges in these steep mountain slopes to spill into the Rhone.

Alpine vegetation. It is summer. The fat soil of the flats along the river is decked with lavish vineyards; fields of potatoes; healthy orchards, well laden with apples, pears, plums, and cherries, and smiling meadows, ready to yield two or three crops a year. Chestnut and walnut trees border the fences and decorate the neat gardens of the Swiss farmers; not a square inch of soil goes to waste. The lower slopes, where uncultivated, are clothed with forests of oak, beech, and maple, but at higher levels these are gradually shouldered out by the gloomier green of pine, fir, and tamarack; these soft woods become smaller, more gnarled, and more scattered as they creep up the cold

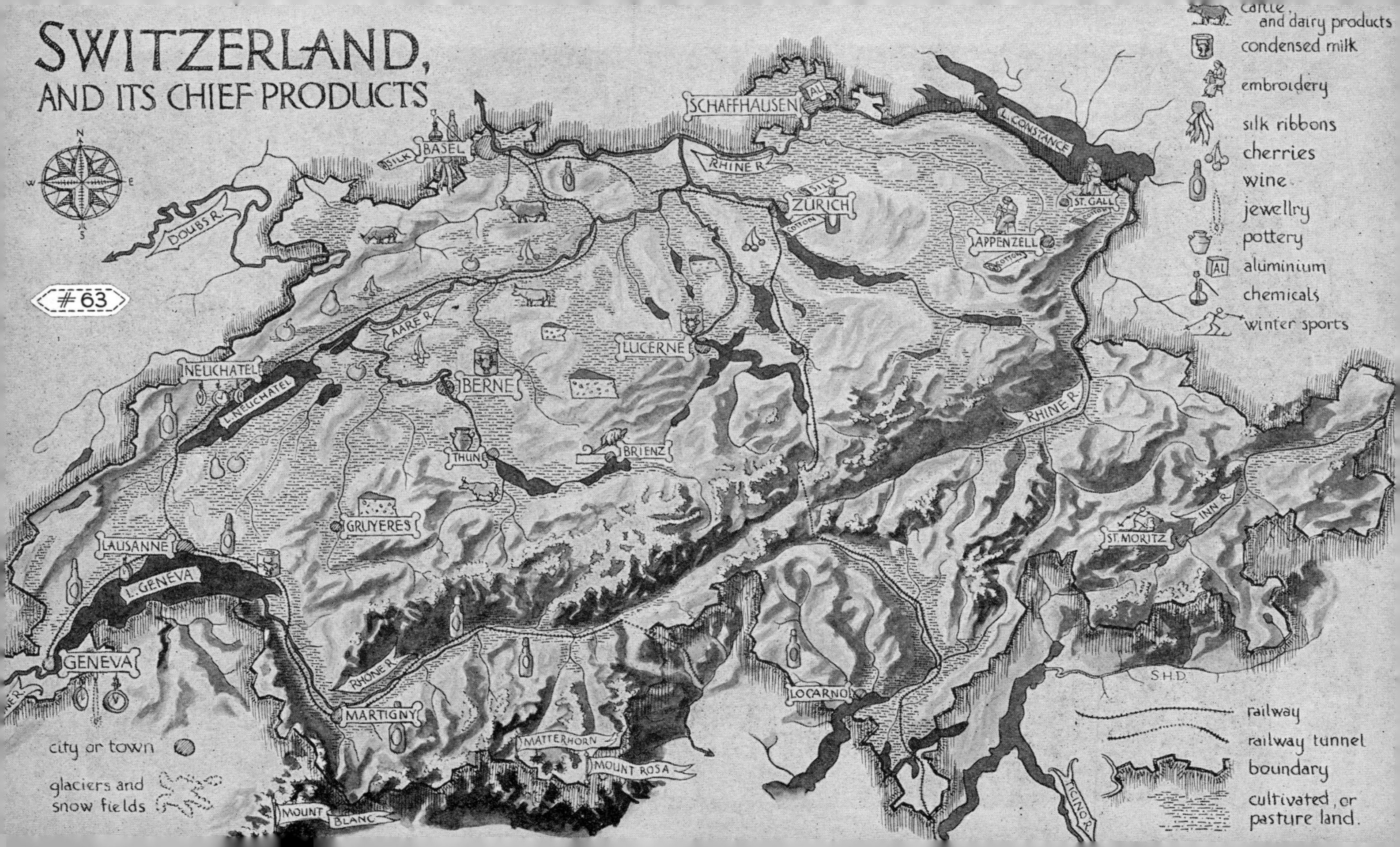

SWITZERLAND,
AND ITS CHIEF PRODUCTS
N
S
E
W
63
cattle, and dairy products
condensed milk
embroidery
silk ribbons
cherries
wine
jewellry
pottery
AL
aluminium
chemicals
winter sports
railway
railway tunnel
boundary
cultivated, or pasture land.
city or town
glaciers and snow fields
BASEL
SILK
SCHAFFHAUSEN
RHINE R.
L. CONSTANCE
ST. GALL
ZÜRICH
SILK
COTTON
APPENZELL
COTTON
DOUBS R.
AARE R.
LUCERNE
NEUCHATEL
L. NEUCHATEL
BERNE
THUN
BRIENZ
RHINE R.
INN R.
ST. MORITZ
GRUYERES
LAUSANNE
L. GENEVA
GENEVA
RHONE R.
MARTIGNY
MATTERHORN
MOUNT ROSA
MOUNT BLANC
LOCARNO
TICINO R.
S.H.D.

slopes, until finally they give way at 6000 feet to perfumed grasslands smothered in flowers. Still higher, the bare ragged rocks begin to push through the green meadows and finally take complete possession of the rugged slopes. A crown of everlasting snow and ice hides all but the steepest slopes of the higher peaks. Little houses can be seen, perched like eagles' nests on the wind-swept terraces, well up the forbidding slopes of these cold, lonesome slopes. It is up to one of these villages (#64) along the valley of one of the torrential streams that we are going, in order to find how it is possible for the mountaineers in these little houses to fight out their lives against the unfriendliness of these bleak, barren, dangerous peaks.

Ascending the mountain gorge. No railway dare climb these roof-like slopes to the south. But postal motor-coaches transfer us upward on wonderful government roads, built by these skilled mountain engineers, where else ascent would be impossible (#64). Upward we wind and zigzag, sometimes grinding the gears desperately against the high grades, groaning cautiously along deep dark cuts in the solid rock, swinging around projecting shoulders, and piercing dark rock tunnels. At last we have gone as far as even the Swiss government can build roads. By mule track we reach our goal, the little group of houses which forms the Alpine village.

An Alpine village. The rough houses scramble over both slopes from the turbulent stream which runs down from the icy peaks to the Rhone valley. The slopes are a checker-board of small fields (#64) separated by rail and stone fences; the houses and barns are the checkers. On some of the steeper slopes, where rain would wash the soil away, successive parallel stone walls one above the other, like a flight of long steps, protect a number of well cultivated terraces. The effort of building such walls and the roads we find everywhere in the Swiss mountains, is colossal, quite comparable to the throwing up of Egyptian pyramids, and requires equal skill. However, as successive generations of mountaineer farmers, each almost fanatically attached to his valley, have century by century contributed their share, the seemingly impossible improvements have gradually taken their present advanced form.

Village houses. The low houses are built of the pine and fir logs (#65), whose odours are wafted daily down the mountain sides with every wind that blows. There is no trouble in obtaining stone for foundations in these rough valleys, where rocks are man's chief

#64 AN ALPINE VILLAGE

enemy. The walls are thick, the windows small and double in winter, and the ceilings low—all qualities necessary to conserve heat in a country where timber for fire-wood is none too abundant, and the fierce mountain blizzards have to be shut out. Indeed, on account of these gales the shingles have to be held down by heavy stones (#68). For economy in heating and saving in building materials, the barn is usually combined with the house, being either below or alongside it. A room open on the south side is well filled with fire-wood, which the summer's sun and breezes will dry.

#65

THE SPRING PROCESSION UP TO THE ALPS

A homely home. The home and garden show that these Swiss farmers are neat, clean, thrifty, and have good taste. The walls are not plastered, but panelled with pine and tamarack, well polished, and suitably carved. Often mottoes are cut on the walls. The furniture is simple, homely, but artistic, and all made by their own hands. Usually well filled book shelves show that although they are far from the world they do not neglect education. Indeed, the school and the church are the most imposing buildings in the village. A big porcelain stove, divided between two rooms, is the most useful article in the house.

The crops. The fields are delightful. As a rule the soil is heavy and rich, often very damp, and therefore particularly suited to growing hay, of which they sometimes cut three crops during the short season. In some parts, however, a niggard soil has to be laboured and belaboured early and late in order to extract from it the barest existence. A small field of wheat or rye will supply bread for the winter; a field of potatoes will supplement the bread; and perhaps the farm may boast of a field of flax, which will be converted into pure glistening linen during the long winter.

Farms in fragments. The farms are all small, seldom more than thirty acres, and their arrangement, or rather disarrangement, extraordinary. A single farm may be composed of twenty very small fields, scattered over the whole valley. The fields of a farm are as badly mixed as if a fierce cyclone had whirled them all aloft and let each settle down where it happened to be tossed. This disorder probably has arisen through the same few families living in the same valley for centuries. Farms were divided and sub-divided many times through inheritance; unions among different families by intermarriage often transferred adjacent fields to different owners, until this strange and inconvenient hodge-podge in the ownership of land has developed.

Farms without men, fields without cattle. In June this village seems to be without men and without cattle. Only women and children are in the houses; the farmyards and stables are empty, and the fields are all in crop. From earliest dawn to the last gleam of twilight the women are at work, for the season is short, and crops must be harvested while the sun shines. The hay is being cut, often by hand, fastened on racks to dry, and often dragged along the ground in bundles or carried to the barn on women's backs, though sometimes the larger farms may boast one horse and a two-wheeled cart. Such strain on these poor women and children is often tragic. The women look old, stooped, and withered while they are still young in years. The pathetic children are often stunted, with their young faces sober and mature, while they should be still gay and youthful.

Carnival of spring. To solve the mystery of the absence of cattle and men let us go back to April. The snow is just disappearing from the village, which is in the forest belt on the side of the mountain. The men are now in the village; the cattle, sheep, and goats have been almost sealed in the stables from the cold for seven long months, eating the hay that was harvested the preceding year. The whole village, fretting under the long, harsh winter, has longed for spring with its warmth, sunshine, greenery, and fresh food. By April all eyes are scanning eagerly the white upper slopes. Here and there dark patches appear in the snow, laid bare by the warm sunshine. Then almost overnight these empty spaces are miraculously changed into as many tapestries, radiant with white anemones, golden crocuses, and pink soldanellos. Of almost unbelievable luxuriance is the April carnival of blossoms right up to the edge of the snow. The glad coming of this delightful season is celebrated in these valleys with processions, games, and many other quaint practices, some, centuries old. In one valley in which black rye bread is exclusively used, the departure of bleak winter and the return of smiling spring is celebrated by every man, woman, and child receiving a generous slice of white bread.

Procession to summer pastures. The great event, however, is the departure of the men, cattle, sheep, and goats for the high Alpine pastures, just freed from snow (#65). The cattle of each farm are arranged in a long procession, with the herdsman in front. The leading cow in each herd is wreathed in flowers, and a huge vase-shaped bell is hung around her neck by a gaily patterned leather collar (#65); smaller bells deck the necks of every other member of the well-groomed black and white cows, and snowy white goats. The herdsmen are dressed in gay costumes, characteristic of the village. Decorated wagons, loaded with simple furniture, supplies, and dairying utensils bring up the rear on their steep march up to the spacious pastures, where they will spend the summer. Usually the women and children are left behind to tend the farms, but it is not uncommon, if they can be spared, for some to ascend and assist with the hard work up on the steep slopes.

In May they ascend to the *low pastures*; by the middle of June these have been cropped close by the hungry herds; as the snow has further retreated, they go higher to the rougher *middle pastures*. July and August finds them so high that the pasture is often bordered by snow, or ice, since the tongue of a white glacier may push its way down between the green slopes of the valley. The more luscious pasture fields are given to the cattle, and the high, stony upper parts, where the rock has not yet been converted into soil, and where bare rocks are projecting upward through the stones, is given over to the goats, which find good grazing on the harsh Alpine shrubs (#66).

Rich pastures. These *upper pastures* are rough but rich. The rock waste of the glaciers is distributed on steep, irregular slopes below the snowline. These slopes are above the clouds in winter but below them in summer. They, therefore, get frost in winter to crumble the rock and are drenched with fog, dew, and rain in summer. Fine soil, abundant moisture, rare air, and bright light produce quick growth and a moist appetizing grassland, knee-deep in flowers, which is the best of cow pastures and too good for sheep or goats.

#66

THE HIGHER PASTURES

Describe the chalet. What marks the middle line of the glacier? What issues from its lower end?

These unfenced Alpine pastures are owned by the whole community, and the number of cattle that each family is allowed to pasture is decided by a meeting of all the men of the village, or *commune*.

The chalet. The men and boys are intensely busy from dawn to dark and suffer hardships accompanied by danger. They live in rough summer houses, called *chalets*. If the chalet is located not too far above the forest, timber is laboriously dragged up the steep slopes and a log cabin built on a stone foundation (#68). In the

#67 BRINGING FIREWOOD DOWN FOR WINTER

highest regions the cabin is made wholly of stone (#66). The roof is low, nearly flat, and heavily weighted with stones to keep the shingles from being blown away by violent gusts of wind. A crude table, with home-made benches, a big iron pot in which to warm the milk for cheese-making, other cheese-making utensils, and a truss of hay to sleep on, completes the scanty furniture. Often a small store-house is raised on four stone stilts, one at each corner, to protect the fragrant cheese from field mice.

A busy day. Every morning at sunrise the cattle and goats have to be milked and taken to pasture. Then the milk has to be warmed and converted into that delicious cheese for which the Swiss are famous and which is on the dining tables in the best hotels of every land. Wood has to be cut in the forest and brought up those back-breaking slopes to the chalet to feed the fires so necessary for cheese-making. On slopes which are too steep for cattle to graze, the hay has to be cut, dried and prepared for taking down later to the village, by sleigh (#67). Often extensive repairs have to be made to

#68

THE CALLING HORN

the chalet on account of damage by snowslides, or *avalanches*. Great masses of snow become loose and sweep down the mountain, destroying everything in their path. The cattle have to be brought back to the chalet at night and again milked. For calling the cattle together, and often for calling the people to vespers down below in the village, they have ridiculously long wooden horns, turned up at the end (#68). Although these are capable of making only a few harsh notes, the sound carries many miles and produces wonderfully sonorous echoes, which, mingled with the tingling cowbells and roaring waterfalls, is a memory which, once heard, is never forgotten.

Each morning some of the milk and cream may be brought down to summer hotels, so that guests can enjoy the freshest dairy dishes for breakfast.

#69

INDUSTRY IN SWISS HOMES

What is each person making?

The Autumn retreat. Before the end of August, winter begins to show his harsh face in these high regions, and the migration through the middle and lower pastures is repeated in reverse order. By October the men and cattle are under shelter in the Alpine village for the winter, which is already showing its cold steel hand. Both houses and stables are now sealed tight for warmth.

Winter's work. The winter is almost as busy as the summer. The cattle have to be fed and milked regularly. The men have to bring the hay from the higher slopes as soon as small sleighs can be used, and they have to cut wood in the forests above, and with

#70

VILLAGE OVEN

great toil and danger, bring it down to be stored (#67). The women are still busier. Besides the ordinary housework and attention to the milk, they spin and weave the sheep's wool, and extract the fine fibres from flax stems and convert it into beautifully glistening white table linen and Sunday garments. The farms are so small and often so barren that, in spite of the strictest thrift, careful cultivation, and unceasing work, men and women must do handiwork throughout the long winter evenings to add to their small earnings. The women work rich embroidery (for which the Swiss are noted) and crochet delicate lace. These are sold at high prices in the cities. The men and boys are also busy. Perhaps they have taken a course at the wood-carving school, and cut out beautifully-finished and artistic toys (#69).

Bread as hard as a rock. The food of these sincere mountain farmers is as simple as their lives and is largely the product of their own hands. Bread, milk, cream, curds, butter, dried and fresh

#71

A SIX WEEKS' SUPPLY OF BREAD

fruits, are the staple products; and coffee is the chief food-stuff which they have to buy. The method of bread making in some villages is interesting. Rye responds best to the harsh soil and is grown on every sunny, sheltered slope. Successive stonewalls support terraces around the slope and every terrace has its rye field. The women cut it with sickles, while the men are still up at the mountain pastures, and carry it to granaries, where it is threshed by hand during the winter. The grain is ground at the village mill into a dark, nutritious flour which contains the bran. The bread is baked in the public oven, a crude affair of stones and bricks covered by a wooden roof (#70). Each family has the use of this oven in turn, once every six weeks. On baking day the women carry the dough to the oven, and when their twenty or thirty loaves are baked, they store them in high racks in the granary (#71). This bread is most wholesome, but it becomes as hard as rock before the six-weeks' batch is all eaten. Though it requires a powerful jaw and strong teeth to chew this rocky loaf, it is largely responsible for the beautifully polished and strong teeth of the unspoiled mountaineers.

The dreaded avalanche. These peaceful looking villages, nestled in among the mountains, often rest under the shadow of a great danger, so that the joyful song needs to be replaced by a prayer that home and all that is dear may not be blotted out in an instant. On the steep slopes above hang great masses of snow, which sometimes become loosened, tear down the valley like whirlwinds, and swallow up the whole village. These *avalanches* are a constant dread; but still more fatal and desolating are landslides, which are great masses of rock and earth that become detached through frost and carry death and destruction as certain as a flood of lava from a volcano. Often the villagers build triangular stone dams to divert these avalanches to both sides. If they have a forest on the slope above it is still better protection, though at times these fearful avalanches will snap trees like pipe-stems and continue their wild career.

Centre of civilization. Life has become more complex for these mountain farmers than for the desert Arabs, or the Amazonian Indians. Switzerland for a thousand years has been at the very centre of the sea of Western civilization, whose culture has beat against her mountains and passed deeply up her valleys. This influence has left its mark on the Swiss, who to-day are skilful in mind and hand, and artistic in taste, and who value the schoolmaster so much that, when they make seasonal migrations up or down the valley, they carry the schoolmaster along, so that the children's education will not suffer.

The mountains are changing. During the last fifty years this mass of mountains with its bracing airs, flower-clad valleys, glistening glaciers, and steep, inaccessible peaks has become a challenge to the adventurous, a magnet to the lover of beauty, and the playground of Europe. In summer it tempts the artist, the mountain climber, and whoever seeks rest and health by means of pure, clear air, simple wholesome food, vigorous exercise, and retreat from the exciting whirl, and rasping noise of cities. In winter it attracts the pleasure seekers from all over Europe who delight in such winter sports as skiing and skating. Many of the Swiss men, if they can be spared from the little farms, act as guides to mountain climbers, and assistants at sports and summer camps. As they receive good wages from tourists, they add, in this way, enough to the family income to make home a little brighter, and allow pleasures otherwise beyond their reach. Indeed, so famous have these sturdy mountaineers become, that Swiss guides form the backbone of almost every expedition that scales high peaks, whether it is the mighty masses of the Rocky and Coast ranges, or the towering giants of the Himalayas.

CHAPTER XI

THE GROWTH OF CANADA

Centres of exploration. The history of Canada, from the discovery until in 1763 it finally became British, is an almost steady strife between France and England for the mastery. The exploration spread out from four focal points, the Atlantic coast, river St. Lawrence, Hudson Bay, and the Pacific coast. The daring French explorers scattered over the interior like a swarm of bees with Quebec as the central hive, they also occupied much of the territory of the Maritime Provinces; while the British, masters of the sea, explored Hudson Bay and the Arctic channels, and from that icy centre sent out their gallant band of men to the harshest and most dangerous parts in the bleak forbidden lands of the north and the fierce mountains of the west. They also attacked the land from the distant Pacific coast, though the exploration of this region began only 150 years after the nearer regions were conquered.

The Vikings. The first explorers were the fiercest bands of dashing sailors and daring pirates that the world has ever known. They were the *Vikings,* who swarmed out from Norway and Denmark around all the coast of Europe, until there was hardly an inlet of the Atlantic, the Mediterranean, and even the Black and Caspian seas, that did not tremble at their bloody raids. They, in their small boats, propelled by oars and one square sail, fought across the storm-tossed waves of the Atlantic Ocean to *Iceland* and *Greenland,* and about the year 1000 A.D., dashed south from Greenland to Labrador, Newfoundland, and Nova Scotia. There is evidence that perhaps they even preceded Henry Hudson by five hundred years in entering Hudson Bay, since recently a Viking sword was found under two feet of soil between Lake Nipigon and James Bay. How it got there is still an unsolved mystery. But the Vikings disappeared from North America and left hardly a trace behind them, unless it is this sword.

The beginning of strife. The real discoverer of Canada was *John Cabot,* who in 1497-8 landed on the Atlantic coast and claimed the whole district for England by right of discovery. In 1524 *Verrazano* made a much more thorough exploration of the coast of the Maritime

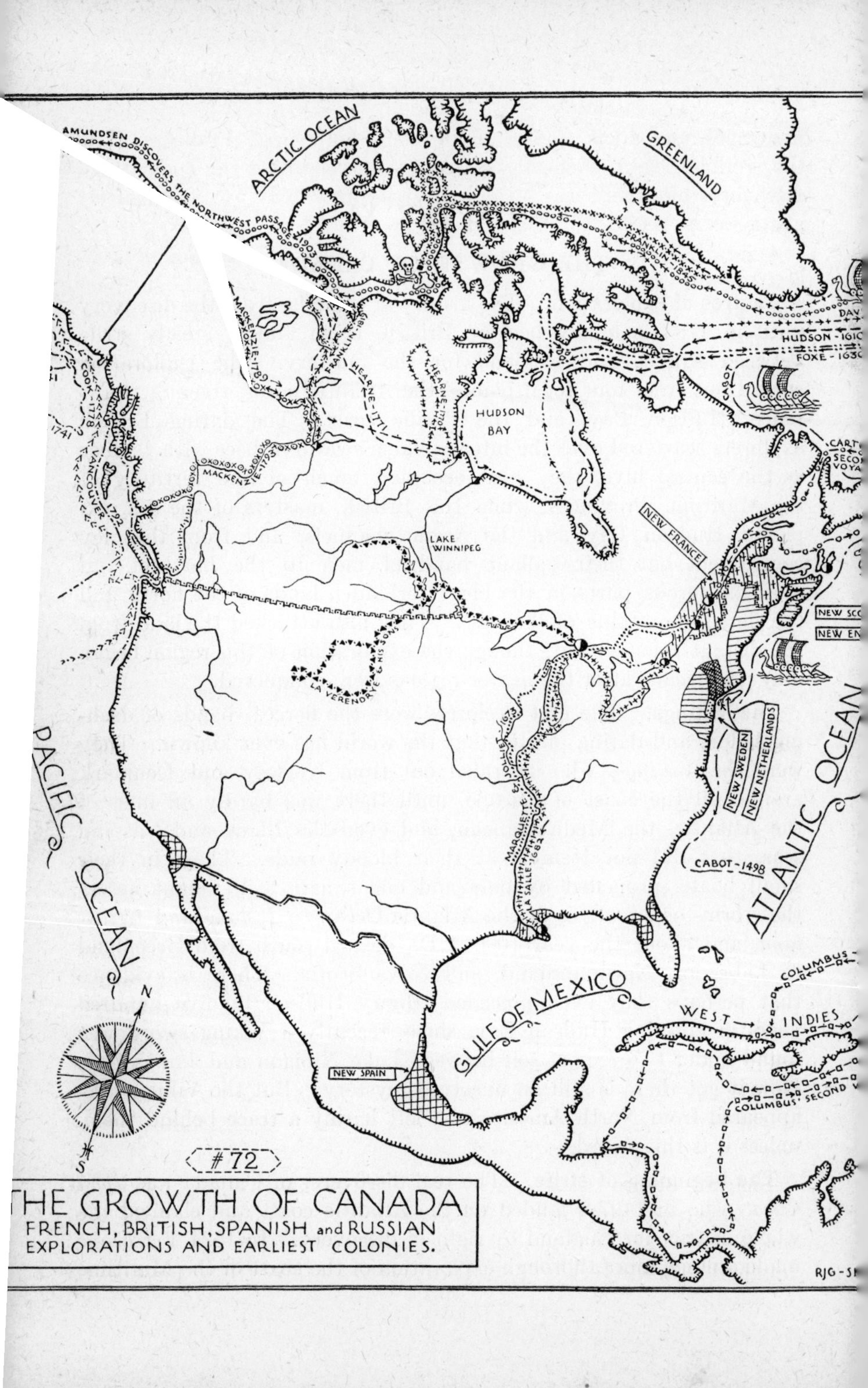
ARCTIC OCEAN
GREENLAND
AMUNDSEN DISCOVERS THE NORTHWEST PASSAGE 1903
FRANKLIN 1845
FRANKLIN 1819
MACKENZIE 1789
HEARNE 1771
HEARNE 1770
HUDSON 1610
FOXE 1630
CABOT
HUDSON BAY
COOK 1778
VANCOUVER 1792
1741
MACKENZIE 1793
LAKE WINNIPEG
NEW FRANCE
LA VERENDRYE & HIS SONS
NEW SCO
NEW EN
MARQUETTE & JOLLIET 1673
LA SALLE 1682
NEW SWEDEN
NEW NETHERLANDS
PACIFIC OCEAN
ATLANTIC OCEAN
CABOT 1498
GULF OF MEXICO
COLUMBUS
WEST INDIES
COLUMBUS' SECOND V
NEW SPAIN
N
S
72
THE GROWTH OF CANADA
FRENCH, BRITISH, SPANISH and RUSSIAN
EXPLORATIONS AND EARLIEST COLONIES.
RJG-S

Provinces, and took possession of the country for France. Thus the trouble began, each of the two nations claiming the country by discovery; but there was no serious strife for nearly a century afterwards, as no settlements were made.

A new Europe. The seventeenth century saw the beginning of settlements all along the east coast of North America. In 1608 *Champlain* planted the first settlers in Quebec and founded New France. In 1620 the Pilgrim Fathers landed at Plymouth Rock in Massachusetts and founded New England (p. 173); in 1621 Sir William Alexander was given possession of Acadia and founded New Scotland, or Nova Scotia; in 1614 the Dutch began to trade at Manhattan (p. 173), now New York, and founded New Holland, or New Netherlands, in 1621; the Swedes settled on Delaware Bay in 1638 and founded New Sweden; and several other English settlements soon throve at Maryland, Virginia, the Carolinas, and Georgia. These colonies, New France, New Scotland, New England, New Netherlands, and New Sweden, were all inclined to take up the quarrels of their parents in Europe and to claim wide areas surrounding their little settlements. They viewed with envy and anger the bullying neighbours, pushing at their side and back doors for room to spread. Wars, or rather raids, began almost at once, and New Netherlands and New Sweden were soon swallowed by the British; so the real struggle settled down to a series of raids by land and sea across the barriers separating English and French.

English farmers against French hunters. From the first the English outnumbered the French. As time went on, the English settlements were strengthened by thousands of colonists, often driven from their home land by persecution. They were a fine type; sober, moral, freedom-loving, and set on succeeding. At once they settled down to the tame, hard work of farming, fishing, trading, and making money. They soon outnumbered the French ten to one. The French settlers were far fewer, more ignorant, knew little of freedom, and were willing to be guided by seigniors, priests, and governors. They were drawn away from the humdrum life of farming by the fascination of the fur-trade, which lured them away from a settled life into the forest to hunt game, trap fur bearers, and explore the country.

Snarling across the barrier. The French along the St. Lawrence were separated from the English along the Atlantic coast by the forest-clad ridge of the Appalachian Mountains (#73); and for two hundred years each stood snarling and ready to spring, the one at

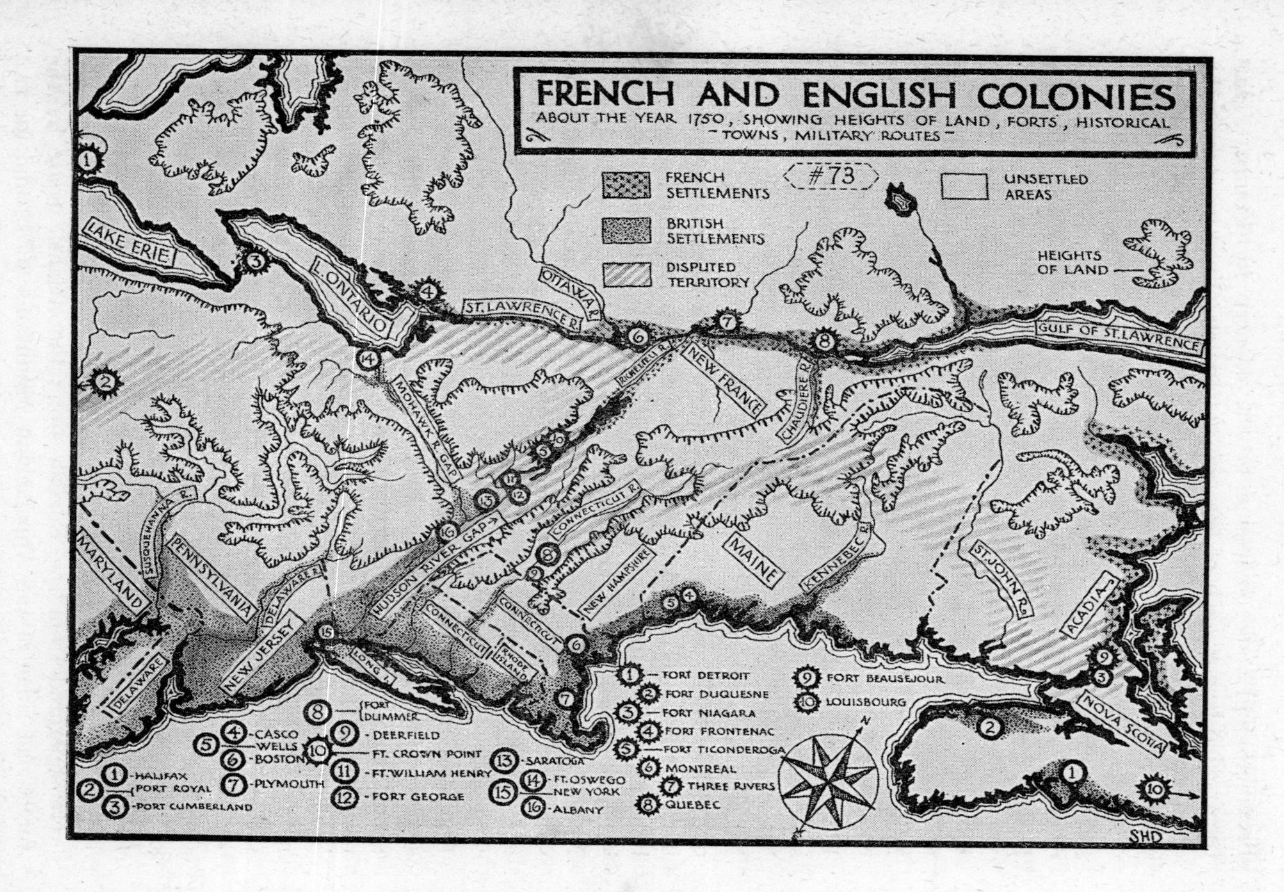
FRENCH AND ENGLISH COLONIES
ABOUT THE YEAR 1750, SHOWING HEIGHTS OF LAND, FORTS, HISTORICAL
– TOWNS, MILITARY ROUTES –
#73
FRENCH SETTLEMENTS
BRITISH SETTLEMENTS
DISPUTED TERRITORY
UNSETTLED AREAS
HEIGHTS OF LAND
LAKE ERIE
L. ONTARIO
ST. LAWRENCE R.
OTTAWA R.
GULF OF ST. LAWRENCE
RICHELIEU R.
NEW FRANCE
CHAUDIERE R.
MOHAWK R. GAP
HUDSON RIVER GAP
CONNECTICUT R.
NEW HAMPSHIRE
MAINE
KENNEBEC R.
ST. JOHN R.
ACADIA
NOVA SCOTIA
MARYLAND
SUSQUEHANNA R.
PENNSYLVANIA
DELAWARE R.
NEW JERSEY
DELAWARE
CONNECTICUT
CONNECTICUT
RHODE ISLAND
LONG I.
1 FORT DETROIT
2 FORT DUQUESNE
3 FORT NIAGARA
4 FORT FRONTENAC
5 FORT TICONDEROGA
6 MONTREAL
7 THREE RIVERS
8 QUEBEC
9 FORT BEAUSEJOUR
10 LOUISBOURG
1 HALIFAX
2 PORT ROYAL
3 PORT CUMBERLAND
4 CASCO
5 WELLS
6 BOSTON
7 PLYMOUTH
8 FORT DUMMER
9 DEERFIELD
10 FT. CROWN POINT
11 FT. WILLIAM HENRY
12 FORT GEORGE
13 SARATOGA
14 FT. OSWEGO
15 NEW YORK
16 ALBANY
N
SHD

the other, across this rugged obstacle. It was in most parts an impossible barrier for armies, but there were two danger gaps through which bands of soldiers could be thrust (#73). One, from Quebec city to the New England States along the *Chaudiere* and *Kennebec Rivers*, was rough and hard to cross; the other along *Hudson River*, *Lake Champlain*, and the *Richelieu River*, navigable for the most part, pointed from the St. Lawrence like a pistol at the heart of the English colonies in New York state. Of course it offered just as good a path of attack for the British colonies to strike at *Montreal* and *Three Rivers*. Indeed, above *Albany* it gave two easy routes for attacking Canada (#73) one north by *Lake Champlain*, the other west through the *Mohawk Valley* to *Oswego* at the foot of Lake Ontario. Soon both these bloody paths bristled with English and French forts to beat back intruders (#73). Besides, both colonies were open to attack by sea, but as the British navy was always strongest, Quebec and Nova Scotia were frequently attacked, their supplies often captured, and more than once taken and destroyed.

An even fight. The French had certain advantages. They were always united, the people obedient to their governors, ready to fight when they were ordered out, and since many were used to the dangers and hardships of hunting and roaming through the wilderness with the Indian savages, they were full of courage, ready for hardship, reckless of life, and skilled in bush warfare. They were friendly with all the Indians except the Iroquois, and usually had them fighting their battles.

The English colonists were spread in eleven or twelve separate colonies along the coast, all jealous of one another, frequently quarrelling, seldom willing to act together, and more intent on farming the lands, fishing the waters, trading on the ocean, and making money, than in fighting the French and extending their colonies. They despised the Indians, treated them rudely, and considered them little better than vermin. They expected regular troops from Great Britain to fight their battles. The French were usually successful in fighting on land, the British more frequently in sea attacks on Nova Scotia and the St. Lawrence.

Kirke takes Quebec. Settlement had hardly started before England and France were at war in the reign of King Charles I of England; *Sir David Kirke,* in 1629, a Britisher, half sailor and half pirate, with a French wife, started out with a fleet to drive the French out of Quebec, for they had already been driven from Acadia, or Nova

Scotia. Since the settlers in Quebec were few, the soldiers still fewer, and all nearly starved to death, he captured the fortress without firing a gun, in 1629, and Canada became British for the first time. England and France soon made peace, and as Canada was not considered to be of much value, both New France and Acadia were returned to France in 1632.

Acadia captured and returned. From 1632 until 1663, the French in Quebec were fighting for their life against the Iroquois Indians and had no thought of attacking the English colonies. However, the New England States always considered the French in Acadia to be intruders, lying like a dog in the manger across their front street to Europe. They attacked them at every turn. In 1654 the whole colony was captured by *Captain Sedgewick* from New England, but was again restored to France in 1667.

The increase in population of both Quebec and Acadia was very slow and mostly due to births in the colonies rather than to the arrival of new settlers. The following figures of population tell the story:

POPULATION

Quebec		Quebec		Acadia		P.E.I.	
1620	20	1688	11,562	1671	441	1720	100
1641	240	1692	12,431	1679	515	1728	330
1663	2,500	1695	13,639	1686	885	1735	542
1666	3,215	1706	16,417	1693	1,009		
1667	3,918	1713	18,119	1710	2,500		
1672	6,705	1719	22,530	1731	6,000		
1675	7,832	1720	24,234	1737	7,598		
1679	9,400	1721	24,951	1748	13,000		
1681	9,677	1727	30,613	1752	4,203	(British and German)	
1683	10,251	1734	37,716	1762	8,104		
1685	12,263	1739	42,701				
1686	12,375	1754	55,009				

French securing the West. Though expansion in population of New France was slow, its expansion in size by exploration was startlingly rapid. By 1658 *Radisson* had reached the Mississippi River (p. **305**), and by 1661 he was on the way to James Bay (p. **305**); by 1670 the dauntless *La Salle* was paddling down the Ohio and by 1670 he had secured for France the route to this river by a fort at Niagara (#73); and three years later *Frontenac*, the French governor, strengthened the hold of the French on the Great Lakes by building a fort at Kingston, first called *Cataraqui* and later *Frontenac* (#73). This also stood as a forbidding barrier across the route of the Indians who dared to divert furs to the English and Dutch colonists in New York.

The French and the English begin to show their teeth. By 1690 the French and English settlements had crept so closely toward each other that the forest wilderness of the Appalachian mountains no longer was able to hold them from flying at each other's throats. A new fighting ground had also arisen in the north by this time. The British *Hudson's Bay Company* entered that bay about 1670, built forts at the mouths of the rivers flowing into it, and the Indians' furs, which the French thought should flow down to Three Rivers and Quebec, were either pouring forth to the Hudson's Bay Company, or stealing past Fort Frontenac along the Mohawk Valley (#73) to the State of New York.

But the fur-trade was the breath of life to New France. By 1690 France and England were about to begin a series of wars in Europe, and their quarrelsome children in America were not slow to follow the bad example of the parent states. *King William's War* dragged along from 1689 to 1697. After a breathing space of three years, *Queen Anne's War* was fought for twelve bloody years, 1701-1713, and the British *Duke of Marlborough*, by a series of brilliant battles, brought Louis XIV, the haughtiest king in Europe, to his knees. Then after thirty years' peace *King George's War* (War of Austrian Succession), raged for four years 1744-1748. After eight years of a troubled peace, the final struggle of the *Seven Year's War* broke out in all its fury. These wars, all hatched in Europe, were fought with as much vigour in Canada as though the French on the St. Lawrence and the English on the Atlantic coast were interested in who should reign in Spain or Austria. Indeed, the colonists, seemed more zealous for the fight than the parent states in Europe, for they not only fought through all four wars but sometimes started before war was declared and continued the struggle after peace had been signed.

King William's War, 1689-1697

Bloody raids. In this, as in most of the struggles, the British struck from the sea, the French through the gaps in the Appalachians (p. 165). The French attacks were bloody raids, in which they and their savage Indian allies skulked stealthily on quiet English villages, pounced on the helpless people, slaughtered and tortured men, women, and children, burned their buildings, and then retreated, dragging away prisoners, only to hand them over to the Indians for future torture. In 1690 they made successful raids up the Champlain gap to *Schenectady*, and up the Chaudiere to small villages, *Salmon Arm* and *Casco*, in Maine (#73).

Port Royal taken for the sixth time. This stirred up the anger of the New England states, and they fitted out a fleet under *William Phipps*, who had the religion of a Puritan but the principles of a pirate. His first prey was poor Port Royal in Acadia, which had already been captured five times. He took it with ease, and his raw soldiers from New England seized everything that could be carried away, from the golden ornaments in the church to the pots and pans in the miserable Acadian kitchens. As soon as he had sailed away, the Acadians returned and in a year had restored their poor homes.

First British possession in Canada. Flushed with an easy victory Phipps returned to Boston to get ready for bigger game, and the next year set sail with forty ships to capture the great fortress of Quebec. But the old veteran soldier, Frontenac, was there to meet him, and after a short struggle, in which Phipps learned he was no match for Frontenac, the ships returned ingloriously to Boston. The French were also successful in Hudson Bay where they captured most of the British ports. After all the struggle, cruelty, and bloodshed, the *Treaty of Ryswick* closed the war and left the English in command of the district around Hudson Bay called *Rupert's Land* and the French in control of New France and Acadia. Rupert's Land is the first part of Canada that became permanently British (#74).

Queen Anne's War 1701-1713

Raids on New England. The British Colonists were farmers, fishermen, and merchants. They wished to be left alone at their work, and had no great desire for conquest or spreading to the west. The French were fur-traders, explorers, and adventurers, with desires to build a big American empire. They began the war by further attacks on New England settlements. In a series of stealthy raids they and their Indian allies spread terror among the English. They would start out from Quebec, often in winter on snowshoes, follow up the Chaudiere Valley (#73), then down the valley of one of the New England rivers to a frontier outpost of scattered little farm houses. Perhaps before daylight they would spring on their sleeping victims, shoot, club, or stab men, women, and children, set fire to the houses, and disappear as suddenly as they had come.

#74

Casco and *Wells,* near the coast of Maine, and *Deerfield* on the Connecticut River, in this way were changed from peaceful, sleeping villages to bloody slaughter-houses.

Victory and defeat. The New Englanders were at last aroused to take revenge on their old enemy, Acadia. After two failures to take Port Royal, they appealed to Queen Anne to send aid. With the assistance of British soldiers, and the British navy, they captured this fort for the last time and with it all Acadia. In honour of the British Queen they changed its name to *Annapolis.* Now this province became permanently British, though when peace came, Prince Edward Island and Cape Breton were still left with the French (#74). Having tasted victory, they decided the next year to strike once more at the heart of New France, by attacking Quebec. But so badly did they sail the St. Lawrence, that soon wreckage strewed the shores of the river, and hundreds of bodies of British soldiers and sailors were floating on the water and being washed up on the shore. The navy did not face the fortress and New France was once more saved for old France.

New France extends her borders. During the thirty years of peace (1713-1744) that followed Queen Ann's War, New France strove to make up for what she had lost in Rupert's Land and Acadia, by exploring westward and southward. But as she got no new settlers from France she had to man her new lands with her all too few settlers from the St. Lawrence. Detroit, a pivot for trade on the Great Lakes, was founded in 1701, and small forts were built right to the mouth of the Mississippi and thinly held (#74). *New Orleans* was founded in 1718, and by 1721 *Louisiana,* a large area near the mouth of the Mississippi with uncertain boundaries, had 5,200 people. The La Vérendryes pushed French exploration right to the Rocky Mountains (#72).

The great fortress. The French, in spite of reverses, went right on building a great empire from the St. Lawrence to the Gulf of Mexico. Though they had lost Acadia, they still owned Cape Breton, and on its south-east corner they built the strongest fortress on the continent, *Louisbourg* (#74). From its land-locked harbour, whose shores bristled with guns, French ships could dart out like a swarm of hornets and pounce on the hundreds of New England merchant ships passing to and fro, and it also stood as a bulwark to guard French ships on their way to and from the St. Lawrence. The

French also prepared for the struggle on land by building a fort, Crown Point, along the Lake Champlain gap (#73).

Disunion among English colonies. The English colonists, though now outnumbering the French twenty to one, were widely scattered, quarrelling among themselves, and for the most so far away from the sting of the French raiding parties that they were prepared to give no help. New England was closest, but was ruled by wealthy traders along the coast who had little to fear. Pennsylvania was settled by sober Quakers whose whole religion was against fighting. New York State was most likely to feel the thrust of the French up the Champlain gap, but they were allies of the Iroquois, who lay across the path of the enemy. New York had already captured most of the fur trade from the French, as they were better traders; they could bring in goods more cheaply than the French and give the Indians better bargains. The most desired article was liquor, and though the Indians liked French brandy better than British rum, the latter was so much cheaper that for one beaver skin traded for rum they could have as long a spell of drunkenness as with brandy bought with six beaver skins. They preferred a long dull carousal on cheap rum to the much shorter, though more pleasing, debauch on costly brandy. To retain their trade the New Yorkers half-heartedly built a number of weak forts along the Hudson-Champlain gap (#73). Fort Dummer (1724) on the Connecticut River, and Fort Saratoga on the Hudson were also placed to guard the easy passage through the Mohawk Valley (#73).

King George's War, 1744-1748

Louisbourg captured by British. When war broke out again in Europe between France and England over the question as to who should be emperor of Austria, the children in North America joined in the fray, as they had grievances of their own. The great fortress, Louisbourg, was a thorn in the flesh of New England fishermen and traders. A thousand fishing schooners had to pass by its frowning guns to reach either the fishing banks of Newfoundland or the English markets. Though Acadia was British, its French inhabitants and Indian allies robbed fishing vessels, murdered fishermen who landed to dry fish, and made life almost unbearable for the weak defenders of Annapolis fortress. The British decided that Louisbourg must be uprooted as though it were a cancer. A combined attack was to be made. Soldiers from New England under *William Pepperell* were to be joined by the British fleet under *Admiral Warren,* and together they would attack this fierce fortress by land and sea. Though the fortress was strong, the defenders were weakened by mutiny and lack of food, and their spirit broken by the depressing dampness of the fog and the bleak gales of winter. In less than two months this band of raw soldiers from New England caused the greatest fortress of the New World to surrender in 1745. Three

#75

WOLFE DIRECTING CAPTURE OF LOUISBOURG

years later the war was over, and when the British government handed back to France this key fortress to the St. Lawrence and New England, the anger of New England knew no bounds. The fighting had been in vain.

Lining up for battle in Acadia. For the next eight years, 1748-1756, England and France were at peace, but their colonies in North America were getting ready for the final struggle, setting their Indian allies at each other, and quarrelling over boundaries all along the line. Though Britain had been given Acadia more than thirty-five years before, the French said its border stopped at the isthmus joining Nova Scotia with New Brunswick, while the British claimed that it included all of New Brunswick and a good part of Maine; soon the French fortresses of *Beausejour* and *Gaspereau* (1755) on one side of the isthmus frowned at the British fortresses *Lawrence* (1750) and *Cumberland* (1754) (#73) on the other. To stand up against the hated Louisbourg, which the French had made nearly as strong as the Rock of Gibraltar, the British had brought out nearly two thousand settlers, mostly soldiers and sailors, founded *Halifax*, and fortified it as a base for the British fleet.

Facing each other along the interior gaps. Along the Hudson and Mohawk gaps (#73), the French, not satisfied with Fort Crown Point, which they had built in 1731, built another at *Ticonderoga* (1755) further up the lake, and the British replied with *Fort William Henry* (1755) right at the head of the lake (#73). The British already had *Fort Oswego* on Lake Ontario, and built several weak forts up the Mohawk River.

The Ohio Valley. But by 1750 a new, more dangerous, and far more valuable no-man's-land had appeared, the ownership of which was sure to decide whether the French or English were going to spread west over the continent. By this date, the British colonies had one million people, were beginning to feel the need to expand, and were determined not to be huddled in between the crest of the Appalachian Mountains and the bleak Atlantic coast. To the west of these mountains spread a vast smiling plain, covered with rich forests and grassy meadows, and with a soil as fertile as that of New England was barren. The Ohio River and its branches spread their gurgling waters like the branches of a giant tree, and served for transport in every direction. This *Ohio Valley* was the prize for which these two young colonies strove. The French first explored its surface, saw that it was needed for their great empire, which was going to stretch from the river St. Lawrence to the Gulf of Mexico, and beat the British in taking possession. Even before 1750 she had a string of fortresses along the Great Lakes and the Mississippi River. In 1753 and 1754 she threw a row similarly across the Ohio Valley with *Fort Duquesne* at the key position (#73, #74).

Braddock's army shot to pieces. It was chiefly the expansion of Virginia, one of the oldest, proudest, and wealthiest of the British colonies, that was blocked by the French forts, along the Ohio River (#73). In 1754 this colony sent *George Washington*, with some untrained troops, to attack Fort Duquesne. He had not got very far before his weak band was defeated by the better trained French troops with their fierce Indian allies. Though England and France were still at peace, their colonies had already started the death struggle. The Virginians pleaded with the other colonies for help and got doubtful promises, few troops and little money. She then petitioned England, who sent out 1500 regulars under *General Braddock.* In the spring of 1755, when this haughty general arrived in Virginia, nothing was ready, though he had to carry all his supplies over the mountains. He had first to beg for wagons, then to cut roads for the wagons and guns; George Washington, with a few hundred Virginians, was his only colonial aid. After dragging guns for two hundred miles along rough roads, hewed out of the

#76

BRADDOCK'S DEFEAT

woods, over two ridges of the Appalachian mountains and across swollen rivers, they were at last within seven miles of Fort Duquesne. Suddenly from the dense forest on both sides of the road a hail of bullets poured on the red-coated ranks from an unseen foe. The volley had a deadly effect. The British regulars had no chance against the French militia, used to bush warfare, and in two hours this fine well-trained army was a terrified fragment, and scattered in a panic. Half of the proud British regulars were dead or wounded, the gallant, even if blundering, Braddock rode calmly through his ranks imploring them to stand firm; but after five horses had been shot from under him, he was wounded to death; only a scattered remnant of his noble company, without wagons or cannon, limped back to Virginia. George Washington and his raw Virginians saved what wreckage was left of this group of British soldiers.

Acadia quieted. Along the isthmus in Nova Scotia, Colonel Lawrence with his soldiers from Fort Cumberland was more successful, and soon had the French fortresses captured (#73). To prevent further fighting among the French, most of the Acadian population was carried away and dispersed among the British colonies (p. 230).

Seven Years' War 1756-1763

Dark days for the British. The slaughter of Braddock's smart and well-trained British red-coats by a mixed band of French soldiers and Indian savages was a severe blow to the British colonies, but a great day for New France. Most of the Indian tribes now joined the victors, and no outlying English settlement along the Appalachian mountains was safe. A band of hideously painted savages would suddenly rush from hiding in the forest and, with a blood-curdling war-whoop, spring on the startled farmers in the fields and on their still more helpless wives and children in the houses, cut them down, remove their scalps, sometimes eat the bodies, and disappear back into the forests. During 1756 and 1757 the French under Montcalm were more than a match for the British and colonial soldiers. His attacks were along the gap of the Mohawk Valley and Lake Champlain. He took Fort Oswego (#73), the key to the former, and burned Fort William Henry, the key to the latter (#73); so that the French were now masters of every fortress east of the Appalachian mountains; held the line along the Mohawk River, which cut off the British from the Great Lakes; held the Lake Champlain gap, which cut them off from attacks on Montreal, and Fort Duquesne, which commanded the Ohio Valley.

Plans of England's great war minister. The year 1758, however, was fatal for the French. *William Pitt*, England's great war minister, was in control. He saw that the war must be won in North America. He put young, alert, thoroughly trained soldiers in command. The British saw the French settlements spread across the continent like a giant snake, with its head at Quebec, its body winding along the St. Lawrence and Ohio Valleys, and its tail wriggling along the

Mississippi to New Orleans. It was fed with soldiers, guns, money, food supplies, and everything else necessary for war from France through the St. Lawrence at Quebec. The proper way to destroy this enemy was not by throwing the chief attack at the body in the

#77

GENERAL JAMES WOLFE

Ohio Valley or even nearer the head through the Champlain Valley to Montreal, but to cut off its head at Quebec, and then all the rest must soon die. England's great war minister planned three expeditions in 1758; the first, and largest, was first to attack Louisbourg and then go on to Quebec to strike the fatal blow; this was to be led by his ablest generals, the cautious *Amherst*, and the daring *Wolfe*. The second was to go up the Hudson River, down Lake Champlain, attack the great French fort of Ticonderoga, then sweep forward to Montreal, and join the first expedition before Quebec (#73). A third and minor expedition was to avenge the unlucky Braddock and take Fort Duquesne.

The sun rises on the British. Wolfe and Amherst found the immensely thick walls of Louisbourg as tough as steel, but Wolfe attacked the French with reckless courage and poured shot and shell into the mighty walls and towers by land and sea until it was finally a jumble of broken rocks. It soon surrendered, and its walls were destroyed almost as completely as Joshua destroyed the walls of Jericho; but the work was so tough and the siege so long, that no attack was possible on Quebec before winter hemmed in that fortress with ice. The expedition to Fort Duquesne, under brave *General Forbes*, who was dying while he marched, was easy. Indeed, they found the fort burnt and empty, as the French had to withdraw their all too few soldiers to defend the St. Lawrence. *Colonel Bradstreet*, in 1758, not only won back Oswego and opened up again a path to the Great Lakes, but crossed over and took Fort Frontenac (#73). The British expedition against the French on Lake Champlain was a ghastly failure. Pitt's biggest mistake was in selecting *General Abercrombie* to lead this expedition. Although his troops outnumbered Montcalm's three to one, his defeat was as complete as that of Braddock three years before.

The French sun sets. The year 1759 was the most fateful in Canada's history. The French no longer controlled the Ohio Valley, as on the ashes of the destroyed Fort Duquesne the British had already erected *Fort Pitt*, soon to be *Pittsburg*; indeed, the British now held the whole of the Mohawk Valley and had destroyed Fort Frontenac on the St. Lawrence. The French still held Lake Champlain, but with such a feeble force that Montcalm ordered that when the enemy arrived they were to blow up Ticonderoga (#73) and make a stand on a safe island at the foot of the lake. New France was not only being strangled by the British, but was being bled white by the corruption of her civic and military officials from the highest to the lowest. While Montcalm, himself, was an able and daring soldier, he was being for ever balked by the fussy and jealous governor, *Vaudreuil*, and his too talkative wife. Supplies were scarce and hard to get past the watch-dogs of the British navy, the crops were poor and food running low. Only the noble courage of Montcalm kept up the spirit of the soldiers and of the people.

Amherst's expedition. William Pitt, decided to leave nothing to chance. He would no longer depend on the uncertain and grudging aid of the British colonies. Though they should have been most zealous in the conquest, they were suspicious of one another, jealous of the British soldier, and mean in money matters. He planned for 1759 two great expeditions, one by land up the Hudson Valley and down through Lake Champlain to make a thrust at Montreal (#73). This was to be led by the chief officer, *General Amherst,* a well-trained cautious soldier, who would be sure to wipe out the stain left by the blundering Abercrombie. He had no fewer than 11,500 well-trained British regulars.

The genius of the war. But the hardest task was left to *General James Wolfe,* only 32 years of age. He had already proved his metal at Louisbourg, where he was the spear point of every thrust, the planner of every move (#75). No better choice was ever made. He was a born soldier. At 16 years of age he fought by the side of King George at the battle of Dettingen. In a time when promotion was slow, he was a major at 21 and a full-fledged colonel at 22. This tall, thin, well-made soldier was noble in his courage, and a sober scholar.

He was so thoughtful of his men and so careful of their training that they would cheerfully follow him through any danger or hardship. Though he had a fiery pride, and was a rigid disciplinarian, he was filled with kindly and noble feelings in a mean and cold age. He was a stern soldier, who loved dogs and flowers, and wrote a tender letter to his widowed mother the night before his most doubtful battle. He found pleasure in noble poetry, and was one of the world's choice gentlemen. He was the man set the impossible task of taking the rocky fortress of Quebec.

Fifty miles of fighting ships. Pitt gave him such equipment for his work as had never been seen on the continent of North America before. On June 1, 1759, a fleet of 250 ships, of which 49 were men-of-war, spread their broad sails to the winds as they crept up the dangerous waters of the St. Lawrence. Hidden rocks, low-lying islands, whirling eddies, and uncertain tidal currents, tormented these uncharted waters; and blinding blankets of fog were not unlikely to settle down on the troubled waters for days at a time. But under the careful guidance of British sailors, all the ships, spread out for fifty miles along the river, came to anchor safely at the head of the Island of Orleans by the end of June (#78).

The strength of Quebec. The west end of the Island of Orleans (#78), which divides the St. Lawrence into two channels, faced to the west the grim fortress of Quebec, perched on a high point of rock jutting out into the St. Lawrence. Point Levis, on the opposite shore, faced the fortress from the south less than a mile away. For six miles to the east of the rocky fortress there was the shelving *Beauport shore,* up and down whose slimy surface the water crept as the tide rose and fell. Behind this was higher land. The river *St. Charles,* winding its course at the foot of the fortress, bounded

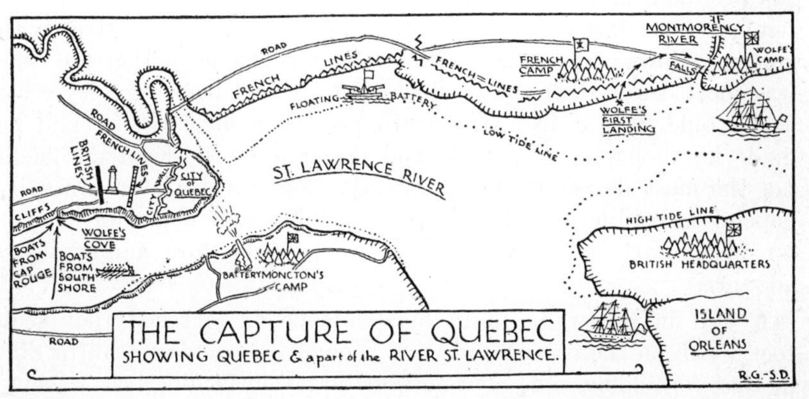

#78

the west of the Beauport shore. Its east border was guarded by the *River Montmorency,* which tumbled down in a high fall a few hundred yards back. The whole of the higher land behind this shore was furrowed with trenches, and bristled with French guns. Behind such protection, strong as steel, lay the most of the French army. West of Quebec the chances of successful attack were still more hopeless. Rising sheer from the water's edge was a steep rocky cliff, which frowned for many miles to the west, and looked as impossible to climb as the side of a stone wall two hundred feet high.

Quebec hemmed in. Such was the redoubtable stronghold that met the keen eye of the delicate, gloomy, young commander, Wolfe, as he first gazed at the fort that had already beat back three British attacks. But undaunted he began his plans. In a very few days British troops were at the point of Orleans, guns at Point Levis poured a ceaseless cannonade into Quebec, and other troops were scattered far up along the south shore. As the British navy, flushed with many a victory, was soon strewed all along the river, both above and below Quebec, he had complete command of the water route, which gave him the choice of a sudden attack at any point. He made his headquarters on the north bank just east of the Beauport shore, with only the River Montmorency between him and the enemy.

Wolfe uncertain. He tried out an attack on this shore, but the over eagerness of his keen soldiers led to disaster, and he had to draw off his men, with heavy loss. His fine mind was too great for his weak frame; he became ill of body and gloomy of mind while the precious summer was fast slipping away. As the touch of September frosts began to splash the whole borders of the river with golden yellows and flaming crimson, he was reminded that the time was short before the navy would have to retreat before the icy crust that would hem it in. His health became a little better, and he made his final plan of attack, and that one of the most daring, if not the most desperate, in all history. But great victories usually depend on bold leaders daring the impossible; and Wolfe was always for dashing attack, just as Amherst was for slow, sure, and cautious, advance.

A dint in the wall. With his telescope he searched that grim rocky wall on the north shore, and thought he could see a little dint in that forbidding cliff. He noticed that women did their washing down along the shore, near the dint, and the garments were soon flying, like flags, on lines at the top of the cliff. There must be a path up from the cove. Soldier's tents at the top showed that this steep gully was guarded. Wolfe made the fateful decision to move his whole army up that narrow timber-strewn path in the darkness of a single night.

Wolfe's soldiers gain the height. On the fateful day of preparation, part of his army was sent up stream and the rest was hidden on the south shore just opposite the cove. Then he began to tease and tire the enemy by having the fleet pretending it was going to attack first at one place then at another. In the darkness of a September night, while the French were kept guessing, the boats, packed with picked soldiers, among whom was Wolfe, drifted down the north shore to the cove. Twenty-five brave volunteers rushed up the path and destroyed the guard at the top. The path was quickly cleared of trees, and all through that crowded night a steady stream of red-coats and Highlanders scrambled up the dangerous slope (#79). Then the troops from the south shore were fed across in feverish haste but in perfect order. By nine o'clock the next morning, the young hero, glowing with pride, and flushed with excited joy, saw the long red lines of Britain's proudest soldiers stretched across the level *Plains of Abraham,* with trenches dug, and everything ready to meet the enemy. Meanwhile, hidden in the forests on each side of the

#79

WOLFE'S MEN DASHING UP THE HEIGHT

plain, Indians and French-Canadian soldiers were sniping off his men as they stood ready for the attack.

Montcalm outwitted. His gallant enemy, Montcalm, had been completely deceived, and when he arrived from Beaufort shore, where he expected an attack, his heart sank when he saw the startling sight of Wolfe's army in possession of the Plains of Abraham, the most commanding position in the whole fortified area of Quebec. No more daring stratagem had ever been attempted than to plant an army in the very heart of the enemy's stronghold before it was discovered. Every action of the sailors, rowers, soldiers, and ships, fitted together like the parts of a jig-saw puzzle. There was not a single slip. Wolfe's keen mind, methodical planning, and rigid order had borne rich fruit; and now the effects of those finer qualities, his inspiring courage, his love for his men, and his noble pride, could be seen filling every soldier along the red lines with dogged determination to support his general with his life.

The battle. In a few minutes the day was decided. Montcalm's equally brave soldiers were hastily gathered together and marched forward to meet the thin red lines. But their motions were irregular, their shooting by fits and starts. The thin red lines, steadily advancing as though on parade, received the bullets of the enemy without flinching, and as one fell out of the ranks another took his place. Not a red coat faltered, not a finger pulled a trigger until the enemy came so close that the whites of their eyes could be seen; then, like a single shot, every rifle blazed out its deadly volley, and almost before the smoke had cleared away they again had reloaded their guns and delivered a second hail of bullets. These two volleys decided the fate of a continent. The whole French army reeled like drunken men before this awful storm of bullets, and the men fell like heads of wheat at the swing of the reaper. Bayonets and claymores swung into action. Soon the French were in hopeless retreat, and Wolfe's brave and well trained soldiers swept everything before them and were complete masters of the field. The ground was a ghastly sight covered with the dead and the agonizing dying. The gallant Wolfe lived long enough to know that his lifework was complete, and died gloriously on the field, pierced by three bullets. His brave rival Montcalm was shot on his horse, carried to a convent, and died a martyr to a hopeless cause.

Wolfe's undying fame. Most men's deeds grow dim with time and are soon forgotten, but for a few of the great geniuses of the world it takes time to reveal to a dull world their true greatness. Wolfe belongs to this class. His fame has grown from age to age, and to-day he is honoured with Marlborough and Wellington among the most distinguished of England's generals.

Though the British still had dangerous times and during the following spring almost lost the city in a stinging attack by the French, who marched down from Montreal; nevertheless, the position of the enemy was hopeless. As the spring sun crumbled the ice of the St. Lawrence, British ships again appeared before Quebec; Amherst closed in on Montreal from both east and west, and before 1760 was past, all Canada was permanently in the hands of the British.

The name Canada. In 1763 the Treaty of Paris was signed, which transferred the whole of New France to England and the name disappeared from Canadian history. The name, Canada, which now stands for the whole country has had a great variety of meanings. Jacques Cartier used it first to indicate a small district on the north shore of the St. Lawrence between Quebec and the Saguenay. It later was used very loosely to indicate the land in the St. Lawrence

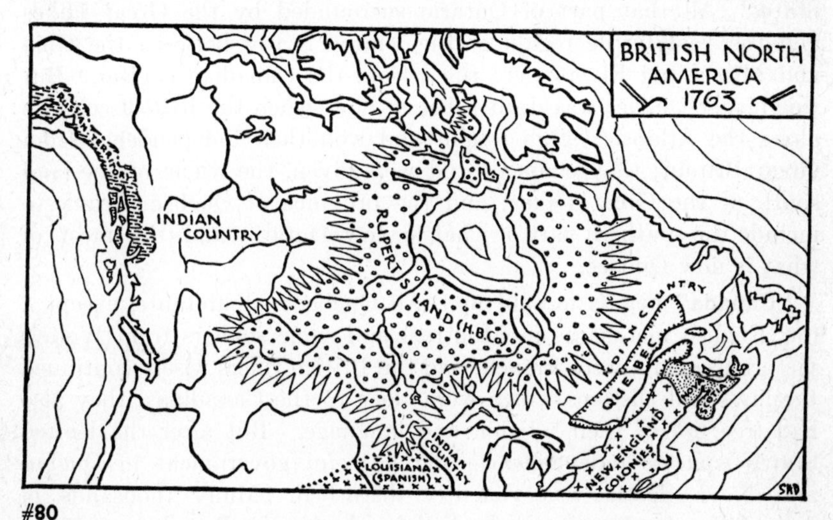

#80

Valley, and the French people were clearly divided into Acadians and Canadians, the former inhabiting the Atlantic coast, the latter the St. Lawrence Valley. The official name, however, for the St. Lawrence settlements, was always *New France.* Indeed, Canada was first used to name a distinct political district in 1791, as we shall soon see (p. 186).

Divisions of Canada. At the conquest "Canada" in the widest sense was divided into five districts (#80): (1) *Nova Scotia* including not only the peninsula and Cape Breton but also New Brunswick and Prince Edward Island; (2) *Newfoundland,* to which was attached Anticosti, Magdalen Islands, and the Coast of Labrador; (3) *Quebec* which covered the settlements along the St. Lawrence and the borders of the Ottawa River; (4) *Rupert's Land*, which embraced all the land as far as it was known, draining into Hudson Bay and Strait; (5) *Indian Lands*, which formed a fringe of uncertain width. They were the hunting grounds of the Indians. It included the region around the Great Lakes and also that vast region in the United States, between the Appalachian Mountains and the Mississippi River (#80).

A shrunken Quebec. In 1774 when by the Quebec Act, a new form of government was given, the province of Quebec was increased to include much of the choicest lands of Canada and the United States. All that part of Ontario surrounded by the Great Lakes and the vast fertile region of the United States between the Ohio and Mississippi Rivers were ruled from the Canadian capital. But this bigger Quebec was dealt a heavy blow when the British colonies along the Atlantic Ocean fought and won their independence from Great Britain; they claimed, and were given, the whole of the land south of the Great Lakes; and the province of Quebec shrunk to include the southern part of what is now Ontario and the most of what is now Quebec.

"Canada" begins to split up. All Quebec's inhabitants, except a few English merchants in Montreal and some soldiers from Wolfe's victorious army, were French; and the British rulers wisely continued the laws and customs which were dear to the Canadians; they also had free use of their religion and language. But after the United States won its independence, the peaceful government in Quebec and Nova Scotia was violently disturbed. Many thousands of United Empire Loyalists, who had fought for the British against the United States, poured into what is now Nova Scotia, New Brunswick, Ontario, and Southern Quebec, in order to live under the British flag. They believed they had suffered grievously for their country and deserved the best treatment. Many of them were highly educated and had been leaders in the English colonies from which they came. They were not slow in demanding their rights. In Quebec, which included most of modern Ontario, they were not satisfied with French laws and French methods of holding land. To correct their complaints, a *Constitutional Act* was passed in 1791, which divided Quebec into two provinces, each with its own government. The Ottawa River was the dividing line for most of the way. The French province was called *Lower Canada*, the English province around the Great Lakes was called *Upper Canada* (#81), and the name Quebec was dropped. Thus the name Canada was officially used for the first time for these two political divisions of our great country. This practice continued until 1840.

Nova Scotia torn to pieces. The loyalists were the first settlers of New Brunswick, and many moved into Nova Scotia, Cape Breton, and Prince Edward Island, all of which up to that time were united in the one province of Nova Scotia, with its capital at Halifax. The loyalists did not get on well with the old settlers, who had come from England, Germany, and the New England states. They were not pleased with the form of government and thought the capital,

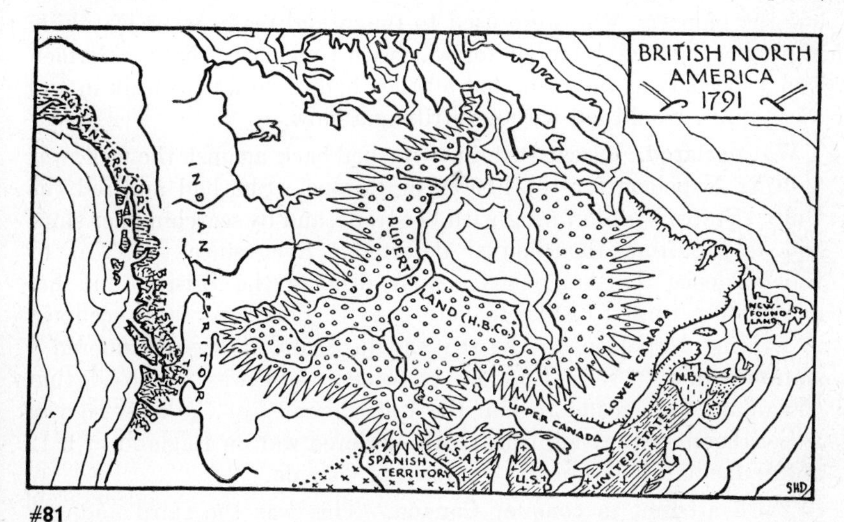

#81

Halifax, was too far away. They were so pressing in their demands on the British government, that by 1784 this small area was split into no less than four separate colonies: Prince Edward Island, split off in 1769, and Cape Breton and New Brunswick, in 1784. While Cape Breton again joined hands with Nova Scotia in 1820, the others have remained apart ever since. Thus by 1791 there were strung loosely, half across a continent, seven struggling British colonies; Upper Canada, Lower Canada, New Brunswick, Nova Scotia, Prince Edward Island, Cape Breton, and Newfoundland.

War of 1812

Covetous eyes on Canada. After the shock of the American war of Independence, nothing disturbed the peaceful growth of the

Canadian colonies until 1812. The numerous settlers of the eastern United States, moving into the Ohio Valley, cast covetous eyes northward at the choice farming land of Canada and felt that all should be united into a single country.

The majority of the population of Upper Canada were Americans who had recently moved in, and many of whom were supposed to be republicans at heart, eagerly waiting for Canada to be absorbed by the United States.

Violent speeches were made in and out of the United States parliament, or congress. The Eastern United States were all for peace; but the pioneers, who were used to rough and ready ways of taking what they wanted, were all for war with Canada. The frontiermen won, especially as they felt that all they had to do was to walk in and the country would fall without striking a blow.

War declared. Great Britain, with her back against the wall, was fighting Napoleon for the liberty of Europe. She had angered the United States by interfering with her trade, and by searching her ships and taking from them naval deserters. Ever since the War of Independence the Americans greatly disliked the British, and her interference turned dislike to hate. As England was fully occupied in fighting Napoleon on the sea and in Spain, it was considered a suitable time to seize Canada, especially as it was supposed that Canadians were ready to join the United States at the least sign of force; therefore they light-heartedly declared war on England in 1812 and immediately began the conquest of Canada.

Third attempt to conquer Canada. This was the third and last attempt on the part of the Americans to conquer Canada. The first attack was made by Sir William Phipps in 1690, before the conquest, when he failed miserably before Quebec; the second was during the War of American Independence in 1775, when *General Montgomery* and *Benedict Arnold* went down to defeat, and the former to death, amid December snows before the invincible fortress of Quebec.

The United States divided. The third and last attempt was on a much larger scale, and continued for three long years, but was no more successful than the other two. Though the United States had a population of eight million and Canada only one-half million, the United States went half-heartedly into the war; the New England States were strongly opposed and would give no help, New York and Pennsylvania were lukewarm, and the southern states showed little interest.

Plan of attack. The United States plan was to attack Canada across the Detroit and Niagara Rivers, so that they would control the lower Great Lakes; along Lake Champlain and St. Lawrence River, so that the armies might converge on Montreal and finally join for a grand attack on the dreaded fortress of Quebec. In not a single year did they make great progress.

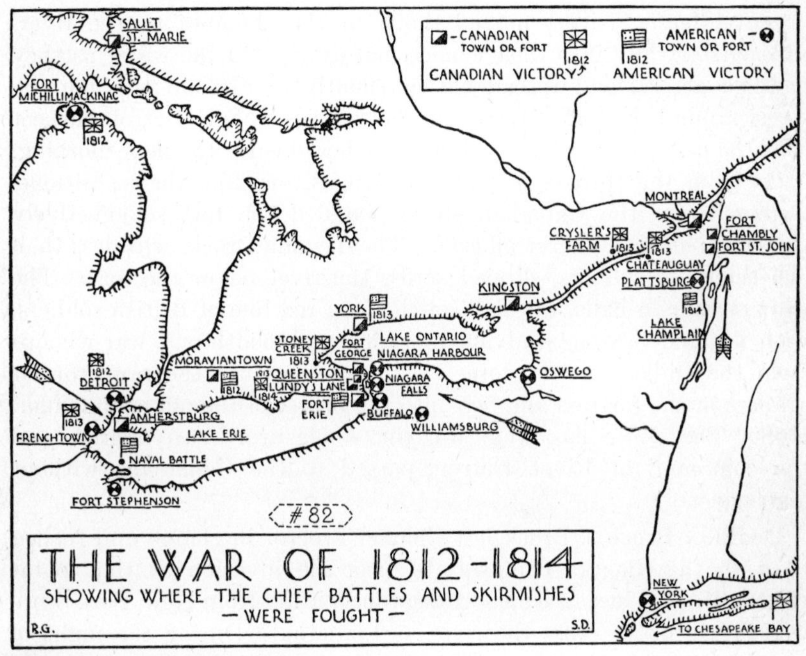

#82

Michillimackinac seized. The Canadians at once sent a small band of soldiers overland who seized *Fort Michillimackinac* on July 17, 1812, and held this key position throughout the war, which gave Canada control of the three upper lakes, and aroused many Indian tribes against the United States.

Brock captures Fort Detroit. *General Hull*, an old and timid veteran of a former war, crossed River Detroit, intending to attack *Fort Malden*, near Amherstburg; but his courage failed and he wasted precious time, while *Sir Isaac Brock*, the real military genius of the war, rushed troops by boat from Toronto and Niagara and attracted to his banner the great Indian chief, *Tecumseh*, with hundreds of fierce and seasoned warriors. Brock decided to strike at once a decisive blow against his timid rival and his trembling troops. A band of his soldiers at once crossed the river to cut Hull's communications with his base, further south. Hull, in a panic recrossed the river, and with all speed shut himself and his two thousand fearful and untrained soldiers in the fortress of Detroit. On the night of August 15th, Tecumseh, with his six hundred dusky fighters from a dozen different tribes, moved down in the shadows to the river, slipped stealthily into their canoes, hardly rippled the water as they paddled across, and spread out as silently as a cloud through the forest around the fort. Little did the sentries at Fort Detroit suspect that the calls of song birds, which they heard early the next morning, came from the throats of painted savages, marking their positions. Cannon from the Canadian shore pounded the fort so effectively that one ball killed three officers. The dashing Brock, with less than one thousand soldiers, slipped across the river below the fort. The whiz of cannon balls, the sight of the long red line of British soldiers, with flashing bayonets advancing, and the bloodthirsty war-whoops from the Indians in the forests struck terror into the raw frontier soldiers in the fortress and fear into the breast of the old commander, Hull. The white flag went up, the whole army surrendered, and the command of River Detroit passed to the Canadians without a struggle.

Death of Brock. Brock left Colonel Proctor in charge and rushed back to the Niagara Peninsula to meet the invasion at that point (#82). He was just in time. A disorderly army from New York State was preparing to cross the river at Lewiston. During the night of October 12, a number succeeded in coming over in boats and, marching in single file on a narrow path between the river and the steep cliff, found a way up the bank of the gorge. With the utmost bravery they drove away the Canadians and won the high point commanding Queenston and the whole country below (#82). Brock arrived in the morning only to see the key position in the hands of the enemy, who were already digging trenches. Gathering what men he could find, he went to the west, clambered up the steep escarpment and then turned east to drive the enemy from the height. He attacked, and seized the British gun, which the enemy had captured, but, suddenly an American stepped out of the trees not forty yards away took steady aim at the hero in his bright coat of red and gold, and shot him dead. The British retired and left the Americans in charge of Queenston Heights.

Americans swept from Queenston Heights. But all was not well in the American army, the most of whom had not yet crossed the river. Many shrank away; many more, as they saw the wounded being brought back across the river, lost heart and refused to fight. As their spirits sank those of the Canadians rose, in spite of the

#83

THE MARQUIS DE MONTCALM

death of their general. Soon Canadian troops again mounted the escarpment, attacked the enemy on Queenston heights, and drove them back pell-mell against the river. Many surrendered, some plunged to death over the height, a few of the more daring rushed down the path by which they had come and swam across the river. The *Battle of Queenston Heights* was won, and the attack at the Niagara River was effectively stopped for that year. The attack by Lake Champlain did not get started in 1812.

Things look bad for Colonel Proctor. *Colonel Proctor*, at Amherstburg, began to attack the Americans on their own ground in January

1813, when he crossed on the ice and took an American fort. But instead of the weak-kneed General Hull, he had a far more energetic man, *General Harrison,* to deal with this year. Nevertheless Proctor did not hesitate to strike the enemy in their own forts at *Sandusky* in May and *Fort Stevenson* in August. But he was getting in great danger. He was hundreds of miles from Niagara and Toronto, from which he must draw his supplies, and a roadless wilderness lay between. Lake Erie must be kept open at all costs or he would be lost. Besides, fresh troops from all the populous states to the south were beginning to pour into Harrison's camp.

Lake Erie lost. Both sides were preparing fleets on Lake Erie to control traffic, and by September, though neither was ready, the young British captain, *Barclay,* decided to fight. Since Proctor's army was short of supplies and their Indian allies were clamouring for food, Lake Erie must be kept open at any cost. He hastily collected what men and guns he could at Fort Malden, and sailed across to attack the renowned *Admiral Perry* with nine new ships. After a desperate and even fight, the British fleet had to surrender; and the Americans gained control of Lake Erie, which they kept for the rest of the war.

Canadians lose west Ontario. Proctor and his army at Amherstburg were in such peril that every day lessened the chances of a successful escape. Yet he lingered. At last in October, he started his fatal retreat through the woods of western Ontario. Harrison with 3,000 men was soon worrying his 500 British and 500 Indians as they rushed along the *River Thames.* They finally overtook Proctor near Thamesville at *Moraviantown,* when with starving, exhausted troops he turned to face the overwhelming number of the enemy. One charge of Harrison's cavalry cut the Canadians to pieces; and though the Indians under *Tecumseh* attacked from the side, the battle was soon over. Tecumseh was killed, Proctor vanished from the scene, and the whole western peninsula of Ontario was at the mercy of the Americans for the rest of the war.

A night charge at Stoney Creek. On Lake Ontario and in the Niagara Peninsula, the year 1813 began badly for the Canadians but ended with the Americans being driven back completely, so that Canadian soldiers had possession of both sides of the Niagara River, and Canadian soldiers were in control of Lake Ontario. An American fleet captured and burned *York,* now *Toronto,* the capital of

Upper Canada in April; took *Fort George,* at the mouth of the Niagara River in May; and began chasing the Canadians along the south shore of Lake Ontario toward Hamilton. But at *Stoney Creek,* five miles from Hamilton, on June 5, the Canadians turned and attacked the enemy at midnight. They charged right through their lines, and by morning the Americans were in full retreat and did not stop till they got to the Niagara River under the protection of Fort George.

Laura Secord's heroic march. When the two armies lined up, the Americans were at Queenston, the Canadians, under *Lieutenant FitzGibbon,* at *Beaver Dams.* Indians in front of FitzGibbon's men gave the Americans no peace. Suddenly they would strike a fatal blow and disappear as completely as if they had sunk into the earth. Any American soldier, who dared to stray into the forest, would suddenly hear a shot or feel the blow of a tomahawk but would never see an Indian. Goaded by these hidden teasing attacks they attempted to clear out the Indians before attacking FitzGibbon, 16 miles away. A United Empire Loyalist farmer's wife, *Laura Secord,* overheard the Americans discussing the attack. She quietly tied her cow, dropped her milk pail, and started off at dawn through the dense dripping woods to bring the news to FitzGibbon. She had to steal her way past the American scouts, and past the hawklike eyes and the keen ears of the Indians. On she went, avoiding paths and open forest, throughout this hot and humid day, pushing through rasping underbrush, stumbling over fallen trees, labouring across slimy swamps, and creeping over narrow logs spanning swollen creeks. At last she staggered up to FitzGibbon, exhausted with her walk of twenty miles, and gave him warning. Though the Indians had already informed him, Laura Secord's heroic walk will never be forgotten by Canadians.

Surrender at Beaver Dams. By the time the Americans had plodded through these dense forests, tormented at every step by the never ending string of these stealthy Indians, who hovered about them like a swarm of hornets, instead of attacking FitzGibbon's men they were ready to surrender, and FitzGibbons won the day without firing a shot. This is the one battle in the war that was entirely a victory by Indians.

Soon this American march into the Niagara Peninsula, which began so jauntily in the spring, became a rapid retreat across the river, and before December was over the Canadians were in possession of all the fortresses on both sides of the river and had burned several villages in revenge for the destruction of 150 houses in the village of *Newark* by the Americans before they retreated across the river.

Chrysler's Farm and Chateauguay. The American advance in the east did not get under way until the autumn of 1813 and finished almost before it had begun with two defeats by much smaller numbers of Canadians. One army was to follow the St. Lawrence from Lake Ontario, the other Lake Champlain, and the two, united, were to attack Montreal. The first was defeated along the St. Lawrence at *Chrysler's Farm*, and the second followed the Chateauguay River until their front of 1,500 came in contact with 500 French-Canadians under *Colonel Charles de Salaberry* and *Colonel George Macdonnell.* The French-Canadians fought so fiercely that the Americans retreated and did not return to the fight but retired to winter quarters on Lake Champlain. Thus after two years, very little progress had been made in the conquest of Canada; the Americans occupied not a foot of Canada in the Champlain region; in the Niagara Peninsula the Canadians held the American side; but the Americans were masters of Lake Erie and the region to the north of this lake.

Chesapeake and the Shannon. When the final campaign opened in 1814, the situation had markedly changed. A naval war had been raging on the Atlantic chiefly as a series of duels between American and British frigates, which were armed sailing vessels. In the first fights the Americans, with better ships, were usually successful; but early in 1813 the tide changed with a naval duel outside of Boston Harbour. The *Shannon* was a British frigate, with a thoroughly trained crew under *Captain Broke*, a very able captain. Though the American vessel, the *Chesapeake*, had a larger crew, both captain and officers were new on the frigate, and many of the men had little experience on fighting ships. The Shannon sailed to Boston Harbour and politely invited the Chesapeake to come out to fight, which she did; an interested audience on the hillside watching the combat. The superior training of the British crew soon settled the affair; every shot raked the Chesapeake from side to side or from bow to stern; in less than fifteen minutes the Shannon came along side the badly crippled American ship, the English rushed on board, and after a short hand to hand fight, the Stars and Stripes was pulled down and the Union Jack unfolded at the mast head of the American prize. When the news spread through Halifax, on Sunday morning, that the Shannon had just brought in this great prize, the congregation of every church emptied and rushed to the harbour, and the whole city celebrated almost as wildly as when the news arrived of the capture of Louisbourg in 1758.

American ships swept from the seas. Soon the British fleet had such complete control of the coast that not an American ship dare show its nose outside a harbour, and United States owners' trade was strangled to death. To avoid this blockade much American goods went down the Richelieu and out from Montreal and Halifax;

#84

THE BATTLE BETWEEN THE SHANNON AND CHESAPEAKE

these cities thrived not only on this trade but by buying and selling captured ships. Indeed, these were the most palmy days in the whole history of Halifax, and many Nova Scotians made great fortunes.

The desperate battle of Lundy's Lane. An American attack in 1814 by way of Lake Champlain was soon stopped with a stinging defeat at *LaColle's Mill* by a mere handful of Canadians. The defeated army retired to Lake Champlain not to advance again. An army of 4,000 well trained soldiers crossed from Buffalo to *Fort Erie* and advanced to *Chippewa,* where they defeated a Canadian force and then began a march across country toward Burlington, near Hamilton. But the Canadians with three thousand faced them on a slight hill, not far from Niagara Falls; here at *Lundy's Lane* was fought the most desperate battle of the war by two armies of trained soldiers. First one side and then the other gained the crest, men falling like autumn leaves before the wind. From 2 o'clock until nearly midnight of a hot summer's day the uncertain strife swayed back and forth. It was nobody's victory, though the Americans, giving up all thought of advance, soon retreated to Fort Erie, and later burned the fort and crossed the river.

Both sides ready to quit. Before 1814 was over, the Americans realized that they could not conquer Canada, for Great Britain's army was now freed from fighting Napoleon in the Peninsular War, and these veterans and the navy were ready for action in America. A large army of well trained veterans was badly led in an attack on the Americans at *Plattsburg* on Lake Champlain and were soundly thrashed by one third their number of American troops; the Canadian navy which took part in the action was also destroyed. But when the British army and navy attacked Washington, burnt the capitol, or parliament buildings, and the president's residence, the United States was anxious to quit. As Great Britain had never been anxious to begin, a peace treaty was soon drawn, and a long period of peace between Canada and the United States began.

Results of war. This war was of the greatest importance to Canada. It left a hatred of the United States so deep, that all chance of the union of the two countries was blotted out forever, and Canada started to develop her own national feeling. The threat from the south gave the first stimulus toward the union of all the British colonies, which was later to bear fruit when the Dominion of Canada was formed. It not only checked but killed any further movement of Americans into Upper Canada (Ontario), and deflected that province from all thought of republicanism to a profound attachment for Great Britain and British forms of government.

The Province of Canada. As Upper and Lower Canada grew in population, education, and industry, the people became more and more dissatisfied with the form of government, which had been planned in 1791 when Upper Canada had few settlers, until finally, in 1837, a small rebellion broke out in both Upper and Lower Canada. Though these risings were easily crushed, *Lord Durham*, a great statesman, was sent out to study what changes should be made. His report was so wise and just that it is considered the greatest document in Canadian history. He recommended that Upper and Lower Canada should again be united to form the *Province of Canada.* This was done in 1840 (#85).

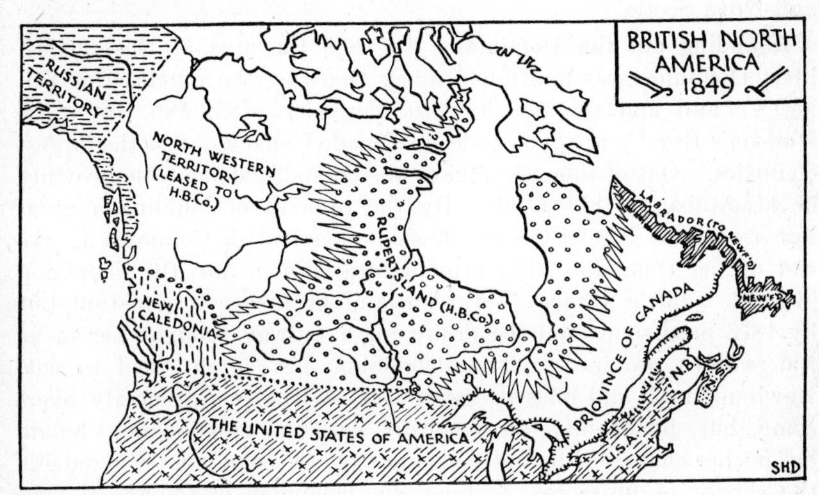

#85

Canada feels growing pains. But as the English part of the province, called *Canada West*, grew in population and wealth, it soon surpassed *Canada East,* which was largely French and Roman Catholic. Each section became more and more jealous lest it should be ruled by the other. It seemed impossible to find a suitable settlement. But events were taking place elsewhere that brought a solving of the problem in sight. The four maritime provinces were growing in population and wealth. New settlements were gathering around the Red River on the Prairies. The discovery of gold had brought thousands of settlers to the Pacific coast, and the vigorous colony of *British Columbia* was thriving.

Birth of Dominion of Canada. Wise statesmen, like Sir John A. Macdonald and George Brown, looked beyond the strife between English and French in the Province of Canada and saw the possibility of a great British country stretching from the Atlantic to the Pacific and from the United States to the north pole, whose problems would be so big that these little racial and religious differences would be lost. They gathered representatives from all the colonies to conferences first at Charlottetown and later at Quebec. As a result of the careful labour and far-seeing statesmanship of these Fathers of Confederation, on July 1, 1867, was born the *Dominion of Canada* embracing the four provinces of Ontario, Quebec, New Brunswick, and Nova Scotia.

Rounding out the Dominion. In 1869 the area of this modest little Dominion was multiplied many times by the purchase of Rupert's Land and the North West Territory (#85, #86) from the Hudson's Bay Company. Indeed, it made Canada one of the largest countries. Out of this vast region was crystallized the little province of Manitoba in 1871 (#86). By the promise of building a great railway to the Pacific Ocean, thus binding British Columbia to the rest of the Dominion, that province was drawn into the Dominion in 1872. Little Prince Edward Island at first was reluctant but by 1873 hard times and the desire for a railway brought her in as the seventh province. The Dominion has often tried to woo Newfoundland, and once or twice her resistance was nearly overcome, but, by dealing too niggardly with this great island, Canada has let her chances slip, and England's oldest colony is still outside the cluster of provinces, making the Dominion of Canada.

New and bigger provinces. After the Canadian Pacific Railway was built, by 1887, settlers from almost every country in Europe began to flow over the fat prairie land in the southern part of the North West Territories. These settlements demanded governing bodies, and the later growth of the Dominion has consisted in dividing parts of the North West Territories, as they got settled, first into districts or territories and later into provinces. In 1882, the districts of Assiniboia, Saskatchewan, Alberta, and Athabaska were carved out of the southern part of North West Territories (#87). In 1898 the District of Yukon, into which had rushed great numbers of gold seekers, was made into a territory (#88); in 1905, the four districts of Assiniboia, Saskatchewan, Alberta, and Athabaska were made into two provinces of Alberta and Saskatchewan (#88), with their capitals at *Edmonton* and *Regina.* Then at last was rounded out the nine provinces of Canada. But from time to time the provinces of Canada, except the Maritimes, have been enlarged by additions from the north. Manitoba at first looked like a postage stamp on the centre of the map of Canada (#86), but she has been enlarged three times (#86, #87, #88), and is now about the same size as the other Prairie Provinces. Ontario and Quebec (#86, #87, #88), have had several large additions and now the former extends to James Bay and Hudson Bay, and the latter right to Hudson Strait.

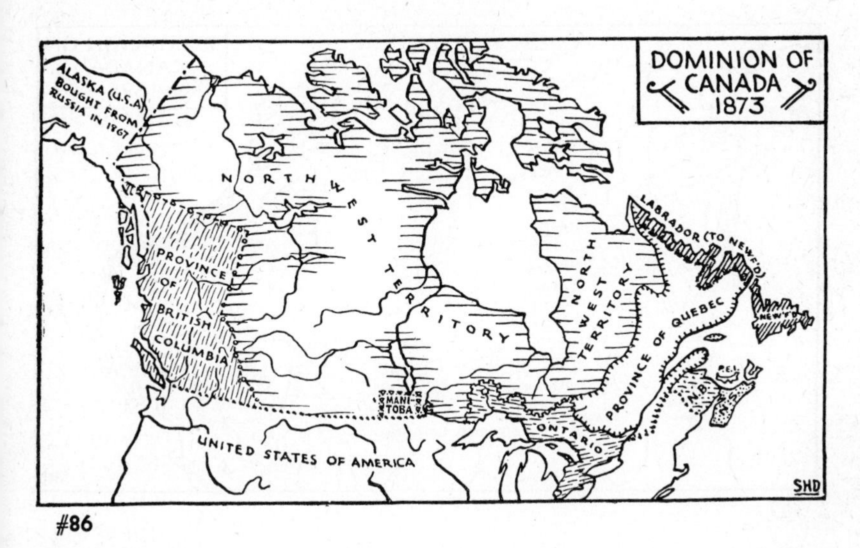

#86

Organizing north Canada. Many changes in north Canada have been made from time to time. In 1880 all the northern islands right to the north pole were annexed to Canada. Royal Canadian Mounted Police have stations scattered over them to keep order, look after the health of the Eskimo, issue licenses, and collect duties. This region has been divided into districts (#87), but the centre of government for the whole North West Territories is at *Smith* on Slave River. Already there is a movement in the western Provinces to push their borders north to the Arctic Ocean.

Fixing boundaries. Canada has had territory nibbled off almost all her borders by arbitrations with the United States. The dint which the state of Maine made in southern Quebec was made still deeper in 1842 by the Ashburton Treaty. In the west the whole of the state of Washington, including the mouth of the Columbia River, was in dispute, but in 1846 it was given to the United States, the 49th parallel of latitude being made the boundary. When the discovery of gold in the Yukon caused a wild rush of tens of thousands of gold-hungry prospectors up the coast of British Columbia, the boundary between Canada and that unfriendly "pan handle" of Alaska became a burning question. In 1903 this was settled by arbitration, not by any means satisfactorily to Canada, since the

#87

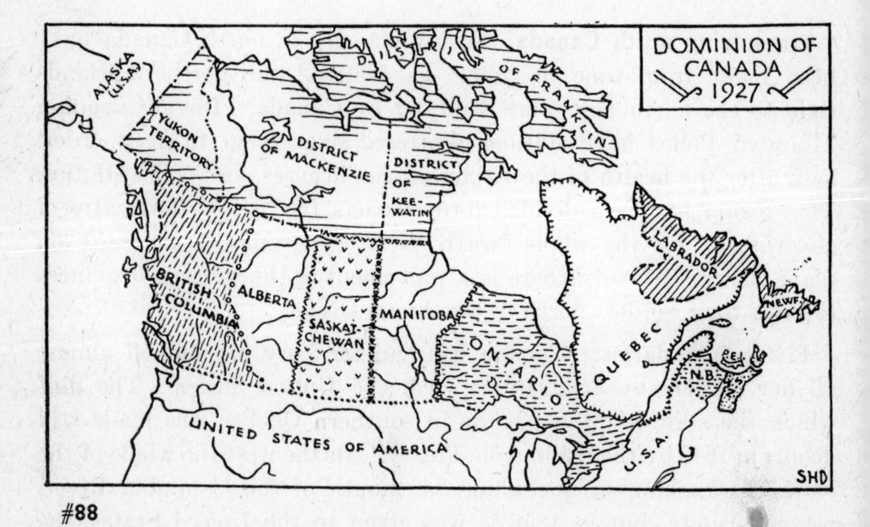

#88

United States got the lion's share, including all the coastal waters, so that to enter the Canadian Yukon it is necessary to cross a strip of Alaska.

Coast of Labrador. The last dispute was a family affair between Canada (chiefly Quebec) and Newfoundland as to the boundary between Quebec and the "Coast of Labrador," a very vague and peculiar expression. This coast had been given to the Government of Newfoundland for the second time in 1825 (#85). Both parties agreed that the boundary should be settled by the Privy Council of Great Britain, the highest court in the Empire. After a study of old maps and documents, in 1927 this body gave Newfoundland an area of 110,000 square miles, which eats deeply into the Peninsula of Ungava (#88). With this decision the last dispute as to boundaries was finally settled. No problems have egged countries on to war more frequently than settling boundaries; and we should be proud that all of these bitter disputes have been arranged in a neighborly way, and thankful that there are no more boundary questions to cause national anger and international quarrels.

CHAPTER XII

THE MARITIME PROVINCES

The Maritime Provinces. Nova Scotia, New Brunswick, and Prince Edward Island, are well named The Maritime Provinces, for they thrust themselves out into the Gulf of St. Lawrence and the Atlantic Ocean; nowhere are they far from the smell of the sea. Prince Edward Island is so long and narrow and so deeply and frequently notched, that the dash of the waves is a familiar sound everywhere. Nova Scotia is a long, narrow peninsula, and an island. Cape Breton, in its fondness of the sea, has actually swallowed a bit of it, which as *Bras d'Or Lake* is one of the most beautiful panoramas, its mystic mazes of salt water being dotted with delightfully green islands. New Brunswick is more solid, and has the sea on only three sides: the *Bay of Fundy*, on the south; *Northumberland Strait* and the Gulf of St. Lawrence on the east; and *Chaleur Bay* on the north; but it has gaping bays gashing its side so deeply that they also bring the back country close to the sea.

In a strait jacket. These Atlantic provinces are sometimes spoken of as the parts of Canada that have been passed by. If that is true, it is not because their location invites neglect. They border the Gulf of St. Lawrence, the finest inlet on the Atlantic coast. Boats entering Canada from Europe first sight the Maritimes, whose ports are hundreds of miles nearer to England than those of Montreal, New York, Boston, or Baltimore. Their geographical position has in some respects hindered their economic development; they have been, from the first, pocketed between the state of Maine and the province of Quebec, and have had no room to expand. All the other provinces had vast, undeveloped lands to the north, whose resources they used, and into which they spread out, but the Maritimes have been confined in a strait jacket from the birth of the Dominion.

Surface

Formation of coal. The Maritime Provinces form a part of the Appalachian Highlands, which will be fully described under Quebec (p. 240), as a series of gigantic parallel wrinkles in the crust, caused by a thrust from under the Atlantic Ocean crushing toward the Canadian Shield. While in east Quebec the worn-down wrinkles

PHYSICAL FEATURES
SOUTHERN UPLANDS 1
COBEQUID MOUNTAINS 2A
CAPE BRETON HIGHLANDS 2B
HIGHLANDS N. OF FUNDY 3
CENTRAL HIGHLANDS 4
NORTH MOUNTAINS 5
ANNAPOLIS VALLEY 6
HEIGHT OF TIDE IN FEET TIDE-45'
OLD. INDIAN TRAILS
ORIGINAL FRENCH ACADIAN SETTLEMENTS
GAME
MOOSE
DEER
BEAR
DUCK
PARTRIDGE
SALMON
WOODCOCK
TROUT
#89
THE MARITIME "PROVINCES"
PHYSICAL FEATURES, GAME, INDIAN TRAILS, EXPLORATION, AND ORIGINAL FRENCH SETTLEMENTS REMOVED BY THE ENGLISH, A.D. 1755.
RESTIGOUCHE R.
NIPISQUIT R.
SAINT JOHN R.
PETITCODIAC R.
KENNEBECASIS R.
JACQUES CARTIER 1534
CARTIER 1535
CAPE NORTH
NORTH PT.
EAST PT.
PRINCE EDWARD I.
CAPE GEORGE
BRAS D'OR L.
CAPE CANSO
DE MONTS AND CHAMPLAIN, 1604-1605

remain exposed at the surface, in the Maritimes these wrinkles, after they had been eroded nearly flat, sank so low that mossy bays, swamps, and shallow water clogged with rank tropical plants, covered a great part of the surface. The rank vegetation died and collected year after year, and century after century. In time it became buried beneath the sea in layers under sand and clay, so that it could not decay. Thousands of feet of such alternating layers gathered on top of the old Appalachian roots and was cemented into rocks by pressure, the red sand becoming sandstone, and the vegetation, coal. At one place, called *Joggins*, on Chignecto Bay, alternating layers of these rocks, 7,000 feet thick, can be seen on an exposed cliff, with no less than thirty seams of coal sticking out. In these seams are still found old hollow trunks of trees standing just as they grew, with fossil lizards trapped at the bottom.

The moulding of the surface. Again this area was raised above the sea and many of these horizontal layers of red sandstone and coal were worn away until the hard Appalachian folds were again exposed to form the chief highlands (#89). These provinces, then, to-day consist of a succession of rolling, rocky, wooded highlands, separated by level, fertile areas, several hundred feet lower, drained by winding rivers.

Highlands. There are four main highlands, all running from south-west to north-east across the Maritime Provinces (#89); the most southern (#89-1) begins at *Cape Sable* and passes under the Atlantic Ocean at *Cape Canso*. At the west its width extends almost across Nova Scotia, *South Mountain*, bordering the Annapolis Valley, being its northern edge. It narrows greatly toward the east. The *Cobequid Mountains*, and their extension eastward through *Cape George*, form the westward part of the second uplands (#89-2A and 2B), which in the north arm of Cape Breton expands to the most massive highlands of the Maritimes. The third highland ranges along the north shore of the Bay of Fundy (#89-3) and is cut in two by the St. John gap. The fourth highland, called the *Central Upland* (#89-4), runs diagonally across New Brunswick from near the south-west border to *Bathurst* and expands into a fairly wide plateau in the centre, from which flow the main rivers of the province in every direction, through deeply cut courses. North Mountains (#89-5), which borders the south shore of the Bay of Fundy is a straight, narrow ridge, that from the bay looks like a rampart. At its north-east end it makes a strange twist to the left to form two examples of the most stately and rugged coast scenery in these beautiful provinces, *Cape Blomidon* (#90) and *Cape Split.* None of these uplands are two thousand feet high, and they decrease in height from north to south. The Central Upland is about 1,700 feet high; the one bordering the north of the Bay of Fundy, 1,000 feet high; the Cobequid Mountains, about 800 to 900 feet high; and the *Southern Upland,* from 600 feet high in the west to 300 feet high in the east. All are composed of

#90 CAPE BLOMIDON

volcanic granite-like rocks, and are the hard projecting parts of a plateau that has worn down gradually, the softer parts being gnawed away by the weather a little faster than the harder.

Red Soils. The most of the lowlands, lying between these uplands, are underlaid by coal-bearing, horizontal, soft rocks, the top one being a red sandstone. This rock, which decays to a striking red soil, brightens cliffs, gullies, river banks, tidal flood plains, and even the soil and the roads in many places. It is especially marked

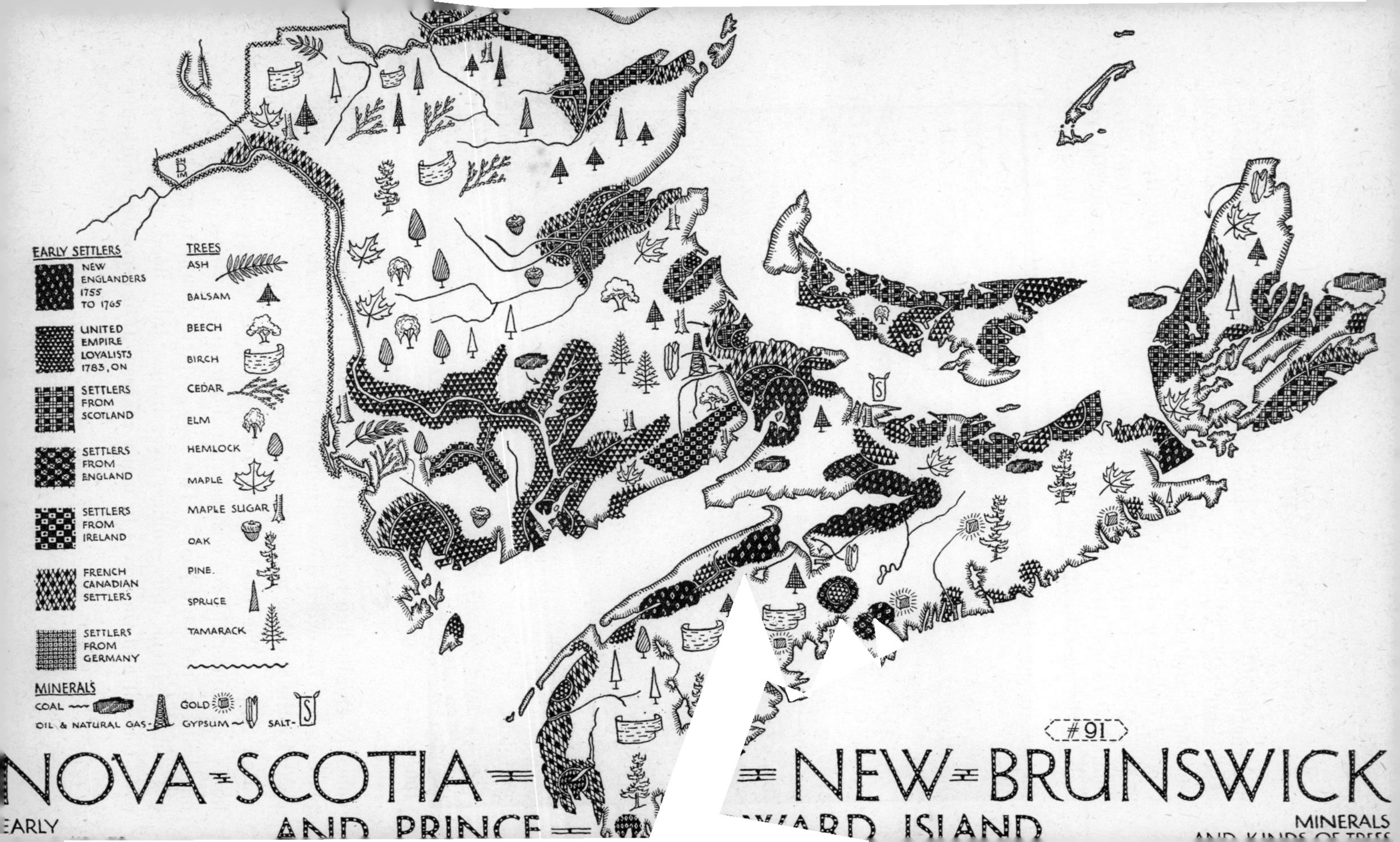
EARLY SETTLERS
NEW ENGLANDERS 1755 TO 1765
UNITED EMPIRE LOYALISTS 1783, ON
SETTLERS FROM SCOTLAND
SETTLERS FROM ENGLAND
SETTLERS FROM IRELAND
FRENCH CANADIAN SETTLERS
SETTLERS FROM GERMANY
TREES
ASH
BALSAM
BEECH
BIRCH
CEDAR
ELM
HEMLOCK
MAPLE
MAPLE SUGAR
OAK
PINE
SPRUCE
TAMARACK
MINERALS
COAL
OIL & NATURAL GAS
GOLD
GYPSUM
SALT
S
#91
NOVA SCOTIA
NEW BRUNSWICK
EARLY
MINERALS

in the red cliffs and soils of Prince Edward Island, into which none of the highlands extend.

Minerals

Gold. Though the Maritime Provinces, as we have seen, have very old volcanic rocks, worn down to their roots, just the place where metallic veins ought to be found, only *gold* has been extensively mined (#91), and it only along the southern upland of Nova Scotia. Dozens, if not hundreds of small mines, like gophers' holes, with piles of rock refuse at the mouth, are scattered from Yarmouth to Canso. Many of these have been long deserted, but in recent years the high price of gold has stimulated new life, and there are at least 20 producing mines.

By far the most valuable mineral is *coal* (#91), which in some places can be seen jutting out of the cliffs. Indeed as early as 1749 supplies were obtained with pickaxe from the cliffs on Cape Breton for the use of the soldiers at Halifax. There are four chief coal districts, one the *Sydney district* on the east coast of Cape Breton, where at *Glace Bay* and *Sydney Mines* immense quantities of coal are mined. Here and there along the west coast of Cape Breton the coal is mined from under the sea. The second district is Pictou County, with *New Glasgow* and *Stellarton* as the chief mining towns. This district has seams of coal thirty feet thick, almost the richest in the world. The third district is just north of Cobequid Mountains from *Joggin* to *Springfield.* The fourth, of much less importance, is at *Minto* near *Grand Lake* in New Brunswick. Natural gas and petroleum are also obtained in New Brunswick south of Moncton.

Nova Scotia's minerals a present to a prince. In 1824 King George IV of England light-heartedly handed over to his brother, the Duke of York the right to exploit, for sixty years, most of the mineral wealth of Nova Scotia. This extravagant gentleman, in order to pay some gambling debts, at once sold the right to an eminent firm of London jewellers, who, under the name of the General Mining Corporation, energetically undertook the mining of coal. They mined the Pictou and Sydney fields according to the best English practice; but in spite of their good work, the people of the province resented their good minerals being lightly handed out to pay gambling debts, and never ceased protesting until the control of this revenue came back to the province in 1858.

Gypsum and salt. *Gypsum* is so widespread, that it forms white seams projecting out of the cliffs in many districts. Large quantities are quarried on Cape Breton and along Minas Basin (#91), and shipped in crude lumps to the United States, there to be converted into plaster of Paris. The only refining plant in the Maritimes is at *Windsor, N. S.*

Nova Scotia has the only salt mine in Canada, at *Malagash* (#91), which should be very valuable, being located in a district where salting fish is so large an industry; but the fishermen, through habits a century old, prefer salt brought back as return cargo from Spain, and *Turk's Island* in the West Indies. However, the Malagash miners are trying to break down these fixed habits by preparing their salt in form suitable to the fishermen.

Coast

A paradise for fish. The inlets and adjacent seas of the Maritime Provinces are more perfectly fashioned for fishing than are any other coastal waters in the world. The whole seaboard has recently sunk, or the sea has risen, and flooded the lower courses of every river valley and gully to make inlets, which worm their way far into the land. Many hills and highlands near the coast have been surrounded, or partly surrounded, by the salt-water to make islands, or irregular peninsulas jutting far into the sea. The whole coast is thus as torn and tattered as the border of a beggar's coat. The fringe of every little inlet is dotted with fishermen's houses, whose humble boats ride at anchor in the protected waters. These inlets also contain ideally smooth, shallow water within which fish may lay their eggs; within their waters also is an abundance of food to attract swarms of small fish, which are valuable as bait. Off the whole coast from Gaspé to Maine the sea is so shallow that the sunlight is able to reach the bottom and stimulate the growth of matted tangles of seaweed, in which swarms of animals live and on which countless fish feed. Nature seems to have neglected nothing, for the cold water gives a firmness and flavour to fish, lacking in the tepid water of the tropics.

Every inlet has its special fish. Every large inlet has feeding grounds favourable to its own particular variety of sea animals (#92). *Chaleur Bay* lying between Gaspé and New Brunswick, with its inflowing river, the *Restigouche*, swarms with aristocratic

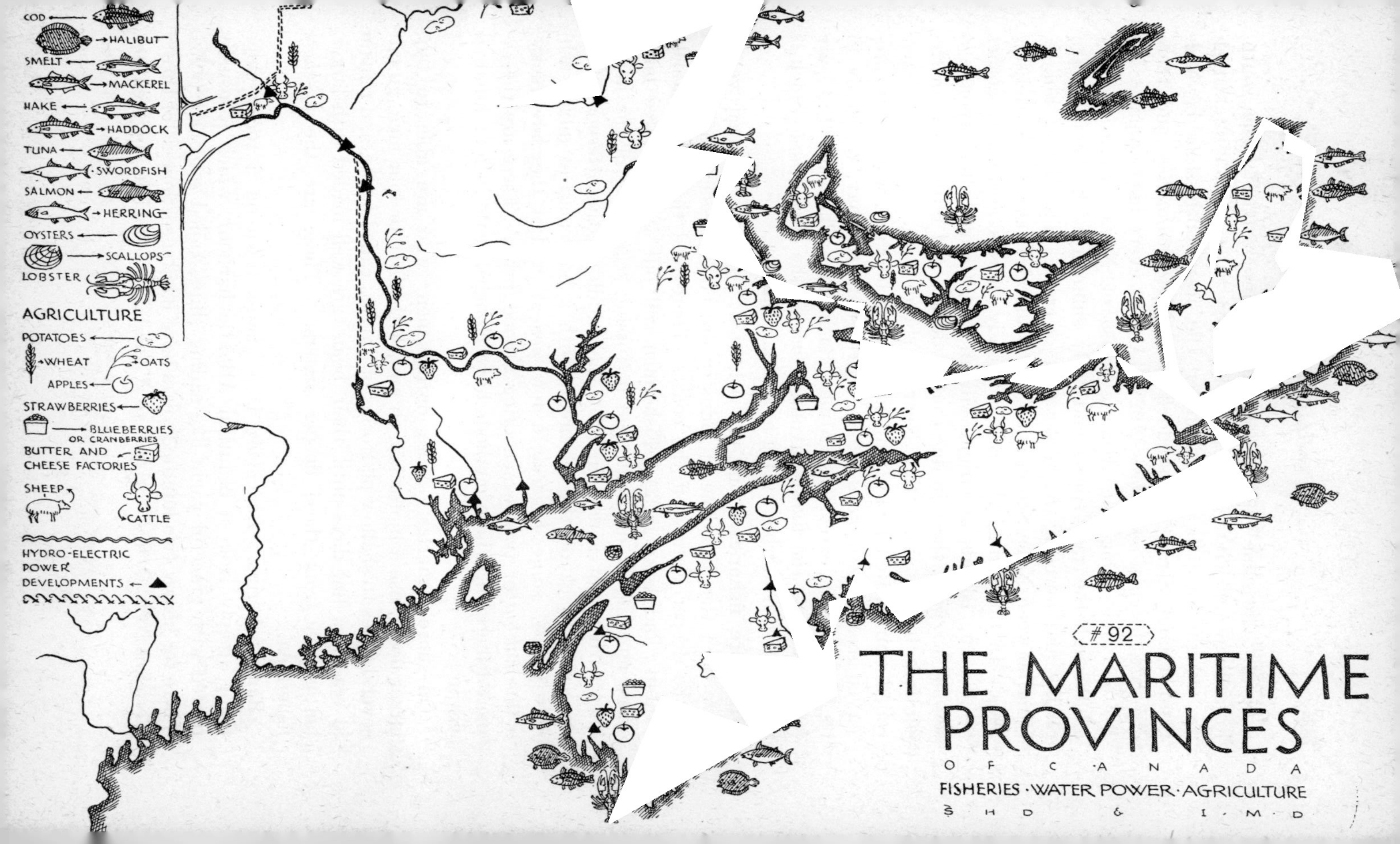
COD
HALIBUT
SMELT
MACKEREL
HAKE
HADDOCK
TUNA
SWORDFISH
SALMON
HERRING
OYSTERS
SCALLOPS
LOBSTER
AGRICULTURE
POTATOES
WHEAT
OATS
APPLES
STRAWBERRIES
BLUEBERRIES OR CRANBERRIES
BUTTER AND CHEESE FACTORIES
SHEEP
CATTLE
HYDRO-ELECTRIC POWER DEVELOPMENTS
#92
THE MARITIME PROVINCES
OF CANADA
FISHERIES · WATER POWER · AGRICULTURE
S H D & I · M · D

salmon. The same Chaleur Bay and Miramichi Bay farther south, in autumn and winter, are alive with tasty little smelt (#92), which are caught in nets placed under the ice. Both sides of Northumberland Strait are studded with *lobster canneries,* and there also, are the chief *oyster beds* (#92). Almost every inlet around Cape Breton and along the Atlantic coast of Nova Scotia is busy all through the raw spring and early summer with the fishermen raising and lowering the *lobster pots,* or traps. The Bay of Fundy, lying between Nova Scotia and New Brunswick, also has its lobster beds. In the spring, the Bay of Fundy, at its north-west corner around the islands of Grand Manan and Campobello, becomes a sea of *herrings;* these in due course, depart and return later in the summer, bigger, and fatter. Near the shore, traps, made of branches, called weirs, catch many millions of baby herrings, which are preserved in oil as sardines (#96). Farther up the bay, especially in its two eastern extensions, Minas Basin and Chignecto Bay, are found perhaps the finest flavoured fish of the coast, *shad,* which is an overgrown herring as big as a small salmon.

In-shore fisheries. But the fish that, for four hundred years, have been the backbone of industry are bottom fish, which swarm in the deeper waters off the coast. There feed the cod, haddock, halibut, and hake (#92). Fishermen sail out two or three miles every morning at daylight, and in tossing seas, drag in the heavy lines with might and main. By the middle of the afternoon they are sailing back on the sea-breeze with a boat load of mixed fish, mostly *cod,* some *haddock,* a few *pollock,* and if they are lucky, a fine *halibut* or two, which bring a good price. These are the *in-shore* fishermen, whose methods have changed little in four hundred years.

Off-shore fisheries. More daring men, who are able to buy larger boats, use a motor launch and, with three or four men, go out several miles to shallow platforms in the sea, called *banks,* and remain at sea for a few days until their boats are well loaded with larger, better fish from these deeper waters. These are the *off-shore* fishermen.

Bank fisheries. *Lunenburg* is a quaint town on a rocky slope rising steeply from a beautiful Atlantic harbour, which is usually dotted with graceful yacht-like sailing ships, called *schooners* (#93). There for nearly two hundred years have lived a settlement of thrifty Germans (#91), first driven to Canada by religious persecution.

Every schooner in the harbour has been made from their native wood, by their own hands, and is usually owned by the builders. The first one was launched about 1870; and since then in increasing numbers, these courageous men and skilful sailors, in crews of about a dozen men, sail out to the Grand Banks just south of Newfoundland and remain for a month or more through dense fogs and fierce gales until the hold of the boat is filled with salted cod, or perhaps their bait being all gone, they are forced to return

#93 LUNENBURG HARBOUR

(pp. 14-26). These ships are the *bankers*, and their crews the *bank fishermen*.

History of Fishing

Fishing from a barrel. We have already seen that almost from the day Cabot reported seeing shoals of fish, so dense that they could be scooped up in baskets, sturdy little schooners from France, England, and Portugal flocked to the banks of Newfoundland every summer to fish. At many of these annual visits, heated disputes over the best fishing grounds arose among the fishermen of these nations. It was not long before they were also fishing off Cape Breton and in the Gulf of St. Lawrence. The fishermen, in barrels lashed to the outside of the ship's railing, fished with two hand lines each, and hauled in the cod as fast as their strength would allow. The fish were cleaned and slitted. Afterwards they were packed in salt in the ship's hold until the boat was loaded or the salt exhausted; then with approximately 40,000 fish on board they arrived back in port after an absence of three to six months. This was called the *wet* or *green fishery*. Before the days of Champlain, there were as many as two hundred British ships and eight thousand men on the Banks in a single season.

Fishing leads to settlement. It was not many years until fishermen found that the valuable cod swam not only on the Banks, but also swarmed along the shore and even crowded into the coves and river mouths. This discovery led to the growth of the *dry fishery*, which began about the time Jacques Cartier sailed up the St. Lawrence (p. 251). Large ships anchored in a harbour in the spring, and the crew of fifty would, in a day or two, build a house to live in with the ship's sail for a roof and sides, and spruce bows for ends. All summer they caught fish by hook and line from small boats, brought the catch ashore each day, and spread the lightly salted fish on racks to dry. These smaller cod were exactly suited for this method of curing, required much less salt, and occupied less space, so that at the end of the season a ship could carry back 200,000 fish instead of 40,000, which was a ship's capacity in the green fishery. These fishermen soon began to marry Indian women, and instead of returning home on the ship in the autumn, would remain with wife and children on the shore and perhaps engage in trapping until the ship returned in the spring. In this way fishing settlements were formed all along the coast during the French régime (#89). One of the first of such settlements was at *Port Rossignol* on Liverpool Bay in 1633.

Later history of fisheries. When in 1763 the Maritime Provinces became permanently British, the *Robins family*, who were also very active on the Gulf coast of Quebec (p. 256), organized the fishing industry in the Maritimes so that it developed rapidly. Many of the United Empire Loyalists who settled near the coast to farm (p. 234), were attracted to the more exciting work of fishing, and the industry developed steadily, but chiefly as in-shore dry fishery. We have seen that not until about 1873 did the Banks green fishery begin to take root at Lunenburg.

History of lobster fisheries. Great changes have taken place in the last fifty years. Such delicacies as lobsters were formerly so numerous that sometimes millions were left stranded on the shores of the Gulf as the tide went out, and were carried off in wagons to fertilize the farms. The same was true of herring. The first three lobster canneries were built in New Brunswick in 1870. By 1910 these had increased to 677; but due to a decrease in the catch, and the tendency for several small canneries to combine into one large establishment, the number has greatly decreased.

#94

UNLOADING A CATCH OF LOBSTERS

Because of reckless fishing, lobsters, which once sold at fifty cents a barrel for the finest, now often bring that price retail for a single one of medium size and quality.

Swordfish and tuna. In recent years two immense fish, formerly cursed as pests that were always breaking fishermen's nets, are now hunted eagerly on account of their great value (#92, #94). These are the swordfish, and tuna or horse mackerel. A fleet of motor boats searches the surface for these giants. A man on the prow shoots one with a harpoon and a wild chase begins, which may last for hours. The great creature finally becomes exhausted and is hauled on board. They are packed in ice and rushed to Boston and New York.

Drainage

Rivers short and tumbling. As there is heavy rainfall in the Maritimes (p. 219), most of the rivers are well filled with water, even during the dry season. Since Nova Scotia is long and narrow, and the highlands run lengthwise, the short rivers, for the most part, tumble down rocks to the sea. Though of no value for navigation, they are the delight of choice salmon (#89), and consequently a delight to sportsmen. Some of these rivers start near one coast, cut right through the highlands, and have their mouth on the opposite coast. For example, it is possible by ascending the *Shubenacadie*, which empties into Cobequid Bay, to paddle within two miles of Halifax harbour. *Annapolis River* is an exception; hemmed in between North and South Mountains, it is forced in the direction of the peninsula and empties into Annapolis Basin.

St. John River. New Brunswick, which is wide enough to have rivers of considerable size, is beyond all other provinces the land of bubbling springs and rippling rivers. Indeed, one can enter this province through any one river and, with few portages, pass in any direction throughout the province. The outstanding river is the *St. John*, one of the most notable in Canada, and the largest that cuts across the Appalachian Highlands between the St. Lawrence and Florida. It rises near the source of the Chaudiere in the highlands of Maine, makes one grand arching sweep north-east, and then cuts diagonally across New Brunswick to empty into the Bay of Fundy (#89). The only serious break to navigation is *Grand Falls*, where the water tumbles down nearly 74 feet. But this grand scene has been destroyed by the engineers, who have deflected all the water through an immense ugly pipe to turn a turbine at its foot. However, this produces the greatest supply of electricity in the Maritimes, giving a flood of electric light to the towns of the north and life to the throbbing machinery of pulp mills and sawmills. The river below this point swings majestically from side to side through beautifully green fields of potatoes and oats, luscious pastures dotted with choice cattle, and rustling forests of maple, birch, spruce, and balsam (#91). Below Fredericton the scene changes, the stream broadens out into lake-like expansions, some of which back far up the branches; soon it begins its course through the southern highlands bordering the Bay of Fundy. Its banks become high, rough land lies on each side, farms are fewer, the stream is deeper, and rises and falls with the tide. It finally breaks through the highland at St. John in a gorge so narrow, that at the outflow of the tide, water tumbles outward in a waterfall and at the inflow water tumbles inward, and we have a reversing fall; but only in exceptional conditions of the tide is the fall of spectacular height.

Uses of rivers. This most notable river of the Maritimes is navigable for good sized ships as far as Fredericton, and during spring and fall, for smaller boats, as far as Grand Falls. Above this break it is still navigable for sixty miles. While the ships on it are few, many millions of logs have been floated down it and its tributaries; but its best lumbering days are past. The *Restigouche* and *Miramichi* (#89), with their complex network of branches cutting right across the province from west to east, are lumbering rivers of the greatest importance. They are a paradise for the angler, whether fishing for lordly salmon, fighting bass, and the saucy speckled trout, or shooting graceful deer and giant moose (#89). All these rivers emptying into the Gulf of St. Lawrence have wide bay-like mouths, kept deep and clean by the movement of tides twice a day. These estuaries have perfect harbours and are fertile fishing grounds.

Water power

Though the rivers of Nova Scotia emptying into the Atlantic Ocean are, for the most part, short (p. 214), they receive heavier rainfall than any other part of Canada east of the Rockies, have many lake reservoirs, and tumble down in rapids and waterfalls to the ocean. Since all good sites for electrical power development are close to the towns which require the power, the province is quickly harnessing the most suitable streams (#92). The two largest water powers, at *Sheet Harbour* and *Liverpool*, are used for running large pulp mills. *Halifax*, *Lunenburg*, and *Yarmouth*, situated far from coal fields, have each developed electricity in convenient rivers. Fortunately, the north side of the province and Cape Breton, whose rivers are small, have abundant coal for power (#91).

Grand Falls on the St. John has the largest hydro-electric development in the province of New Brunswick, but there are several others of importance (#92). Power developed a few miles west of St. John

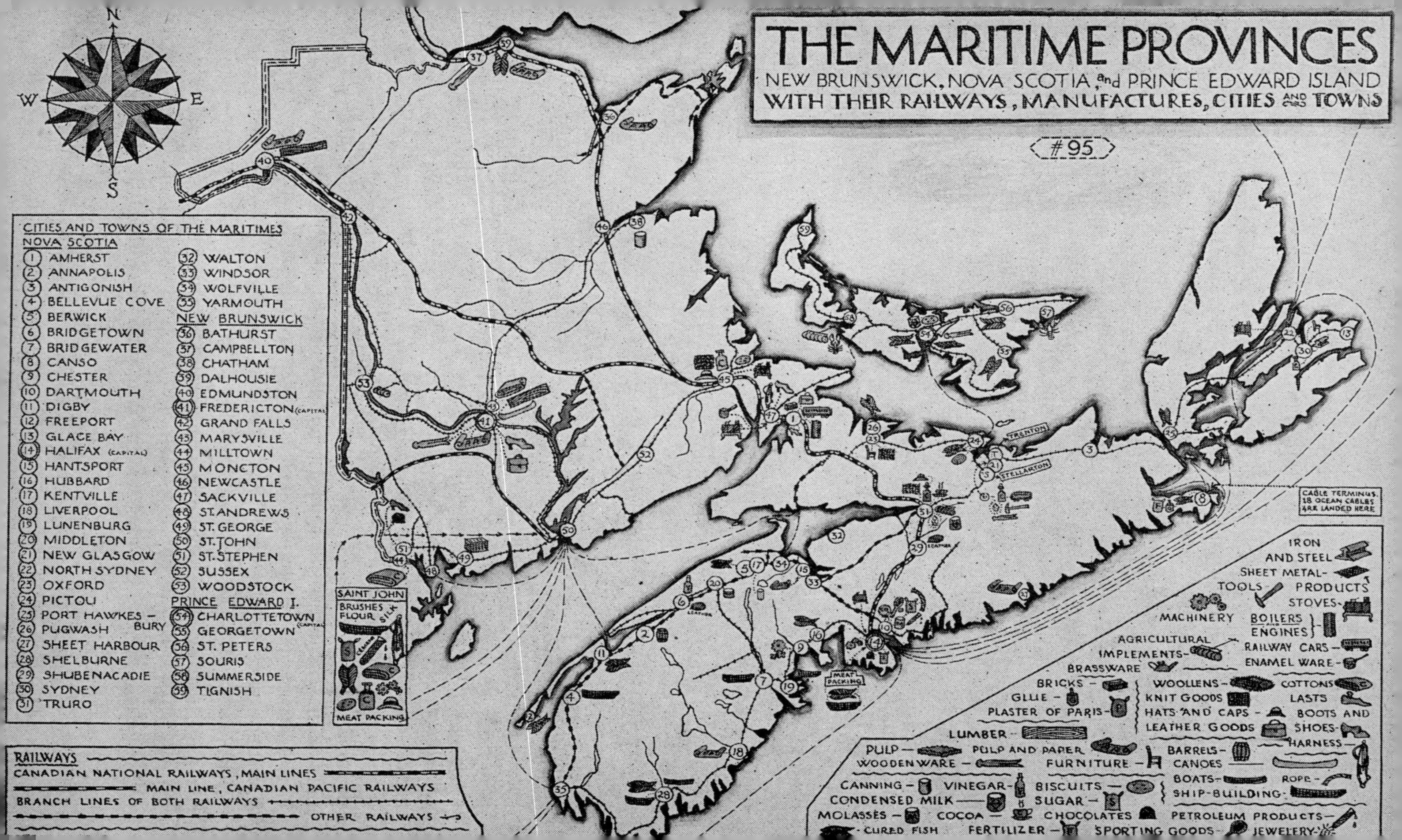
THE MARITIME PROVINCES
NEW BRUNSWICK, NOVA SCOTIA, and PRINCE EDWARD ISLAND
WITH THEIR RAILWAYS, MANUFACTURES, CITIES and TOWNS
#95
N
S
E
W
CITIES AND TOWNS OF THE MARITIMES
NOVA SCOTIA
1 AMHERST
2 ANNAPOLIS
3 ANTIGONISH
4 BELLEVUE COVE
5 BERWICK
6 BRIDGETOWN
7 BRIDGEWATER
8 CANSO
9 CHESTER
10 DARTMOUTH
11 DIGBY
12 FREEPORT
13 GLACE BAY
14 HALIFAX (CAPITAL)
15 HANTSPORT
16 HUBBARD
17 KENTVILLE
18 LIVERPOOL
19 LUNENBURG
20 MIDDLETON
21 NEW GLASGOW
22 NORTH SYDNEY
23 OXFORD
24 PICTOU
25 PORT HAWKESBURY
26 PUGWASH
27 SHEET HARBOUR
28 SHELBURNE
29 SHUBENACADIE
30 SYDNEY
31 TRURO
32 WALTON
33 WINDSOR
34 WOLFVILLE
35 YARMOUTH
NEW BRUNSWICK
36 BATHURST
37 CAMPBELLTON
38 CHATHAM
39 DALHOUSIE
40 EDMUNDSTON
41 FREDERICTON (CAPITAL)
42 GRAND FALLS
43 MARYSVILLE
44 MILLTOWN
45 MONCTON
46 NEWCASTLE
47 SACKVILLE
48 ST. ANDREWS
49 ST. GEORGE
50 ST. JOHN
51 ST. STEPHEN
52 SUSSEX
53 WOODSTOCK
PRINCE EDWARD I.
54 CHARLOTTETOWN (CAPITAL)
55 GEORGETOWN
56 ST. PETERS
57 SOURIS
58 SUMMERSIDE
59 TIGNISH
RAILWAYS
CANADIAN NATIONAL RAILWAYS, MAIN LINES
MAIN LINE, CANADIAN PACIFIC RAILWAYS
BRANCH LINES OF BOTH RAILWAYS
OTHER RAILWAYS
SAINT JOHN
BRUSHES
FLOUR
SILK
MEAT PACKING
TRENTON
STELLARTON
LEATHER
MEAT PACKING
CABLE TERMINUS
18 OCEAN CABLES
ARE LANDED HERE
IRON AND STEEL
SHEET METAL PRODUCTS
TOOLS
STOVES
MACHINERY
BOILERS
ENGINES
AGRICULTURAL IMPLEMENTS
RAILWAY CARS
BRASSWARE
ENAMEL WARE
BRICKS
GLUE
PLASTER OF PARIS
WOOLLENS
COTTONS
KNIT GOODS
LASTS
HATS AND CAPS
BOOTS AND SHOES
LEATHER GOODS
HARNESS
LUMBER
PULP
PULP AND PAPER
WOODENWARE
FURNITURE
BARRELS
CANOES
BOATS
ROPE
SHIP-BUILDING
CANNING
VINEGAR
BISCUITS
CONDENSED MILK
SUGAR
MOLASSES
COCOA
CHOCOLATES
PETROLEUM PRODUCTS
CURED FISH
FERTILIZER
SPORTING GOODS
JEWELERY

is distributed by cable as far east as Moncton, and electricity from the power stations on the St. John River is transmitted to all important towns along the river.

Exploration

Champlain's first settlements. The fearless men, who first explored the rocky coast, picked their way through the inlets studded with hidden reefs. They never knew they were brave, and would have laughed at being called explorers. Their names are unknown, for they were humble fishermen, who pushed ever farther forward, but wrote no books nor letters; indeed, the most of these lowly French fisherfolk could neither read nor write. From the days of the Cabots, these men, in frail ships, came yearly to fish, first south of Newfoundland, then off Cape Breton; and it was not long before they were drying their fish on the sloping shores of all the inlets. Hence, when *De Monts* and *Champlain* worked their way along the Atlantic coast of Nova Scotia in the spring of 1604 (#89), every inlet they explored had no doubt been well known to the coast fishermen for many years. These explorers had two ships and over one hundred workmen and settlers. The scene along the shore was a varied and endless panorama; birches, oaks, maples, and spruces (#91), charmed their eyes at every turn; wild ducks, wild geese, snipe, and plover, furnished thrilling sport for guns and savoury food for the table (#89). On a cliff they gathered a barrel of meaty eggs of the cormorant; gannets, as big as geese, they killed with sticks, and some rocks were literally covered with seals so tame that the sailors took as many as they wanted. The ships rounded the western end of the province (#89), and soon entered the Bay of Fundy. At last *Digby Gut*, with the towering walls of *North Mountain* on each side, was traversed, and *Annapolis Basin* in all its loveliness spread before their astonished and admiring eyes. Little did they think how often this peaceful spot was to hear the boom of guns, see the land drenched in blood, and witness destruction of humble cottages, fertile fields of wheat and oats, and helpless cattle. They did not settle in this charming valley, but foolishly sailed east nearly to the head of the Bay of Fundy, whose high tides they saw. Then they returned along the north shore, passed St. John's noble harbour and settled down at a most unsuitable place on a barren, woodless, waterless island at the mouth of the St. Croix River (#89). The settlers built houses, even planted seeds in autumn, and waited for winter, which this year came early and was exceptionally severe. On October 4th the first snow fell, and soon these settlers from sunny France were shut in freezing huts, with the whole island under four feet of snow. Even the wine froze solid, and cider was cut with an axe and served by the pound. The dreaded scurvy settled down like a black shroud on this helpless little group, and when in spring its ugly form lifted, thirty-five out of sixty-nine were dead.

Founding of Annapolis. Often during that dark, icy, horrible winter, they must have longed for the peace, sunshine, and beauty of Annapolis Basin, which they had reached the summer before. As soon as the dull and snail-like spring was passed, they moved all that was left of their belongings back to that spot; and there they founded the first permanent settlement in Canada.

During these times young Samuel de Champlain was very busy exploring and drawing maps of every coast and inlet. Once in 1604 he pushed south along the New England coast as far as Cape Cod (#89). Next year he repeated the journey, rounded Cape Cod, and went as far as Nantucket. If he had gone a little farther he would have reached the alluring harbour of New York. Its attractions might have drawn the first settlement to that superb situation instead of to Annapolis, and the whole history of North America might have been changed.

Climate

Seasonal changes. The Maritime provinces are children of the sea. Great masses of air moving eastward from the land are masters of the temperature, while moist masses of air moving in from the Atlantic Ocean bring most of the rain, snow, and fog. The temperature of Nova Scotia and the south part of New Brunswick is much like that of south Ontario, the Atlantic Ocean more than making up for the Great Lakes; but Ontario is hotter in summer. The northern, western, and higher parts of New Brunswick have the more rigorous winters of western Quebec. Summer throughout these provinces is mild and comfortable, with cool nights, because of the evening breezes from the sea. September is a pleasant autumn month with a threat of frost toward the end. Though there are fine days in October, the weather toward the end of the month begins to get cold, and severe frosts bring growth to an end. November is unpleasant with bleak, cold winds. Winters in Nova Scotia and south New Brunswick are milder than in south Ontario, because of the moderating effect of air masses coming in from the Atlantic. All the harbours along the Atlantic coast and Bay of Fundy are open throughout winter, but those of the colder gulf water are sealed with ice. The icy waters of the Gulf of St. Lawrence and Atlantic make winter linger too long, so that even June is unpleasantly cool and raw along all the coastal regions. By the end of June, masses of warm air moving eastward obtain the mastery and bring balmy summer again. In the higher, interior parts of New Brunswick, where the influence of the ocean is least, summers are warmest and winters coldest. In Prince Edward Island, summer, with clear sky, sunny days and cool nights, is delightful; and the long autumn with mellow sunshine, pleasing coolness, dazzling landscape splashed with gold, green, and crimson, and air as vigorous as a tonic, is nearly perfect. Winter is steady and bracing, but the less said about the raw, prolonged spring, the better; it can get no warmer than the frigid waters of the gulf.

Rainfall. Rainfall is heavier than in any other part of Canada except the coast of British Columbia. It is mostly caused by warm, moist masses of air, moving in from the Atlantic Ocean, being pushed up above the cooler masses coming from the interior. Rainfall is heaviest on the coast of Nova Scotia, and becomes less in north-west New Brunswick. Everywhere it is sufficient to give a bright greenness throughout the growing season; and in winter, precipitation is heaviest of all. Most of this is snow, which is very deep in north New Brunswick and usually covers the whole area for several months. Along the mild Atlantic coast much rainfall is mixed with the snowfall.

Fogs. The cold Labrador current creeps south-west along the whole Atlantic coast of Canada, while one hundred miles from shore the warm Gulf Stream moves slowly forward in the opposite direction. Though this latter current has little effect on the climate of the Maritimes, it breeds fogs all along the coast, which are very dangerous to shipping. The masses of warm, damp air over the Gulf Stream, as they move toward the coast with every east and south-east wind, have their bottom layers so chilled in passing over the icy Labrador current, that the moisture is condensed to a thick bank of fog, which drifts in, like a huge curtain of wool, over the land. These fogs are especially frequent in summer on the Atlantic coast and on the Bay of Fundy. Prince Edward Island is almost free of fog.

Forest Industry

Masts for the British navy. Heavy rainfall in all parts of the Maritimes stirs tree growth so rapidly that parts which have been disfigured by fires or gashed by the axe, soon cover over their wounds with a fresh forest growth. When the early discoverers arrived all but the barest rock and the sandiest stretches were clothed by dense forests of mixed hardwoods, such as beech, birch, and maple, and softwoods, such as pine, spruce, balsam, and hemlock (#91). As early as 1700 a French ship, which brought supplies to a fort near *Marysville* on the St. John River, returned with a load of magnificent, white-pine masts, some three feet across at the butt and one hundred feet long. After the English conquest, the white pine along the St. John and other convenient rivers was set aside to supply masts for the British navy and merchant ships; and since a good mast sold for five hundred dollars, some snug fortunes were made by lumbermen at Fredericton and St. John in the early days. Then a little later the square timber trade with Great Britain filled the St. John and Miramichi rivers with rafts, and the harbours at their mouths with ships.

To-day the best of the white pine is gone in both provinces, and spruce and balsam fir are coming to the front.

The sawmills and the rivers. Both provinces, with their complex network of streams reaching back into the forests, are well equipped for moving logs down to the mills. The short streams of Nova Scotia, serving small areas, have produced many small sawmills; while the longer and more richly branching rivers of New Brunswick, draining much larger areas of dense forests, and served by larger mills, especially along the St. John, Miramichi, and Restigouche have a much larger output of lumber. A great stretch of forest in north New Brunswick between the Miramichi and Restigouche is almost unbroken by a single settlement (#91). In Nova Scotia, the broad stretches of forests are dotted with little settlements, like a lake with islands, so that the settlers who work lean farms in summer to get bare necessities, can earn a little extra by cutting logs in winter and add some comforts to life. This is a great advantage to lumbering, since skilled lumbermen are everywhere obtainable.

Methods changing. Since the best timber has already been cut, lumbermen are being forced farther and farther back to get good logs. Streams are getting shallower on account of the clearing of forests, and it is not so easy to float logs down. To add to the logger's troubles, rivalry from British Columbia is getting keener, since the digging of the Panama Canal gives that province cheap freight to the eastern United States and to Great Britain. In order not to fall behind in the race, New Brunswick is now using more efficient methods. Logging railways are being pushed back from the main railways and rivers into the forests; caterpillar motor-trucks, which can be used the whole year round, are replacing sleighs (p. 49). Sawmills are being taken from the mouths of rivers back to the timber in the forests; and machines are replacing horses for loading.

Shoe pegs and toothpicks. New Brunswick, and to a less extent Nova Scotia, have fine stands of firm hardwood, which have not been extensively used. Large quantities of this timber are required for making such little things as spools for thread, pegs for shoes, toothpicks, lasts for boots, sticks for candy "suckers," etc. Much birch for such purposes is cut in New Brunswick, shaped into suitable sizes, and shipped to Great Britain and the United States for manufacture (#91).

Pulp and paper mills. Both New Brunswick and Nova Scotia are now using their second-growth spruce for pulpwood, and pulp and paper mills are springing up along the rivers near falls and rapids, where hydro-electric power is easily obtained (#92, #95). The largest one in Nova Scotia is near *Liverpool*, and the largest in New Brunswick are at *Edmundston*, *Bathurst*, and *Dalhousie*. #95 shows the position of these mills.

Both New Brunswick and Nova Scotia have splendid groves of sugar and rock maple (#91), from which are obtained large quantities of maple sugar and also some maple syrup.

Agriculture

Bad springs, bountiful autumns. Though the fisheries of Nova Scotia and the forests of New Brunswick catch the eye, nevertheless, the products of the farm surpass those of the sea, the forest, and the mine in each of the provinces on the Atlantic. The bountiful rains throughout the growing season drop down fatness, and the balmy summers and prolonged, gentle autumns give a greenness and richness of growth surpassed nowhere in Canada. The raw winters and the too long, harsh springs are serious drawbacks, since animals have to be fed in stables for a large part of the year, and early planting is impossible.

Hungry granite rules the desolate scene. The southern highlands of Nova Scotia (#89), composed mostly of granite rocks, are, for the most part, bare and barren, though here and there, pockets of soil left by glacial ice, and small river valleys, supply feeble settlements with scant farm lands, and with pasture for a few cattle. Islands of soft and crumbly slate in the granite have decayed and these support further, little scattered farm settlements along part of the coast; but from Halifax to Canso the forbidding, hungry granite rules the desolate land for the most part. The higher parts of the Cobequid ridge are almost as barren as the southern highlands.

Three classes of farm lands. All the lowlands along the north side of the province, including west Cape Breton, are covered with moderately fertile soil and are farmed. These lands are of three classes (a) *uplands*, (b) *intervales*, and (c) *salt marshes.*

The uplands are rougher lands, rising between the river valleys; they are most suitable for pasture and hay. The valleys between these uplands are flat and fertile, with fine silt spread out by the streams in flood. These, called *intervales*, have rich soils and are the choicest lands for fruit, turnips, potatoes, and vegetables. The famous *Annapolis* and *Cornwallis* valleys and the middle course of the St. John are the richest of the intervales, but there is one in the lower course of almost every little stream.

The salt-marsh land. It was on the bosom of the salt marshes around the Bay of Fundy that the first settlements of Canada were nourished. The tides of this bay are probably the highest in the world. The bay opens to the south-west like a funnel. As the tidal wave, twice a day, enters this throat from the Gulf of Maine (#89) it is only 10 to 12 feet high, but as it advances with a rush it gains in height what it loses in width; at Yarmouth and St. John it has swelled to 24 feet in height, at Digby it has mounted to 27½ feet, and at Annapolis to 29 feet. By the time the tide reaches Minas Basin it is a wave fifty feet high climbing up the shelving shore. This great surge drowns every inlet and rushes up the rivers with a roar, flooding the land for miles. With every advance of this turbid tide, brown as chocolate, it washes up and leaves on the flats great quantities of red, slimy silt. As the tide drops, the water creeps and trickles back, and leaves miles of perfectly flat shore, covered with an ugly layer of fertile, red slime. Year after year, and century after century, this deposit grows thicker and thicker.

From slimy ooze to luscious grass. When the Acadians first landed in Annapolis Basin, they had the choice of clearing the forest

#96
Find the following: a silver fox, woman packing sardines, dyked land with marsh hay, a potato field, pulp mill, orchard, sheep farm, apples ready for shipment, a steel mill, and a wagon load of lumber. In what parts of the Maritimes can each of these industries be seen?

or of shutting out the tides from these broad, rich flood-flats. As along the Bay of Biscay, in their home land, the French peasants had already learned how to build dikes, or mounds of earth, to shut out the sea, they selected the salt marshes in preference to the forests, and settled all along the shores of Minas Basin and Chignecto Bay (#89) as well as along the flood plains of the streams emptying into them. There they dammed back the tides by earthen walls, strengthened by branches of trees; planted their gardens, laid out their apple orchards, and above all harvested rich crops of marsh hay to fatten herds of sleek cattle during the winter and long spring (#96). These marshes are still, after three hundred years, the scene of some of the most profitable farming in both New Brunswick and Nova Scotia. So deep and rich is the soil that, in spite of neglect to fertilize it, it continues for long periods to give bountiful crops. When at last it shows signs of barrenness, the farmer opens the dike, allows the rich, red silt from the high tides to settle down on it for a year or two, closes the dike, and again it begins a new period of vigorous growth.

The intervales. The farms of New Brunswick are largely on the intervales of the River St. John and its tributaries, and of the rivers emptying into the Gulf of St. Lawrence. These intervales are much wider and larger than in Nova Scotia, so that far fewer of the settlements are shut off in the bush.

The million acre farm. Prince Edward Island's farms, with fertile, bright coloured soil, cover almost the whole surface like a red rash. In no other province of Canada is there so little waste land. Juicy hay, plump oats, perfect potatoes, and richly-flavoured turnips are grown on nearly every farm, and support flocks of choice dairy cattle, whose milk is turned into butter at the creameries (#92).

Dairying. Since hay, oats, and turnips are three of the chief field crops, Maritime farmers are rapidly learning that it is most profitable to themselves and most beneficial to the land not to sell these crops and starve the land, but to feed them to stock, especially dairy cattle. They have found it better to sell the meat, milk, butter, cheese, and wool, and to put the manure back on to the land. Consequently, mixed farming is general (#92). The horses, especially of Prince Edward Island, are noted for their vigour, sturdiness, and perfection, but the motor car has decreased the demand. Dairy cattle are plentiful and of good quality in all the best farming districts along the marsh lands of the Bay of Fundy and the intervales of the St. John River (#92). About 1890, cheese factories were first started in New Brunswick and soon became numerous, especially in the south. In all three provinces the cheese factories have been driven out by butter factories (#92); though it is more work for the farmer to separate cream, and greater care is necessary to get the product quickly to the creamery, profits on cream seem to be higher; and, besides, the farmers have the skimmed milk to feed the hogs. Indeed, creameries and hogs seem always to thrive together.

Fifty miles of apple blossoms. The two farm specialties of these provinces are apples and potatoes (#92). While apples are grown along the St. John River and around Minas Basin, and could be grown successfully in most sections of these provinces, the *Annapolis valley* surpasses all other apple regions of Canada, except the Okanagan valley in British Columbia, in the production of apples. Protected from cold winds by the wall of the North Mountains, this shielded intervale, with rich soil and sunny skies, puts size, colour, and flavour into the best varieties of apples. During the last forty years, British people, settled from the New England states (#91), have added the skilled intelligence and careful co-operation necessary to build up a great apple industry. The location, nearest to the British market, the ability to grow the varieties of apples the British people prefer, the satisfying flavour which the Nova Scotian soil and climate diffuse into the fruit, and the long bearing life of the trees (often a hundred years) have all played a part in making the Annapolis valley in May a glorious spectacle of fifty miles of apple blossoms.

Strawberries and blueberries. While all the small fruits, such as raspberries, gooseberries, and currants are cultivated everywhere; strawberries are widely grown, and bring a ready and profitable sale. Their period of ripening is thrust so far forward by the long harsh springs that they appear on the market just after the supplies from other districts have failed (#92). The canning of wild, or semi-wild, blueberries is now a growing industry in west Nova Scotia and along the gulf coast of New Brunswick (#92), and can be much more widely extended as the demand for this fruit grows.

Prize potatoes. Abundant summer rains in New Brunswick and Prince Edward Island give potatoes a vigorous start; the moderate warmth of summer is just right for steady growth, but not sufficiently hot to encourage blight and other plant diseases. Another favourable factor is the long, mild autumn with warm rains which allows the potato plant to coax all its numerous tubers to full size and graceful oval shape. Not only are the people of Ontario and Quebec willing to pay higher prices for Maritime potatoes for table use, but they, and even more the growers of the southern United States, pay very fancy prices for these potatoes to use as seed (#92).

Fur farming. A branch of farming, which started in Prince Edward Island fifty years ago, was nursed to success through the skill and patience of a few ingenious men. It has already spread over the rest of Canada and the United States, is now taking root in every other civilized country, and nobody knows to what importance it may

#97

A FOX FARM

develop. It is fox farming (#97). Occasionally an elegant silver-black pup may appear in a litter of the ordinary red fox. *Charles Dalton* at Tignish and *R. T. Oulton* near Alberton, by patient selection, got the silver-black strain to breed true, that is, produce nothing but silver-black pups. Then they had to find the conditions under which these delicate animals could be reared and kept in captivity. These problems have now been solved, and it has been found that each pair of foxes should be kept in a small yard, enclosed by wire, which must be high enough, and arched inward enough, to prevent these active creatures from climbing over. The wire must be buried deep

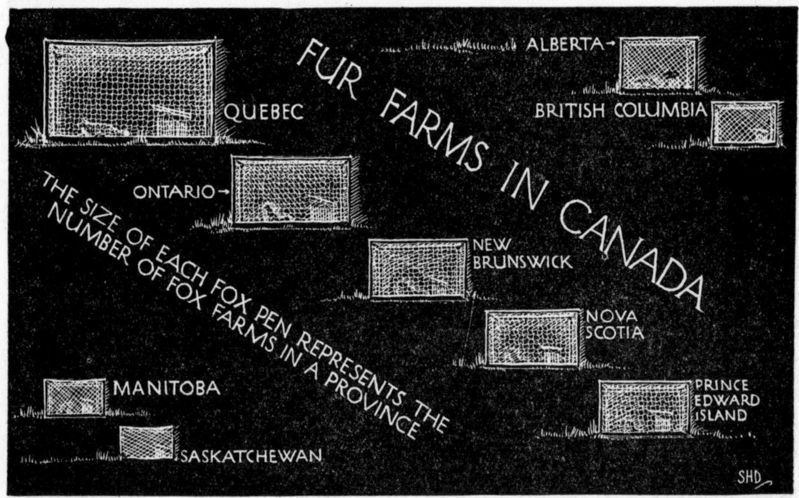

#98

enough in the earth to prevent them from digging their way out; often the whole floor is lined with wire netting. In each yard there should

#99

be a covered den to which the foxes may retire (#97). Everything must be kept very clean, and the quantity and quality of food must be carefully watched. Many thousands of beautiful foxes are obtained each year from these farms (#98). Success with the silver-black foxes has led to the establishment of fur farms for rearing mink, muskrat, raccoon, fitch, and other fur-bearers. This new kind of farming is spreading, and perhaps in a short time the world's supply of fur will be largely obtained from the farm rather than from the fast-vanishing forest (#99).

Settlement

Settlers on the graveyard of the Atlantic. One of the most desolate spots on the face of the earth is a long, low, narrow strip of shifting sand called, *Sable Island*, which barely lifts its dangerous face above the surface of the stormy Atlantic, over 100 miles south-east of Nova Scotia. It is without a harbour and is usually hidden under a dense fog or swept by storms. There are long treacherous sandbars at each end, running out for miles just under the water. It lies in wait, like an ugly shark, for the unsuspecting ship, and is so low that its dull colour cannot be seen by ships for any distance, even in clear weather. It is no wonder that this ship's trap is called the "graveyard of the Atlantic." Yet, unbelievable as it may seem, the first settlers brought to Canada were brutally landed in 1598 on this island of doom by the *Marquis de la Roche*. He returned to France and left them to their fate. They were chiefly criminals and beggars, snapped up from jails and slums in France. Cattle and pigs, put off on the island many years before, had managed to multiply, though their only food was rank grass and coarse weeds, which grew in the barren sand. With these semi-wild animals, with fish, seals, and walruses for food, sixty of these unlucky men fought a losing battle with starvation and death for five years. Finally a fishing boat arrived and took a dozen half-dead wretches — all that were left — back to France.

Birth and death of Port Royal. The next attempt at settlement was in 1604 when the ill-fated settlement at St. Croix Island was started (p. 217). The transfer of the ghastly remnant of these people to the comfortable climate, pleasant surroundings, and productive soil of Port Royal was the first planting of French Acadians on the soil of the Maritime Provinces. In 1606 a group of carpenters, blacksmiths, and other workmen arrived, and this budding settlement began to sprout and spread. They built the first American boat and erected the first water-wheel, which turned the first grist-mill. At this time

Great Britain and France were at war; and as the New England states, just to the south of Acadia, were colonies of Great Britain, the colonists fought as well. In less than ten years a New England fleet burnt the buildings, slaughtered the cattle, destroyed the crops, and carried off the Acadians. Thus, a prey to the savage beast of war, this quiet, prospering settlement of peasant farmers at Port Royal (the first north of Florida) was nearly wiped off the map. A few of the settlers escaped into the woods.

Birth of Nova Scotia. In 1621 Charles I of England granted the whole territory to a Scottish nobleman, who called it *New Scotland*; so that now there was a New France (Quebec), a New England, and a New Scotland. But in those days charters were written in Latin, and as the Latin for New Scotland is Nova Scotia, that was its name in the document and that has been its name ever since. Though this Scottish owner professed to intend great things, he did scarcely anything, and only a handful of settlers had arrived, when in 1632, Charles I treacherously returned Nova Scotia to France, so that it once more became Acadia.

The Acadians settle on the salt marshes. In 1632 Port Royal was founded for the third and last time. Two or three hundred hardy farmers from the shores of the Bay of Biscay in France were planted in this beautiful valley. From this little acorn has grown the mighty oak of the French settlements, whose many branches spread over the three Maritime Provinces, and whose people now number more than two hundred thousand. They preferred the tidal meadows (p. 224) to the forests, because there were no trees to be cut down, and they were more like home to men who had lived on diked land on the edge of the Bay of Biscay all their lives. Indeed, as they multiplied, which they did very rapidly, they spread out and soon formed a fringe all around the blind ends of the Bay of Fundy (#89), for there they found broad flat, tidal flood-plains, covered with fertile, red slime and ooze. Then they began creeping up the flood-plains that border the rivers. #89 shows that the first settlements, in 1632, were around *Grand Pré* at the west end of Minas Basin; by 1680 they had reached east to *Truro*; and by 1700 they had crossed the *Cobequid Mountains*, were fast taking up land along *Chignecto Bay*, and were moving up the *Avon River* to *Windsor*, and the *Petitcodiac* toward Moncton.

Acadian farming. Their peaceful villages fringed the shores of all these inlets by 1700. They harvested marsh grass, which they fed to sleek French cattle, cut fields of wheat, barley, and oats with sickle, and pulled flax and cut hemp, the fibres of which they wove into cloth and twisted into cord and rope. They also planted the first apple orchards.

Few Acadian fishing villages. It is true that along the south coast of Nova Scotia a few scattered fishing villages grew up (#89). At *Canso*, as early as 1510, there were huts on the shore and probably some fishermen remained through the winter. In 1653 *Nicholas Denys* was given the whole coast along the gulf from Canso to Gaspé, and he formed settlements of French fishermen. But fishing did not thrive, and nothing but a remnant was left when the British finally conquered the country.

The loves and hates of the Acadian farmer. The Acadian farmers were the backbone of the colony. They were different from the French who settled along the St. Lawrence, who loved the forest, followed the fur trade, neglected the farms, and looked up to the seignior (p. 271). The Acadians loved the meadows, were zealous farmers, not greatly interested in the fur trade, and each owned his own farm and had no seignior, and hence paid no seignioral dues. Both groups had three characteristics in common: first, they loved their home and village more than they loved life itself; secondly, they loved France; and thirdly, they disliked and did not trust the British, with whom they were often at war.

The Acadians' terrible fate. This fact proved the undoing of these simple, peasant farmers. By 1710 Acadia was conquered for the last time by the British, and the French farmers became citizens of a British colony, which was engaged in a death struggle with their mother country, France. Though the British left them their farms, gave them freedom to worship in their Catholic churches, and dealt justly with them, their love for France left them suspicious, and they refused to take the required oath. As the sprinkling of British in Halifax (about 3000) had the French of Quebec on one side, with guns bristling in the great fortress on the St. Lawrence, and the French of Cape Breton on the other, with Louisburg's guns threatening every British ship, and twelve thousand Acadians almost in their midst, they felt a real danger. From time to time they had reason to believe that the Acadians were helping their enemies. The suspicion grew as the grim struggle became acute. War inflames the passion and deadens the reason. At last the British decided on a terrible cure. This was no other than to remove the whole Acadian population from the beautiful homes, which they had been building about them for one hundred and twenty years, and which they loved as a mother loves a child. In the autumn of 1755, the Acadians were seized in their settlements, rushed aboard ships, and distributed among

#100

REMOVAL OF ACADIANS

all the states along the Atlantic coast (#100). About three thousand were landed in South Carolina. Such large numbers were put off at the ports of Virginia that she refused finally to receive any more. A boat load for Philadelphia took smallpox and were held for months on the ship until only a remnant was left, and they, almost naked and starving, were received kindly by some Godly Quakers. It was a cruel and terrible punishment; but it was a war measure, and all war is cruel and terrible.

Settlers from New England. The Acadians' homes and barns were burned so that they could not return; their cattle, sheep, pigs, and horses were driven to the British forts; and their fields were left waste, though not for long. These salt-marsh lands were the best farm lands along the whole Atlantic coast. An invitation was sent to the English settlers along the bleak New England coast to come in and take possession of these fertile farms. A regular "gold rush" began. Between 1755 and 1762 thousands of these New Englanders poured into all the Acadian settlements: into the *Annapolis* valley; around *Minas Basin* and *Chignecto Bay*, into *Yarmouth*, *Shelburne*, *Liverpool*, and *Chester*. A large group settled as far away as *Guysborough* county (#91). These thousands from New England were the first English settlers in the province, with the exception of those at Halifax, and their descendants still till the marsh lands around the Bay of Fundy.

British and German settlers. The first British settlement in the Maritime Provinces was composed of a group of about two thousand soldiers and sailors, who selected *Halifax*, a district, rocky and barren, but with an excellent harbour. It was chosen for military purposes. It could be strongly fortified and was in a good position to cope with the French fortress of Louisburg in the defense of shipping from England to the New England States. A year or two later over one thousand German Protestants were brought over from Hanover, of which the king of England at that time was ruler. They settled on the slope of a rocky hill looking down into a harbour as beautiful as a mountain lake. This became the flourishing town of *Lunenburg*. It was quite natural that thirty years later (1785), when more Germans arrived, they should settle near their kin. They moved back into the country and became farmers, and called their new home *New Germany* (#91). These thriving German settlements still retain many of their old customs, and in this part of Nova Scotia, cabbage is an important crop in every garden, and a barrel or two of sauerkraut finds its place in most cellars.

First settlement of New Brunswick and Prince Edward Island. Up to this time New Brunswick and Prince Edward Island were almost empty, except for *Micmac Indians*. A few, feeble fishing villages sprawled along the Gulf of St. Lawrence (p. 212), and the rich intervales along the lower St. John River had from time to time lured a few Acadians (#89). Indeed, this region had been laid out in seigniories, and seigniors were given possession; but the Acadians were too independent, the seigniors soon disappeared, and the settlements did not prosper.

Some Acadians escaped. Of eleven thousand Acadians, the British soldiers seized and deported only about six thousand. The others escaped into the woods and fled to New Brunswick, Prince Edward Island, and Cape Breton (#91). Between 1755 and 1800 these, with Acadians that returned from the Atlantic states, made settlements all along the east and north coast of New Brunswick, on the upper St. John, and on the coasts of the two islands, as shown in #91.

Return of the Acadians. We must not forget the poor Acadians strewed along the hostile Atlantic coast among unfriendly peoples, who spoke a strange language and had a different religion. These humble farmers, although they had seen their homes burned before their eyes, their sleek cattle driven away, and their sacred churches occupied by soldiers, never forgot their beloved marsh farms beside the roaring tides of the Bay of Fundy. They gathered in pathetic groups, they walked hundreds of miles through rocky forests along the rugged coasts; they hired creaking old boats, ready to fall to pieces, and many of them returned only to see their farms in the hands of New Englanders. It was these returned exiles who formed many of the French settlements of New Brunswick, Prince Edward Island, and Cape Breton.

A thousand mile march. The wandering of one group is filled with pathos and suffering. Nearly one thousand of the scattered Acadians gathered at Boston in 1766, their hearts still longing for the salt-marsh farms. They decided to walk back over the rocky coast to the spot, which they still called home. For four months this miserable band of men, women and children, footsore and starving, dragged along. All were hungry, numbers took sick, and many died on the way. At last they crossed New Brunswick and, nearly dead, they arrived at the head of the Bay of Fundy only to see their beloved farms in possession of new owners. Though their spirits sank in their tired bodies, these iron-hearted peasants pushed on westward through the Annapolis valley, till they found unoccupied land. There, between *Digby* and *Yarmouth*, they settled down on the shore, and the sons and daughters of these undaunted people live to-day in neat snow-white cottages, speak antique French, till their well-kept farms, and drive their ox-carts to their beautiful churches, as French as when they lived in the "Basin of Minas."

United Empire Loyalists in Nova Scotia. We shall soon see that, as the result of the United States winning its independence in 1783, great numbers of United Empire Loyalists from New York and Pennsylvania flocked into Ontario (p. 322), and from Vermont and New Hampshire into Quebec (p. 273). There were also larger numbers of Loyalists in the New England States. They soon went through the same agony that they had helped to deal out to the Acadians. Their homes were seized and their lives made so uncomfortable that they were forced to become exiles. Many of these were highly educated, graduates of universities, the best of citizens. They chose the Maritime Provinces as their new home under the British flag. The best spots in Nova Scotia had already been occupied by men from Massachusetts, Connecticut, and Rhode Island, who eagerly sought the Acadian farms (p. 232). Nevertheless, large numbers entered this province. They occupied the wooded part of Annapolis valley (#91); great numbers swelled Shelburne almost to a city, but soon had to disperse along the coast; a whole necklace of small settlements was strewed along the Atlantic coast from Shelburne to Guysborough and Sydney (#91).

The making of New Brunswick. But if the United Empire Loyalists rounded out the population of Nova Scotia, they made New Brunswick. We have seen (p. 233) that before their arrival, there was merely a thin thread, with a few village knots of Acadians along the coast of the Gulf of St. Lawrence and of Chaleur Bay. Now many thousands of splendid men and women tramped into the forested wilderness along the St. John and other rivers emptying into the Bay of Fundy (#91). With axe in hand, they hewed out homes, cleared fields, and turned the wastes of the intervales into smiling farms, thriving villages, and growing towns. *St. John*, *Fredericton*, *Woodstock*, *St. Andrews*, *St. George*, and *St. Stephen* were all cut out of the woods by these sturdy men, who have given to this province so many of its political and scholarly leaders.

Arrival of the highlanders. Even before the United Empire Loyalists were well settled, Scotland began to contribute some of its best sons to these provinces. While the New Englanders came in from the south through the Bay of Fundy and along the Atlantic coast, the Scots came in through the Gulf of St. Lawrence. From

1770 to 1825 they founded or refounded such important towns as *Campbellton, Dalhousie, Newcastle, Chatham* and *Shediac* in New Brunswick. About the same time, 1773, the ship *Hector*, brought its first consignment of Scots to *Pictou*. A little later the Scots founded *New Glasgow*; and from 1791 to 1825 Scottish Highlanders settled in great numbers in *Antigonish*, and in both the north and south of *Cape Breton*.

Proprietors of Prince Edward Island. About the same time the whole of Prince Edward Island was handed over to some sixty proprietors. These men continued to live in Scotland, and while they brought out many Scottish settlers, there was no prosperity while great tracts of land were in the hands of absentee owners. The most notable settlement was one by Lord Selkirk at *Belfast* in 1803.

Irish immigrants. After the stream of Scottish immigration had exhausted itself, famine and discontent started that emigration from Ireland that emptied it of half its population. Wave after wave of Irish Settlers entered New Brunswick, fewer Nova Scotia and Prince Edward Island (#91). These settled along the new roads and railways, and to-day in New Brunswick they surpass in numbers the Scots.

The Prairies drain the Maritimes. After 1850 the attractions of the great fertile areas of the United States were too strong to be resisted; and emigration from Europe was deflected to the south, so that the stream to the Maritime Provinces began to dry up. With the opening of the Canadian prairies by the building of the Canadian Pacific Railway, fifty years ago, the lure of the promising west started a discomforting movement, which for thirty years has been draining the Maritime Provinces of their brightest and most forceful young people. Increase in population in New Brunswick has been at a snail's pace. Nova Scotia has barely been able to hold her own; and Prince Edward Island has had an almost alarming decrease.

Manufacturing

History. Nova Scotia built the first water wheel, the first gristmill, and the first ship, on this continent. But these were small, as they were made by the workmen brought out by Champlain to Fort Royal in 1606 (p. 228). Until long after the conquest by the British in 1763, the Acadians devoted themselves to farming the salt marshes, and fishing along the coast for cod. After the conquest, the progressive

New Englanders and United Empire Loyalists started manufacturing in a small way. Fishermen built their own boats, and lumbermen set up sawmills along the St. John and other streams to cut lumber for building houses and barns. As the population increased, these industries expanded. Soon skilful carpenters along the coasts began to use the excellent timber, which the forests supplied, to build larger and larger ships. In 1850 a splendid wooden schooner was built at New Glasgow, which sailed across to Scotland and was the largest ship that had ever entered the great harbour of old Glasgow up to that time. Then all around the coast the port towns buzzed with busy shipbuilders, and no less than 300 sailing ships were launched in Nova Scotia in the single year 1874, and sold to all the maritime nations of the world. Some of these were the largest, full-rigged ships afloat, running up to 2500 tons. But alas for Nova Scotia, the steel age appeared! It was found that ships could be made of steel; and so successful was this new type of ship that by 1880 the industry in Nova Scotia was dwindling. It has never recovered.

No large home market. The Maritime Provinces, standing at the gateway of Canada, are convenient to receive raw products of every continent, and have snug, deep harbours of ample size indenting the Atlantic coast. Nova Scotia is well supplied with coal, the seams of which lie almost beneath the harbours; and while the Maritimes have no Niagaras, they have numerous sources of water-power, easily harnessed, and close to transport routes. Nevertheless, the position of these provinces is against them for manufacturing. There are no dense groups of free buyers near at hand as in west Ontario or the Montreal district. They are shut out from New England markets by crushing duties; and the state of Maine, thrust up like a wedge between them and the populous St. Lawrence valley, makes railways so roundabout and freight rates so high that the value of their markets is lessened.

Local manufactures. Still there are factories of great importance in these provinces (#95). Foremost are numerous sawmills, and large pulp and paper mills, which are turning the wood of the forest into finished products. These find a convenient market in Great Britain and the United States. Almost every town has its planing mill, sash and door factory, and in some towns a stave mill. These further work up the lumber and hardwood, largely for local use

(#95). Almost every town and village around the coast has its fish-curing factories (#95): many can lobster; some, especially in New Brunswick, can clams; large numbers extract cod-liver oil. Smoke-houses for herring and haddock are in operation in goodly numbers; and along the shores of *Passamaquoddy Bay* at the south-west of New Brunswick are several factories, where small herring are converted into canned sardines (#96).

Steel and sugar. Coal mines in east Nova Scotia have attracted several very important industries. To make iron and steel, not only vast quantities of cheap coal, but also iron-ore and limestone, close at hand, are necessary. *Sydney* is nearly surrounded with coal mines, immense quantities of high grade iron ore can be lifted from mines at Belle Isle and Newfoundland, placed aboard boats, and in less than two days be unloading in Sydney harbour. Limestone is abundant within a few miles of the city. In consequence, the great steel works at Sydney is one of the biggest industrial plants in Canada. Since raw sugar, brought in for refining, comes from the West Indies, South Africa, and Brazil, there are large refineries (#95) at both Halifax and St. John, the chief ocean ports.

Halifax. *Halifax*, the capital of Nova Scotia, was the first settlement made by the British, and it was selected because of its almost perfect harbour and its key position on the Atlantic. Its natural harbour has been perfected for commerce, with wharves, warehouses, loading machinery, and cold storage plants. It has lines of ships sailing to every continent; and as its harbour is open during winter, when the St. Lawrence is gripped by ice, its traffic is particularly heavy at this season.

Halifax was selected by the British for a fortress; and though it has never seen fighting, many expeditions have sailed from its harbour to victory. Stone-walled towers, masked batteries, and frowning fortresses, some old and decaying, none very vigorous, surround the harbour; and the citadel, the largest in Canada, stands out very plainly in the middle of the city. At one time it was an important British naval station, and its decaying dockyard is a grim reminder of those busy times.

Halifax is justly proud of its old streets, old parliament building, old monuments, and old churches. No city on the continent, except Quebec, can compare with it for quaintness and interest.

St. John. *St. John*, situated at the mouth of the river St. John, and on a deep-water harbour connected with the Bay of Fundy, is a busy seaport with an ice-free harbour in winter. As the terminus of the Canadian Pacific Railway, it has extensive shipping. Steamers of twenty oceanic and fifteen coastal lines enter and leave its ports regularly (#95). While it lacks the historic and military associations of Halifax it has more numerous factories. It was for long the hub of the lumbering industry on the St. John, but the sawmills have now moved farther back. Nevertheless, it still handles large quantities of lumber. The chief industries of other cities and towns are shown in #95.

CHAPTER XIII

QUEBEC

A distinguished province. Quebec is in many ways a distinguished province. It is by far the largest (#101) — almost half as large again as Ontario, more than twice as large as any of the prairie provinces; three areas the size of France, and nearly four Germanies could be carved from within its borders, and out of the fragments left, Denmark, Belgium, Holland, or the Maritime Provinces could

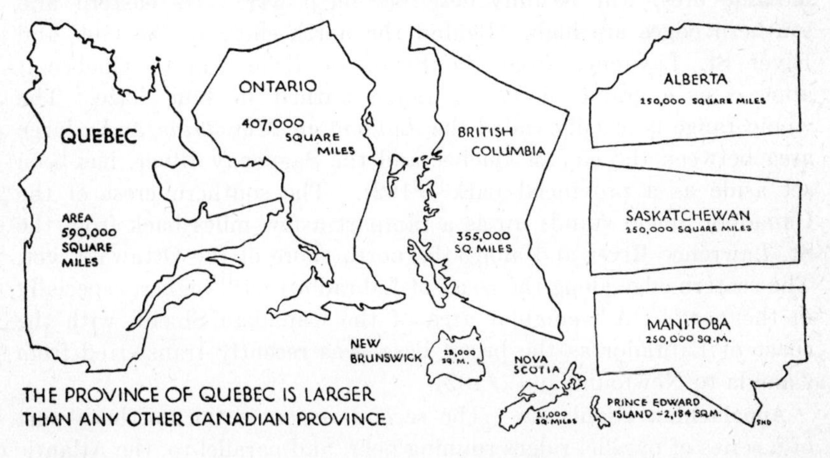

#101 RELATIVE SIZES OF PROVINCES

be made. Though compact, it is the longest and widest of the provinces; it extends farthest north and farthest east, the sea bathes it on three sides, and sends the gaping estuary of the St. Lawrence into its very heart. It was the acorn from which the mighty oak of Canada has gradually grown, until to-day it spreads its broad branches over half a continent.

A new picture in an old frame. This vast province, stretching twelve hundred miles, from Hudson Strait to the New England States, is divided by nature into three distinct regions: (1) the *Canadian Shield* on the north, occupying nine-tenths of the province, (2) the *Appalachian Highlands* on the south-east; and, framed in between these ancient relics, (3) the more modern *St. Lawrence lowlands.*

A gigantic ancient crack, called the *Champlain fault*, running along the south shore of the St. Lawrence from Gaspé to Quebec and then south-east to the head of Lake Champlain, divides the St. Lawrence lowlands from the Appalachian Highlands; and a cliff, which looks like a mountain from the south, running along the St. Lawrence from the Strait of Belle Isle nearly to Quebec, then sweeping overland from Quebec to Ottawa, separates the St. Lawrence lowlands from the Canadian Shield (#102).

Canadian Shield. The Canadian Shield, bespangled with lovely lakes, networked with writhing, pulsating streams, and prolific with metallic ores, will be fully described on p. 341. Its eastern and southern edges are high. Behind the north shore of the Gulf and River St. Lawrence from the Strait of Belle Isle to Quebec it appears as a grand mountain range bathed in blue haze. The whole range is usually called the *Laurentian Mountains*, and a large area between the city of Quebec and the Saguenay River, has been set aside as a provincial park (#102). The southern crest of the Canadian Shield stands up as a rampart a few miles back from the St. Lawrence River and along the north shore of the Ottawa River. The eastern edge along the coast of Labrador is still higher, especially in the north. A triangular area of the Canadian Shield, with the coast of Labrador as the broad base, was recently transferred from Canada to Newfoundland (#102).

Appalachian Highlands. The second region is the northern part of a series of parallel ridges running near, and parallel to, the Atlantic coast from the southern United States to Newfoundland. In eastern Quebec these ridges run north-east from the south border to near Levis and continue parallel to the south shore of the River St. Lawrence. They pass under the Gulf of St. Lawrence beyond the rugged *Gaspé Peninsula.* As they enter from the south they are in four distinct ridges (#102) and have peaks almost 3,000 feet high. Before they reach Quebec, these ridges have melted into a single, broad, confused, broken highland, called the *Notre Dame Mountains.* As this range reaches Gaspé Peninsula it becomes higher and is sorted out again into several ridges, which pass out into the Gulf of St. Lawrence as a number of peninsulas ending in wild, rugged capes. The chief range in Gaspé is the *Shickshock Mountains,* some of whose peaks are the highest in the province.

Making of mountains. In the far distant past the Atlantic waves beat against the high and frowning front of the Canadian Shield.

HE PROVINCE OF QUEBEC

OWING ITS PHYSICAL FEATURES, EARLY EXPLORATIONS, FORESTS, PARKS, GAME, ETC..

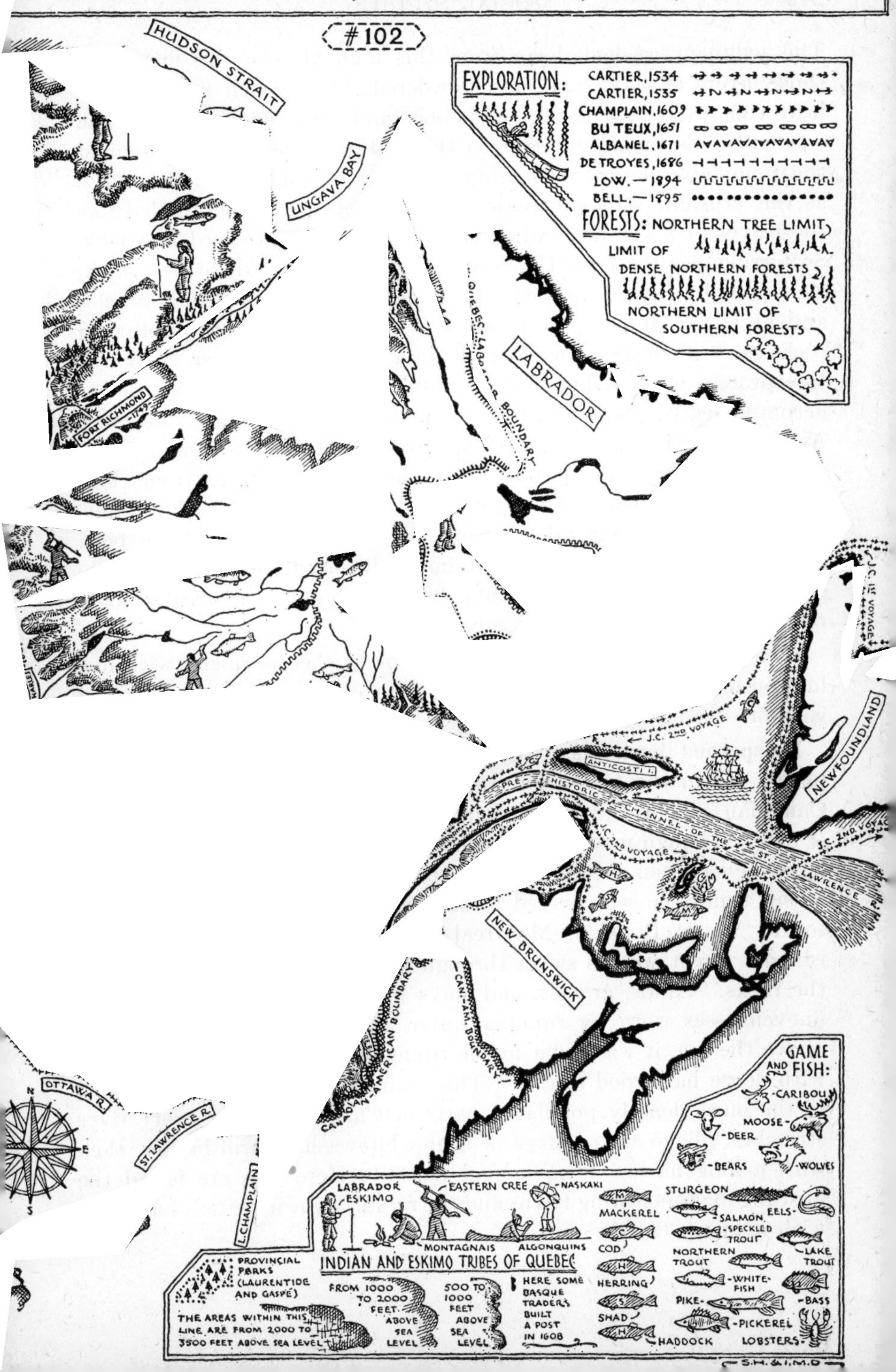

The sediment washed down from this high plateau was spread in thick layers of clay, sand, and powdered shells, which pressure and heat converted into shale, sandstone, and limestone layers. Then a tremendous pressure began to thrust these layered rocks against the unmovable mass of the mighty Canadian Shield. Slowly, steadily, as this crushing strain increased, these layers became crumpled like the leaves of a big book when the edge is pushed toward the back. So terrific was the force that layers of rock, thousands of feet thick, were actually turned upsidedown at the edge of the Canadian Shield, and at Gaspé and on the Island of Orleans to-day these upturned rocks can still be seen, the older piled on top of the younger. Though this process was so slow that it could scarcely be observed, it was accompanied by many violent earthquakes, and desolating volcanoes. At last, when it was completed, a series of high ranges of mountains, massive as the Rockies, beat back the Atlantic waves from their outer edge. This was the beginning of the Appalachian Mountains. Ever since that tremendous event, rivers, ice, and air have been gnawing away those giant mountains until they have shrunk to mere shadows of their former mighty bulk, and only the harder roots remain as comparatively low ridges.

Consequently the Appalachian area of Quebec is ridged with low hills on which are good pastures, and the valleys between are carpeted with prosperous farms.

Prosperous lowlands. The St. Lawrence lowlands lie flat and placid, set in the triangular frame between the rugged edge of the Canadian Shield and the crumpled remnants of the Appalachians (#103). Though made of the same layers as the Appalachians, these have never been folded or disturbed. But within comparatively recent times the sea covered these lowlands so deeply that only the top of Mount Royal at Montreal projected as a dome-shaped rock island. Great whales swam through the seas and shellfish clung to the rocks. Sands, gravels, and clays were deposited to level up the unevennesses worn by running water, so that when finally it rose above the sea, it was a flat fertile triangle that soon became covered with dense hardwood forests. This triangular lowland to-day is one of the most densely populated parts of Canada. Along every river stretches village after village of neat whitewashed farm houses, each with a long narrow farm behind (#103). Here also are found the hustling manufacturing towns and the rapidly growing cities, humming with industry (#103).

Volcanic roots. A remarkable row of dome-shaped mountains, which stand out like sentinels above the peaceful monotony of this level plain, tell a story of violent volcanoes in the far distant past. Though this region escaped the foldings of the Appalachian, it

#103

THE ST. LAWRENCE VALLEY

Notice the Laurentian Highland away back and to the left. The flat St. Lawrence plain is in the foreground.

evidently felt the strain, which caused a crack to form along the line occupied by these mountains. The pressures from beneath forced up molten rock along this line of weakness, and a row of volcanoes belched forth streams of molten rock until volcanic cones were probably built. Below these cones were pipes, through the layered rocks, to the supplies of lava beneath. As the strain was gradually lessened, the lava cones cooled; the molten rock, filling the pipes, hardened; and the rain, water, and air began to wear away both the lava cone and the layered rock of the plain. So long is the time since these far-off volcanic outbursts, that the cones are all worn away, and thousands of feet of the softer layered rock have been crumbled to soil by wind and weather and carried to the sea by water; but the harder lava rocks in the pipes have been left projecting as mountains. Fragments of the layered rocks are found still adhering to sheltered portions of these pipes, evidence of the time when the layered rock stood at that level throughout the plain.

Minerals

Early mining. While the early Spanish explorers were loading their ships with gold and silver, which they seized from the Indians of Mexico and Peru, the French in Quebec were satisfied to trade trinkets for valuable furs, and devoted themselves to bringing the joys of Christianity to the savage Indian. Yet the traders listened eagerly to every story the Indians had to tell about precious metals, and they sent explorers on long and dangerous journeys to Lake Superior to seek the mines from which came the lumps of copper, that were brought in by Indian traders. But no gold, silver, nor even copper mines were found. Indeed, no great progress throughout the French period was made in the mining industry except in the smelting of iron. As early as 1733 the first smelter on the continent was built near *Three Rivers* to extract iron from ore found in the bogs of the district. Wood charcoal was used for extraction. It continued to turn out a high quality of iron for 150 years, when the supply of iron ore became exhausted.

Noranda gold mines. The Canadian Shield, which is, beyond doubt, rich in minerals, covers over ninety per cent. of Quebec, and as the greater part of the south is occupied by the roots of the worn-down Appalachian plateau, this province should be noted for its mineral production. Though most of the northern peninsula, Ungava, has never been seen by the white man, its southern fringe is now

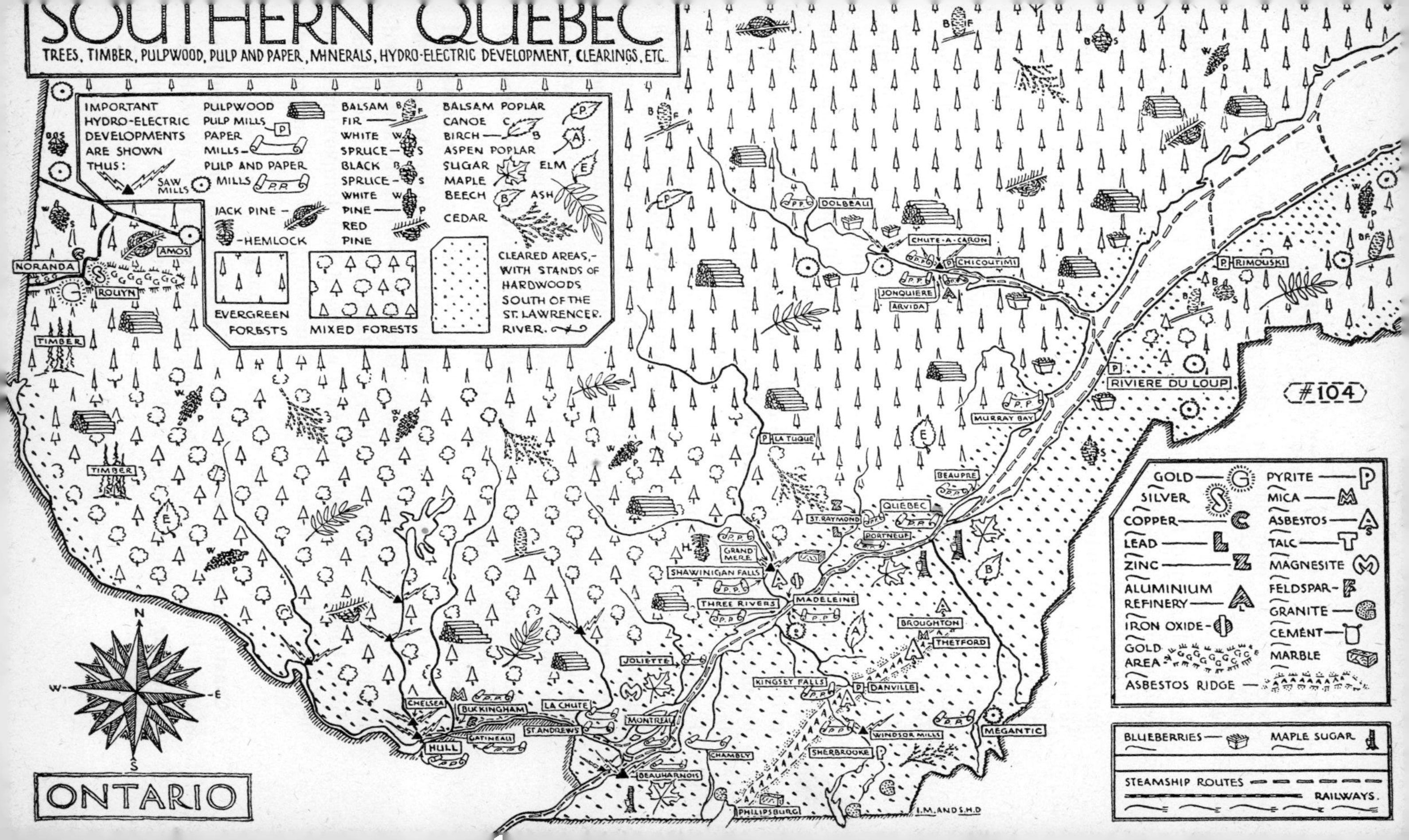
SOUTHERN QUEBEC
TREES, TIMBER, PULPWOOD, PULP AND PAPER, MINERALS, HYDRO-ELECTRIC DEVELOPMENT, CLEARINGS, ETC.
IMPORTANT HYDRO-ELECTRIC DEVELOPMENTS ARE SHOWN THUS:
SAW MILLS
PULPWOOD
PULP MILLS
PAPER MILLS
PULP AND PAPER MILLS
JACK PINE
HEMLOCK
BALSAM FIR
WHITE SPRUCE
BLACK SPRUCE
WHITE PINE
RED PINE
BALSAM POPLAR
CANOE BIRCH
ASPEN POPLAR
SUGAR MAPLE
BEECH
ELM
ASH
CEDAR
EVERGREEN FORESTS
MIXED FORESTS
CLEARED AREAS,- WITH STANDS OF HARDWOODS SOUTH OF THE ST. LAWRENCE R. RIVER.
NORANDA
ROUYN
AMOS
TIMBER
TIMBER
DOLBEAU
CHUTE-A-CARON
CHICOUTIMI
JONQUIERE
ARVIDA
RIMOUSKI
RIVIERE DU LOUP
#104
MURRAY BAY
LA TUQUE
BEAUPRE
QUEBEC
ST. RAYMOND
PORTNEUF
GRAND MERE
SHAWINIGAN FALLS
THREE RIVERS
MADELEINE
BROUGHTON
THETFORD
JOLIETTE
KINGSEY FALLS
DANVILLE
CHELSEA
BUCKINGHAM
LA CHUTE
MONTREAL
ST. ANDREWS
GATINEAU
HULL
WINDSOR MILLS
MEGANTIC
CHAMBLY
SHERBROOKE
BEAUHARNOIS
PHILIPSBURG
ONTARIO
N
S
W
E
GOLD
SILVER
COPPER
LEAD
ZINC
ALUMINIUM REFINERY
IRON OXIDE
GOLD AREA
ASBESTOS RIDGE
PYRITE
MICA
ASBESTOS
TALC
MAGNESITE
FELDSPAR
GRANITE
CEMENT
MARBLE
BLUEBERRIES
MAPLE SUGAR
STEAMSHIP ROUTES
RAILWAYS.
I.M. AND S.H.D

beginning to pour out its riches. The discovery of the famous *gold-copper-silver* Noranda mines at Rouyn, south-east of Lake Abitibi (#104), has given gold mining a fresh start, and new mines are springing up all around this district. It may soon be a rival to the rich gold mines in the adjoining fields of Ontario, since already Quebec stands next to Ontario in production of both gold and copper. The metals are extracted at blast furnaces in *Rouyn* and shipped to a refinery at *East Montreal*, where gold, silver, copper, selenium, and tellurium are separated.

Cloth made of rock. Quebec has long supplied the world with most of its *asbestos*, which is obtained from a ridge in the Appalachian region, running from the southern boundary to near the Chaudiere River (#104). The three chief mines and mills are at *Thetford, Danville*, and *East Broughton*. This remarkable rock flakes into satin shreds, and its value varies from 350 dollars a ton for fibres an inch long, to 15 dollars a ton for fibres one-quarter inch long. The long fibres are woven into coarse cloth, from which are made fire-proof garments for firemen, theatre curtains, gloves, etc. The short fibres are matted into felt, pressed into wall board and shingles, woven into brake lining, and cemented into covers for furnaces and steam pipes. Formerly Quebec produced three-fourths of the world's supply, but now that Rhodesia is quarrying asbestos of the highest quality, and Russia is increasing her output at a furious pace, Quebec's production is reduced to a bare one-half. Unfortunately most of Quebec's asbestos is shipped crude to the United States, there to be worked into cloth, felt, gloves, and other manufactured products.

Other minerals. Other important minerals of Quebec are *felspar*, *magnesite* and *mica* (#104) in the Canadian Shield north of the Ottawa River, and *granite* and *marble* for monuments and building stones, quarried near the southern border of the Appalachian area. The amber mica of Quebec is of the best quality and is used largely in the electrical industry. Magnesite is made into bricks, which, because they will not melt at the highest temperatures, are used for lining furnaces. Felspar is used in paint and in pottery.

Coast waters and drainage

Coastal waters. The eastern coast of Hudson Bay and the southern coast of Hudson Strait, besides being blocked with ice for a good part of the year, are high, rugged, and with few inlets suitable for harbours, though the mouths of the numerous rivers will be found to afford shelter if it is ever needed in this bleak, harsh land. The splendid array of long, branched, deep, narrow inlets on the east coast of Ungava, worming their way through the high ramparts of the Canadian Shield, supply unlimited shelter for ships in the deepest water, but this coast now belongs to Newfoundland, not to Quebec.

The main artery. But what ports are missing to Quebec in the rigorous north are more than made up for in the fruitful south by the mighty St. Lawrence, passing like a main artery to the heart of the province and opening like a wide gateway towards Europe. This river with the Great Lake expansions is 1,900 miles long. Near its mouth it is so wide that one shore can hardly be seen from the other. Only at Quebec does it become narrow enough for guns to defend the passage; and for this reason, from the earliest settlement, this city became the key to Canada, and has seen more attacks and defences than any other spot on the continent.

Navigation of St. Lawrence. The wide, deep channel of the St. Lawrence gave entrance to the largest ships as far as Quebec, which for two centuries was the centre of shipping. Between Quebec and Montreal numerous shallows, especially in Lake St. Peter, had to be blasted and dredged, before a wide, safe passage thirty feet deep, was obtained. This allows all but the very largest ships to pass. These improvements have transferred commercial leadership from Quebec to Montreal.

The Gulf of St. Lawrence. The river opens by two mouths around *Anticosti Island* into the Gulf of St. Lawrence, which has three passages to the ocean: a northern one through the *Strait of Belle Isle* is the shortest to Europe; however, swift tidal currents sweep in icebergs during the winter and spring, and winds often blind it with thick fogs, so that ships cannot use it until June. A narrow passage, the *Strait of Canso*, treacherous with drifting ice and baffling currents in winter, is little used except for local traffic. The third, between Newfoundland and Cape Breton, is wide and deep, though often masked in fog. Through it steam the great ships of the ocean. The lower St. Lawrence River is also much vexed with fog. However, Canada has devoted such attention to this passage that now it is lighted almost like a street by night, and well marked with buoys and horns by day.

The estuary. Although this great blue river pours two million gallons of fresh water into the wide gulf every minute, twenty miles below Quebec the water is brackish, and a few miles further down it is salty. This carrying of salt-water far up this stream is due to swift currents that accompany the tides. These, in two great waves, rush each day at a rapid rate up this estuary, giving a rise of tide water as high as 18 feet at Quebec, but are finally exhausted at Three Rivers. The kindly currents, swishing back and forth with the tides, scour the St. Lawrence channel and keep it free from bars and delta formations. The water of the St. Lawrence is beautifully blue, since it has several expansions — lakes *St. Francis*, *St. Louis*, and *St. Peter* — in whose still waters all discolouring particles settle.

A splendid scene. There is no more impressive beauty, on the grand scale, than this majestic river, especially below Quebec. The south shore is garnished with its unending line of snug white houses nestling on the lower land, a splendid church with its graceful spire breaking the evenness of the line every few miles; behind the houses are long narrow fields, parted by rail fences, stretching back to melt into the waves of higher and higher woodland, which make the background. On the north shore the stern granite cliffs rise abruptly into blue mountains almost from the shore, except where a stream has gnawed its way through the barrier; clumps of fishermen's huts cling like nests of sea-fowls to the foot of the cliff, wherever space can be found. Between two such clear-cut splendid shores flow the limpid, blue waters of the river itself, with rocky islands scattered now on one side and now on the other. The stately ocean liners, steaming silently up the channel, with the stillness of a sunny evening broken by the melodious vesper call of the church bells across the glowing water, make the picture complete.

Tributaries from the Canadian Shield. All the other rivers of southern Quebec are tributaries of the St. Lawrence. Those emptying in from the north are the largest and longest, as they come from far back on the height of land. The *Ottawa*, the longest and most important, and the *St. Maurice*, which empties by three mouths at *Three Rivers* (hence the name) are typical Canadian Shield streams, with lake expansions, rapids, and waterfalls (#102). The *Saguenay* is one of the most remarkable of rivers. Its channel is like a cleft through a great mountain, and for 112 miles it takes its gloomy, almost ghastly course deep down along this black chasm, six hundred feet deeper than the bed of the St. Lawrence. Its source is Lake St. John, placed in a hollow in the Canadian Shield. Into this lake tumble streams from all directions, some of which rise hundreds of miles back in the rocky wilderness and flow through gloomy spruce and pine forests. As far as *Port Alfred* the Saguenay is navigable for the largest ships, and for moderate sized ships as far as the fine commercial city of *Chicoutimi.* With ocean ports and immense quantities of hydro-electric power right at hand it is no wonder that the whole district, which only yesterday was a wilderness, is to-day lined with thriving, rapidly-growing factory towns.

Richelieu River. The tributaries south of the St. Lawrence are shorter, less turbulent. They rise in glacial lakes in the Appalachian plateau. Only the *Richelieu,* which flows over the St. Lawrence Lowlands, is navigable. It rises in delightful *Lake Champlain* and empties at *Sorel.* A canal overcomes the rapids near *St. John;* this route is of considerable importance, since a canal connects Lake Champlain with the Hudson River and New York City; there is at present, considerable traffic from Montreal to New York, and a century ago much freight from New York state passed out through the Richelieu and the St. Lawrence. If its least depth, 6½ feet, were increased (and a commission is at present studying the practicability of doing this), it might in future become of the greatest importance. It has played a stern part in Canadian history, since along this route there is the lowest gap across the Appalachian mountains between the two countries. It was the chief route of invasion in the wars between the French and English, and during the war of 1812. This gap is studded with old and crumbling fortresses, and decorated with bronze tablets telling of the heroic struggles of the past.

Rivers and logs. The other two important tributaries from the south run their course far too much across the hard ridges of the Appalachians to be of use for navigation. All of these rivers, with dozens of others, both north and south, even when useless for ships, have been invaluable to the lumbering industry. Every spring, for more than one hundred years, their waters, flooded with melting snows, have floated countless numbers of logs down to sawmills and pulp mills.

A river bed in the ocean. On the bed of the St. Lawrence estuary, and along the Gulf, through *Cabot Strait,* and for 250 miles across a shallow shelf of the Atlantic is a deep trench (#102), found only by sounding the sea. This must be the bed of an earlier River St. Lawrence, and in order that such a trench could be cut, the whole region must have been at one time above sea-level. A marvellous story is indicated by the presence of this old river channel, a story telling of a time when the Gulf of St. Lawrence was dry land; when Canada and the United States reached 250 miles farther out into the Atlantic Ocean than at present.

Exploration

The only Atlantic gap. The Canadian Shield and the Appalachian Highlands form fierce barriers along the Atlantic front, and not a single easy passage pierces either of them. But the great river St. Lawrence, lying between the two, makes the only sure and safe route to the very heart of the continent. The first explorer to enter this gateway was *Jacques Cartier*, a skilled seaman, who hailed from the rugged port of *St. Malo*, the home of many hardy sailors.

Cabot discovers Canada. The startling discoveries made by Columbus filled the courts of Europe with excitement and their rulers with greed. As flotillas of ships, laden with silver and gold, in a steady procession began to sail across the Atlantic to Spain, and as the belief grew that China and Japan, with their fabulous wealth were near at hand, Europe ran wild to reach these new, rich lands across the Atlantic Ocean. England and France, especially, were jealous of Spain, the patron of Columbus, and vied with each other to snatch a share of the plunder from the ignorant Indians. England's energetic king, Henry VII, sent out *John Cabot* to search for these rich lands and bring back their wealth. But the king was cautious as well as covetous, and warned them to sail north and west in order not to enter the region already claimed by Columbus for Spain. These skilled seamen reached the bleak coast of Newfoundland, probably viewed the barren highlands of Cape Breton through the dense, dank fogs; and may have discerned the still more forbidding rocky ramparts and yawning inlets of Labrador. They saw no people, much less Chinese rulers, dressed in silks. The best Cabot had to report was harsh headlands, dense fogs, rocky lands, thickly covered with gloomy pines and spruce; but as for gold and silver, pleasant gardens, and spicy lands of China, there were none.

Sailors of sterner stuff for the north. Exploration north and west was a far more serious matter than the exploit by Columbus in more temperate regions. He had steady, favourable trade-winds, ocean currents, clear weather, blue skies, and strewed along the latter part of his path, archipelagos of small islands clothed with luscious vegetation, and loaded with fruits and nuts. In the north it required sailors of sterner stuff. The sailing ships had to fight against the opposing and uncertain westerlies every inch of the way; the cross movements of the Gulf Stream and Labrador current confused and stealthily carried them out of their courses; icebergs, great masses of field ice, and treacherous fogs dogged their steps. The icy shores were often hard to separate from the icy seas; and islands, such as Newfoundland, Cape Breton, and Prince Edward Island, were so large, and torn, and tattered with bays and peninsulas, that they might sail for days or weeks without knowing whether they were following the coast of an island or the mainland. The sight of greasy, poverty-stricken Eskimos, half-naked, savage Indians instead of Chinese Mandarins clothed in rustling robes, and basking in gardens scented with spices and flowers under peaceful sunny skies, was not very encouraging.

Jacques Cartier discovers Quebec. It was to face such perils, confusion, and disappointment that *Jacques Cartier* on 20th April, 1534, with two small ships and sixty hardy seamen, sailed forth, amidst the cheers and fears of his townsmen from St. Malo to search north-westward for China, Japan, and the Spice islands. In less than a month he reached the west coast of Newfoundland (#89, #102); slaughtered with clubs two boat-loads of the great auk, a bird that swarmed on the coast but could not fly. He entered the Strait of Belle Isle, and was so disgusted with the barrenness of Labrador, that he considered it "the land God had cursed before he gave it to Cain, the murderer." How his heart must have sunk as he sailed south along the whole western rocky coast of Newfoundland amidst wet fogs, raw cutting winds, and fields of ice in June. At last his heart was cheered as in July he skirted the flat, fertile land of Prince Edward Island. He next entered the beautiful *Chaleur Bay*, perhaps the entrance to the Western Sea! But alas! he found that it ended blindly, and did not give him access to China. He landed at Gaspé on July 25, the first white man to set foot in Quebec, and erected a huge wooden cross on its rocky brow, taking possession of the land for France. John Cabot had, almost forty years before, raised the flag of England and taken possession of the land for Henry VII. Thus began the bitter and bloody strife, which continued its desolate path for two hundred years, and only ended when the heroic Wolfe and the no less brave Montcalm gave their lives on the Plains of Abraham.

Cartier's return. But Cartier pushed forward, as he was looking for a western entrance that would lead to China. He crossed to Anticosti, skirted its north coast, found there were two entrances to the west, one north and one south of this land, which of course he did not know to be an island. It was August. Fishermen had told him of furious autumn storms and icy seas in this region; he seized two Indians and was soon back in St. Malo. But instead of gold and silver, instead of stories of great chiefs in silks and ornaments, he had nothing to show but two startled Indian savages.

Cartier discovers the St. Lawrence. This St. Malo sailor was not discouraged, for he had been told of a great river, which pierced further back into these unknown lands. He was able to report that a bountiful fishery existed in the gulf, that water-fowl covered the rocks, and splendid trees densely mantled the land and wafted their pine odours to the sea. The next year, with three fine ships, he crossed the stormy Atlantic safely and was soon sailing past Anticosti through that entrance to the unknown, which even to-day, though marked out with buoys by day and lighted at night, with projecting rocks blasted away, and shallows removed, is still so difficult to navigate that boats take on pilots who have spent their lives on these dangerous waters. Cartier eagerly guided his three sailing-vessels up its uncharted waters against its confusing currents, among the many islands, with no better pilots than the two Indians whom he had captured the year before. He was much impressed with the sombre Saguenay, taking its gloomy course between high cliffs of naked rock. At last the river narrowed. As he approached Quebec, hundreds of naked and half-naked Indians swarmed about his ships and begged him to visit their chief *Donnacona*. He marched through fields of ripening corn and tobacco along the *St. Charles River*, then examined the foul, bark wigwams of the Indians, and conferred with their uncouth and lying leader. They soon informed him that farther up the river, which they called the *Hochelaga*, was a far grander and more impressive Indian fortress.

Cartier at Montreal. Early in September Cartier eagerly set out with his smallest ship and two rowing-boats to reach this interesting village. He soon passed that famous, rugged rock, *Cape Diamond*, thrusting its fierce front out into the surging stream. Little did he dream that on its haughty brow was soon to bristle the most powerful fort on the continent, and that it was to be the key to Canada. For thirteen delightful days the sailors followed the windings of the river between green banks of beautiful trees draped with wild grape vines, loaded with their pungent purple fruit. The stream swarmed with wild ducks and other waterfowl, and the songs of the thrush reminded them of the nightingale. Then at last a thousand savages rushed wildly to the shore, danced and shouted a joyous welcome, and tossed on their ships putrid fish and mouldy corn as gifts. After examining the huts and visiting the Indians in their triple-walled fortress, Cartier ascended Mount Royal, and there viewed a scene which made him forget the filthy, festering huts, the offensive smells, the foul talk, and the low life of the half-naked savages in the village below. The broad, level, fertile plain (#103), covered with frost-touched trees in all their autumn glory of gold, scarlet, and orange, spread like a carpet before him. The noble river took its silver course through the midst of the plain. Most enticing of all was the stream, as far as the eye could see, winding its way from the wistful west. This was not China, far from it; but did that river lead into its gay gardens of spices, its prosperous cities, and its untold riches?

The bite of a Quebec winter. Cartier hoped so, and lingered on the scene; soon the cold north-west winds and the frosty nights warned him only too truly that he must hasten back to Quebec to get ready for the winter. Even his brawny sailors from St. Malo went down before the fury of a Quebec winter. Soon their houses were smothered in snow, four or five feet deep. Lack of fresh food brought down the curse of scurvy like a dark pall, the sickly party melted away, and twenty-five frozen corpses were soon lying under the snow outside their huts, waiting for the ground to thaw out in the spring in order that graves could be dug. They marvelled how the nearly naked savages could wade through the deep snow in the depth of this Arctic winter, as they came to visit their encampment. To make matters worse, the Indians became sullen and threatening. At last delightful spring broke through the dismal gloom; the harsh ice crumbled and vanished before the warm sun; their ships were freed, and the miserable remnant of the party departed for St. Malo, but not before they had done a shameful act. They seized chief Donnacona and some of his men, whom they took to Paris. There these disgusting savages told tall stories about a land called the Saguenay, where lumps of gold and silver were as common as stones, where people could live without eating, and where the natives had only one leg, instead of two. Soon, in their new surroundings, they sickened and died.

Though Cartier made a third voyage he added nothing new to his explorations.

The discovery of Lake Champlain. *Samuel de Champlain,* the great explorer of the Ottawa River and Georgian Bay, as well as of the Atlantic coast, placed and housed the first small group of French

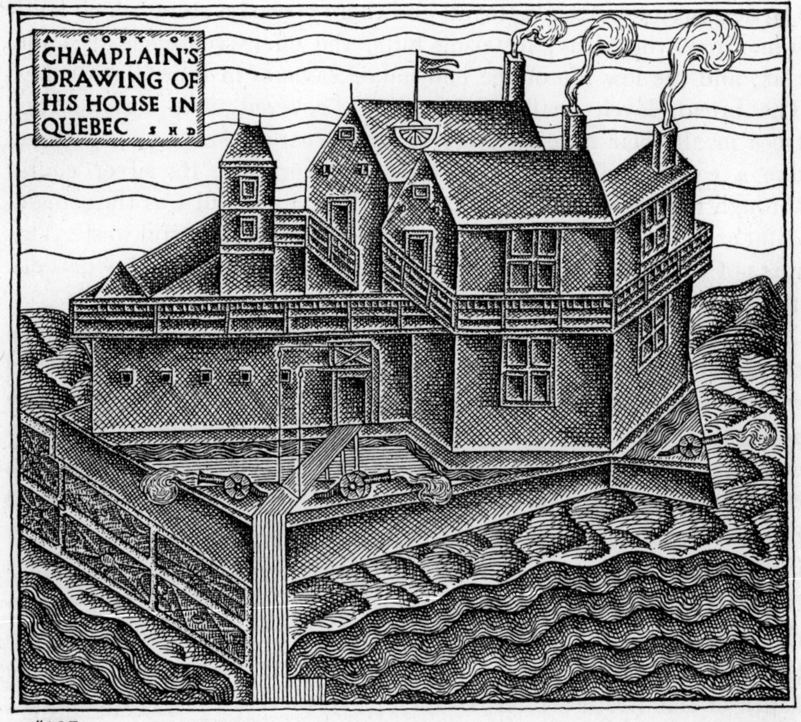

#105

settlers in Quebec in 1608. After cheering this little group through their first hard winter, he turned westward in the spring to search for China and the Western Sea (#102). But he soon met seven hundred *Huron Indian* warriors, encamped above Quebec, who had more ghastly work for him to do. He, unluckily, had promised a Huron, the year before, to assist his tribe in their war against the *Iroquois.* They were so persuasive that he soon turned south with his sailing-boat on to the *Richelieu River* and followed its clear waters through a delightfully fertile country. Yet not a hut, not a garden, not an Indian did they see; for this was already a deadly "no man's land" along the fighting route of the Hurons and Iroquois. At the *Chambly Rapids*, now traversed by a canal, Champlain was loath to leave his fine boat for the poor protection of Indian canoes. Soon the river opened into that beautiful lake, whose polished surface Champlain was the first white man ever to see and which still bears its discoverer's name. His exploration of the lake was soon cut short by

#106

CHAMPLAIN'S DRAWING OF HIS FIGHT WITH IROQUOIS

Which is Champlain? What protection had the Iroquois? What clothing did the warriors' wear? How did the warriors reach the battlefield? How did Champlain's gun differ from a modern one?

a savage battle with the fierce Iroquois (#106); the Indians, however, were unable to stand for long, with bow and arrow, against the deadly bullets of Champlain's rifle, which brought down the three dashing Iroquois chiefs in no time. After witnessing the sickening torture of the prisoners, Champlain was soon back in Quebec, sadly convinced that these savages, with hearts of tigers, needed most of all, Christian missionaries to teach them the gospel of love.

Later explorations. Ungava even to-day is largely a blank space on the map. #102 shows the routes of a few of those fearless men, who have traversed its forbidding tracts, beginning with *Buteux*, who in 1656 reached far up the St. Maurice, and ending with

Low, who had such a love for this harsh bare country that he attacked it from almost every corner.

Fisheries

Fishing in Quebec. When John Cabot returned in 1498, and made his graphic report that fish in the gulf were so numerous that they could be scooped up in buckets weighted with stones; it started the exodus of many Basque and Breton fishermen from France who came flocking to the Gulf of St. Lawrence. They settled for the summer on the Magdalen Islands and along Gaspé long before there was any permanent settlement in the province. These two regions, with scattered villages along the north shore of the Gulf, right to the Strait of Belle Isle, are still the centres of fishing (#102). This quiet industry was long neglected by the French settlements for the more exciting and more profitable fur trade. Almost immediately after the British conquest in 1763, however, the energetic *Robins family* from Jersey, in the Channel Islands, south of England, settled at Gaspé Peninsula, began buying cod fish, equipping fishermen, and soon had control of almost the whole fishing industry along the Quebec coast. This, the first great fish company in Canada, is still master of a large share of the Quebec fishing industry. It has changed its name, has moved its head office to Halifax, but from the little town of *Paspebiac* in Gaspé it still controls curing stations all around the coast. The chief sea fish are cod, lobster, and herring; and the chief river and inshore fish are salmon, smelt, and eels. #102 shows the distribution of these. The sea fishing has not changed in four hundred years and is still done by hook and line from small boats, but numerous gill and trap nets border the shore of the St. Lawrence River below Quebec.

Fishing for sport. Quebec is the delight of the angler who sets his heart on big salmon and is willing to pay for his fun. More than two hundred rivers and five thousand lakes are rented at high prices to sportsmen, who keep them well stocked and have the exclusive right to fish them. Though this renting prevents the waters from being denuded of fish and brings large sums in taxes to the province, it unfortunately excludes all but the wealthy from the nearest and best fishing spots.

Hydro-electric Power

The fountain of industry. Every important river in Quebec tumbles down to the St. Lawrence from the Canadian Shield or the Appalachian mountains in a tumult of rapids and waterfalls. As the St. Lawrence is the centre of industry, it is therefore not surprising that water-power is Quebec's biggest natural resource, that she has nearly half the total installed hydro-electric power in Canada, and that electricity is the milk that nourishes all her industries. Her pulp and paper, mining, metal extraction, chemical and textile industries would be helpless without its energy. She has already working, immense electricity generators on the Ottawa, St. Maurice, Saguenay, St. Francis, and the St. Lawrence above Montreal (#104), and factories have sprung up around these generating stations like date-palms around a spring in the desert.

Harnessing the streams. Though the quantity of electricity should be fairly steady throughout the year, this is hard to obtain because the flow of streams slackens during the dry summer and autumn but rushes forth in torrents during the melting of the snow and after the spring rains. To equalize the flow throughout the year, the electrical companies and the provincial government have flung more than twenty great walls, or weirs, across the headwaters of the main streams. During the spring freshets these weirs are closed, and millions of tons of water are held behind them in huge artificial lakes; this water is gradually fed out as required to keep a steady flow throughout the year. The *Gouin Dam* on the St. Maurice River is one of the largest in the world.

Climate

A comfortable climate. A province, which extends 1200 miles from north to south, is bound to have a wide range of temperatures. The south-west district, in which most of the people live, has a comfortable climate, very similar to that of south-west Ontario, except that it is not affected by any large body of water. While summer in both provinces is much alike, winter is more steadily cold in Quebec, with little melting of snow. Autumn comes earlier and is shorter than in Ontario. By the middle of September the maples on the hills fly flaming signals that frost has come; the trees are soon bare, and November is cold and gloomy, with the low temperature, high humidity, and frequent sleet storms making outdoor life uncomfortable. The great piles of wood in every farm woodshed is a good indication that winter is long and cold. In most districts snow falls during about fifty days, and since little of it thaws in the cold weather, there is often three or four feet of snow on the ground, which not only forms a warm blanket over the winter wheat, but makes good roads into the lumber camps, and attracts thousands of tourists to take part in the winter sports. Rainfall is greater by ten inches than in southern Ontario.

Summer comes suddenly. If harsh winter rushes in suddenly from the north in November, it is driven out still more quickly in the very short spring. Indeed, winter almost meets summer. The ice on the river near Quebec may make a good road for horse sleighs to-day, to-morrow be restless, and in less than a week be full of holes. In eight or ten days the boat will have replaced the bobsleigh. The sudden change from winter to warm spring, carpets, like magic, the forest and thicket with gayly coloured flowers and clothes the trees with leaves in a few weeks.

Climate of north and east. The city of Quebec and districts farther east have much colder winters, heavier snowfall and shorter summers with a far shorter growing period between the last spring and the first autumn frost. The part of Quebec along the Canadian National Railway has a climate very similar to that of the Clay Belt in Ontario (p. 310). Still farther north the summer shrinks to a couple of months, all but a few inches of soil is forever frozen, and winters are extremely severe.

Forests

The forests of Quebec. When Cartier and Champlain explored the St. Lawrence, dense forests of big trees covered the whole southern part of the province. In the St. Lawrence Lowlands and the Appalachian Highlands, hardwoods — maple, birch, elm, oak, ash, beech, and basswood—with their broad leaves, rule the forest (#104), but even here they are mixed with pines, spruce, hemlock, cedars, and balsam, which have played, and continue to play, the chief part in the lumbering industry. Over the southern part of the Canadian Shield the hardwoods are dethroned by the narrow-leaved evergreens (#104). Still farther north, white and red pine and hemlock drop out, and spruce, balsam, jack pine, poplar, and canoe birch are left to struggle with the short growing season, the fierce winds, and the biting cold of this region. Trees become smaller, and gnarled, with growth so slow that a scrubby tree, a few feet high, may be fifty years old. At last near Hudson Strait even the tenacious spruce, tamarack, and scrubby birch surrender and leave a naked fringe of barren lands along that desolate coast.

The noble hardwoods gone forever. The hardwood forests of the St. Lawrence plain have been eaten into by the farmer's axe in the clearing of the land as in Ontario (p. 311), until now that splendid forest is gone forever, except for wood lots connected with farms, from which firewood, fence posts, and rails are obtained.

Lumbering. Because lumber from white pine has a clear white colour, a distinct, beautiful grain, is durable, easily worked with tools, and takes a nail readily, this tree is prized above all others. In Quebec, however, the best of the white pine has been cut, and spruce and balsam are her chief timber trees, followed by jack pine, and birch. Formerly the chief sawmills of the province were studded along the Ottawa and its tributaries, but the timber has been cut so far back that this region has to take second place to Gaspé Peninsula, where at *Rimouski*, *Matane*, *Priceville*, and other mill towns, some of the largest sawmills in the province are found (#113). Small sawmills for local use are widely scattered along all the streams, more than 1300 were cutting lumber during 1933. On account of the great expense of floating logs down long distances, through streams shrunken by the clearing of the forests, the tendency now is to put the sawmills back as near as possible to the timber, and to use hydro-electric power for running the saws. The British Columbia plan of using logging-railways to bring the logs to the mill (p. 400) is used in some lumber camps.

Preparation of pulp wood. Nowhere else are pulp and paper mills as big as those that cluster around the lower Ottawa, St. Maurice, and upper Saguenay, where electrical power is cheap and abundant, and small spruce and balsam grow thick in the forest (#104). The pulpwood trunks are sawed into eight foot lengths and floated down to the mills, where they are further cut into two or four foot lengths and barked (#107). A good deal of pulpwood is also cut by farmers and others and shipped to the mills by train or truck. For convenience and cheapness in shipping, this is first barked and cut into standard lengths.

Mechanical pulp. These sticks are to be made into pulp. Two chief processes are used, one *mechanical*, the other *chemical*. The first is simple; the side of the stick is kept pressed firmly against the rough rasping surface of a rotating grindstone, and a spray of water,

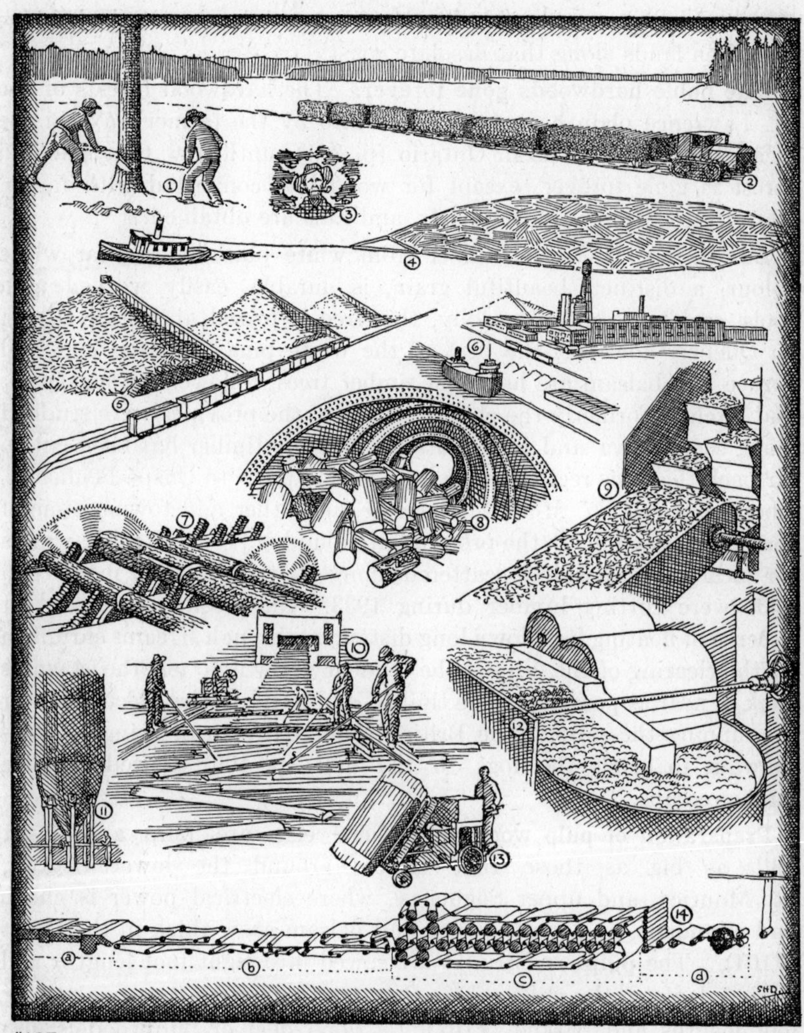

#107

A TREE TURNED INTO A ROLL OF PAPER

1. Cutting the tree. 2. Tractor train towing-pulp logs to railway. 3. A pulp-wood cutter. 4. Towing a "boom" of pulp-logs. 5. Piles of pulp wood at mill. 6. A typical pulp and paper mill. 7. Saws cutting up logs for barking and grinding. 8. Logs in rotating drums of the barker. 9. Chips on way to digester. 10. Feeding logs to the saws. 11. Huge digesting tank. 12. Pulp on the way to the paper machine 13. An electric truck carrying a roll of newsprint. 14. Paper machine (a) pulp is screened and fed on belts of felt (b) pulp on felt is dried and pressed (c) paper picked off felts and thoroughly dried (d) finished paper.

playing against the stone, washes the pulp away as a thin soup. In a big mill there are dozens of these mighty, noisy grinders pulping log after log for twenty-four hours in the day. Every cord of wood (worth 6-7 dollars) delivers a ton of pulp, worth 16-18 dollars.

A chip of wood falls to pieces. A stick of wood is composed of millions of very minute fibres, each as thin as a cobweb. These are stuck together by a cementing material to make the stick. The fibres themselves are tough, white, not acted on by chemicals, and will not decay even after hundreds of years, but the cement has not these properties and is injurious to the paper. The mechanically produced pulp does not separate the fibre from the cement but grinds both together. By the *chemical process*, on the other hand, the wood, in fine chips, is boiled with suitable corrosive chemicals, which dissolve the cement until the solid chips fall to pieces into millions of delicate fibres. All the chemical is washed away, so that nothing but a thin soup of pure fibre is left. This pulp is so much better than mechanical pulp that the latter always has chemical pulp added to make it mat and to give paper strength. Often pulp is bleached or coloured before it is made into paper.

A mile of paper in five minutes. The making of paper from this mixture of pulp and water is an interesting process (#107). A wide, horizontal moving belt of wire-netting has a steady stream of the mixture fall across one end. As the belt moves forward, the water runs away and leaves a sheet of matted pulp, which is whipped from the belt by a rotating cylinder and squeezed so tightly between two steel cylinders (like clothes passing through a wringer) that it comes out like thick, wet paper. Then it is drawn around a whole series of horizontal, rotating, steel cylinders, filled with steam. This dries it. Finally it is pressed, under tremendous force, between solid steel cylinders to smooth it. At the wet end of a paper-making machine enters thin soup containing millions upon millions of fibres, thinner than the finest wire; and at the other end comes out a sheet of white, smooth, tough paper twenty feet wide. Every five minutes these swift machines roll up a mile of paper, and this process goes on hour after hour and day after day without a break.

Uses of pulp. Most of the pulp is made into newsprint, which is used in newspapers. But much of it is used to make other paper, cardboard, roofing paper, shingles, millboard, beaverboard, etc. Paper is rapidly replacing wood for containers; and most of the cans, packages, bars of soap, etc., that come to the grocer are packed in cardboard boxes. Paper towels, napkins, caps, plates, and even spoons, are products of pulp machines. Nearly all artificial silk or rayon,

#108

#109

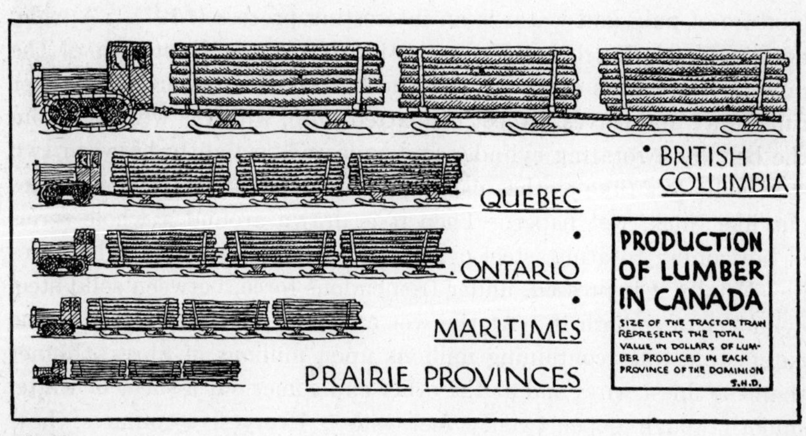

#110

which allows even poor people to have beautiful clothes, is made from a very pure chemical pulp.

The graphs (#108, 109, 110) show how outstanding are the forest industries of Quebec. She is first of the provinces in production of pulp and paper, second in production of lumber (#110) and first in cutting of fire-wood, fence posts, and rails, and second in preparing railway ties.

Making maple-sugar. Despite the general slaughter of hardwood trees in the St. Lawrence plain before the axe of the settler, fortunately

#111

BOILING DOWN MAPLE SAP

How are the sap buckets carried? How is the pot fastened? What is the use of the paddle? What is the man doing with the dipper? What is the boy thinking about?

many beautiful groves of stately spreading sugar-maples have survived (#104). Each spring these trees are tapped, and the clear frothing liquid that flows out is evaporated to maple-syrup, or still further to maple-sugar. The centre of this industry is the Eastern Townships (#104), and model factories for the making of the two products are at Plessisville and Quebec. The days of romance, when the boiling down was done in big iron pots over open fires in the bush, and the product was an inky liquid or a dark brown solid, containing (besides sugar) soot, ashes, sticks, and charcoal, are now almost gone (#111). To-day the sparkling sap is caught in covered pails, transported often through long tubing to an up-to-date factory, where the evaporation is rapidly completed by live steam and vacuum pumps (#112). The light amber, transparent maple-syrup and the odorous, light-brown maple-sugar have a much finer flavour;

#112

NEW METHOD OF MAKING MAPLE SAP PRODUCTS

What do the bottles contain? What is being poured into the moulds?

but it is more than doubtful whether they have half the tang of the nondescript product obtained in the good old days at the sugaring-off parties in the evenings, when neighbours and friends gathered together from far and near, and the seething syrup was poured sizzling out on the snow to harden into sugar.

Quebec produces about two-thirds of this typically Canadian product, Ontario produces most of the remainder.

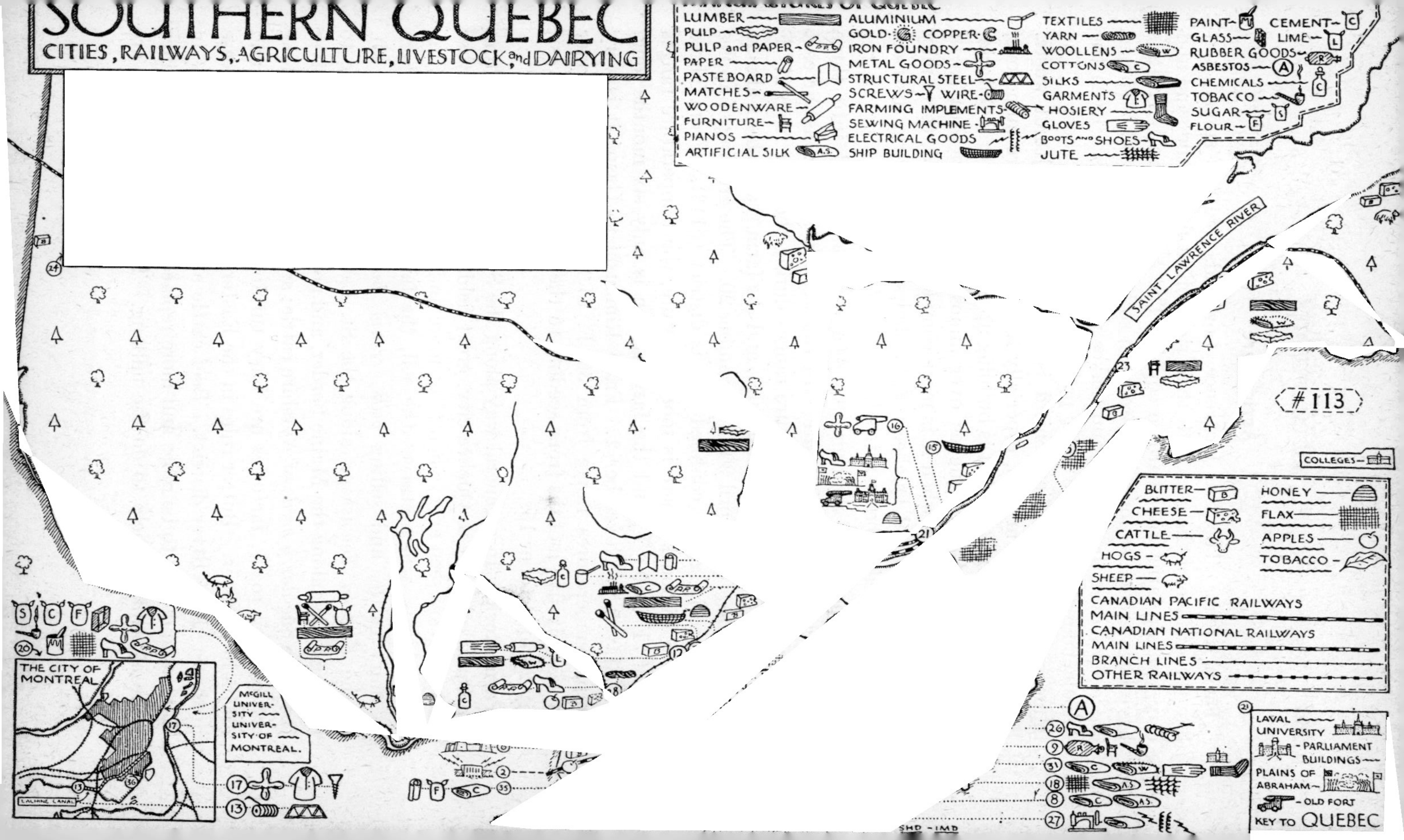
SOUTHERN QUEBEC
CITIES, RAILWAYS, AGRICULTURE, LIVESTOCK and DAIRYING
LUMBER
PULP
PULP and PAPER
PAPER
PASTEBOARD
MATCHES
WOODENWARE
FURNITURE
PIANOS
ARTIFICIAL SILK
ALUMINIUM
GOLD
COPPER
IRON FOUNDRY
METAL GOODS
STRUCTURAL STEEL
SCREWS
WIRE
FARMING IMPLEMENTS
SEWING MACHINE
ELECTRICAL GOODS
SHIP BUILDING
TEXTILES
YARN
WOOLLENS
COTTONS
SILKS
GARMENTS
HOSIERY
GLOVES
BOOTS AND SHOES
JUTE
PAINT
GLASS
RUBBER GOODS
ASBESTOS
CHEMICALS
TOBACCO
SUGAR
FLOUR
CEMENT
LIME
SAINT LAWRENCE RIVER
113
COLLEGES
BUTTER
CHEESE
CATTLE
HOGS
SHEEP
HONEY
FLAX
APPLES
TOBACCO
CANADIAN PACIFIC RAILWAYS
MAIN LINES
CANADIAN NATIONAL RAILWAYS
MAIN LINES
BRANCH LINES
OTHER RAILWAYS
THE CITY OF MONTREAL
LACHINE CANAL
McGILL UNIVERSITY
UNIVERSITY OF MONTREAL.
LAVAL UNIVERSITY
PARLIAMENT BUILDINGS
PLAINS OF ABRAHAM
OLD FORT
KEY TO QUEBEC
SHD - IMD

Agriculture

Farm areas. The St. Lawrence Lowlands (#103), spread out like the surface of the sea, with their rich soil of fine fertile sand and clay, contain farm lands hard to equal. The Appalachian Highlands have been eroded so long and are worn so low that the soil is spread out deeply over the valleys and makes prosperous farms. Even the Appalachian ridges make good pastures for cattle and sheep. The immense, rocky Canadian Shield is not all an agricultural waste. Pockets of good farm land, especially in river valleys, are numerous, and the great *Clay Belt*, that is building up a new agricultural Ontario, spreads eastward into Quebec over almost as large an area as in Ontario, and perhaps when it is more completely explored, there may be a new agricultural Quebec. #114 shows the chief farm districts already occupied or being settled.

Growing periods. Southern Quebec can be divided into three climatic regions for agriculture. In the western section, as far east as Three Rivers, the temperature ranges during the year from 27° F to 93° F, there is sufficient moisture, and the farm season lasts almost seven months, from April 20 to November 20. The growing of apples and tobacco is largely confined to this district (#113). From Three Rivers to Rimouski there is more moisture, the temperature ranges from 30° F to 90° F, and the farm season is only six months long, from May 5 to November 1. From Rimouski to Gaspé it is very damp, temperatures range from 30° F. to 80° F., and the farmer has scarcely five months from seeding to the completion of harvest, May 20 to October 15.

Farm stock. In a general way along the St. Lawrence River from the Ontario border to Quebec city, great fields of *hay*, *oats*, and other field crops are seen everywhere; but more and more, instead of selling hay and oats, which starves the soil, the farmers are feeding them to dairy cattle, and selling *milk*, *cream*, *butter* and *cheese* (#113). In the *Eastern Townships*, along the St. Lawrence from Quebec to the Saguenay, along the Maine border, and north of the Ottawa River there are fine *dairy herds* of Ayrshire cattle; and every few miles along the country roads, factories are busy turning milk into cheese and cream into butter. Butter rules in the Eastern Townships and cheese in the St. John River district. Beef cattle are not so common as in Ontario and tend to be more and more replaced by dairy herds, since there is plenty of help to do the milking. The best beef cattle are found north of the Ottawa River. Sheep are most common in the Eastern Townships, where they find suitable pasture on the more barren ridges.

Other crops. Quebec has its specialties in agriculture. A strong *tobacco* has long been grown for home use, but more careful attention to the quality has developed an export trade in the western end of the province. The growth of *apples* is centred in the same district as tobacco, which is noted for its beautiful *Fameuse*, or *Snow apples.* The island of Montreal is as famous for its *musk melons,* as is the Annapolis valley of Nova Scotia for its apples. Flax has been grown since earliest times, and for over 200 years French-Canadian farmers went to mass every Sunday with white shirts which were made from flax fibres, grown, extracted, spun, and woven by their own families on their own farms.

Quebec's first farmers. *Louis Hébert,* brought to Canada by Champlain, was Canada's first farmer. By 1620 every spring he was ploughing and seeding the land where now stands the stately cathedral and great, gray seminary of the city of Quebec, and every autumn harvesting good crops of wheat, corn, and peas. He soon had apples as well. *Abraham Martin* in 1643 was cultivating what is now the historical Plains of Abraham. *Pierre Boucher* from 1653 onward did much for agriculture in the fertile district around Three Rivers, and had a greater variety of crops on his farm than perhaps could be found on many Canadian farms of to-day. He grew wheat, barley, oats, rye, corn, peas, beans, tares, vetch, hemp, flax, timothy, buckwheat, sunflower, horse-beans, turnips, beets, carrots, parsnips, oyster plant, and cabbage. Though no monuments are erected to most of these pioneer farmers, their names are not forgotten, as many of the most cultured families of the province are the descendants of such worthy stock and bear their names.

Quebec's first live stock. Champlain introduced the first cattle from Normandy and from that hardy stock is descended the cattle seen grazing peacefully today in the fields along the St. Lawrence. They have developed into a special breed, the only one that has been originated on this continent. By 1663, oxen, cows, pigs, sheep, hens, turkeys, pigeons, dogs, and cats were introduced, and they even brought along live rats to make food for the cats. How these farm animals adapted themselves to a changed climate can be gathered from the following: A herd of twelve horses, introduced in 1665, had increased to 145 in 1679, to 218 in 1688, and to 684 in 1698.

In spite of rich soil and hard-working farmers, agriculture did not prosper in Quebec for many years, especially between 1763 and 1850. War had destroyed their buildings, played havoc with the crops, slaughtered many of the best men, and caused the migration to France of many of the leaders. The French settlers were left cut off not only from France but from the rest of the world, and in 1850 they were still farming by the methods they had brought over from France two hundred years before. The long farms were divided into two strips; on one was grown hay and oats for three successive years; the other was left to be covered by nature with wild grasses, which were used for hay and pasture for the all too few stock. Then the use of the two strips was reversed for the next three years. The fields were never seeded with grass and clover, but depended on nature. There were too few cattle to keep the land fertile; the agricultural implements were home-made, heavy, and awkward. The French-Canadian farmer, even since 1850, has been slow to change his methods, but through the efforts of such leaders as *William Evans*, *J. X. Perrault*, and especially *Bernard*, who went through every county lecturing to farmers on improved methods, the whole industry has been born again, and fine stock, large farm buildings, the latest implements, and intelligent cultivation are seen everywhere.

Settlement and Transportation

Earliest settlements. Fur trade and farming no more mix than oil and water, and as farming means settlers and fur trade waste land, the story of early settlement of New France, as Quebec was called, is sad and disappointing. Companies, like the *Hundred Associates*, pledged to bring out large numbers of settlers for a monopoly of the fur trade, were so intent on profits from furs that they almost forgot settlers. In 1608 Champlain erected buildings at Quebec (#105), and with about twenty clerks and missionaries spent the winter there. Quebec was soon fortified, had a few real settlers who tilled the soil, but had to start all over again, for in 1629 it was captured by the British and all the French were taken prisoners. The real settlement began about 1635 when a few farmers settled along the north shore of the St. Lawrence at and below Quebec (#114). At about the same time a second settlement at Three Rivers became the key to the fur trade. But the settlement at Montreal in 1642 was of still greater importance. It was founded by *Maisonneuve*, and at its start no less than 60 settlers were busy erecting houses,

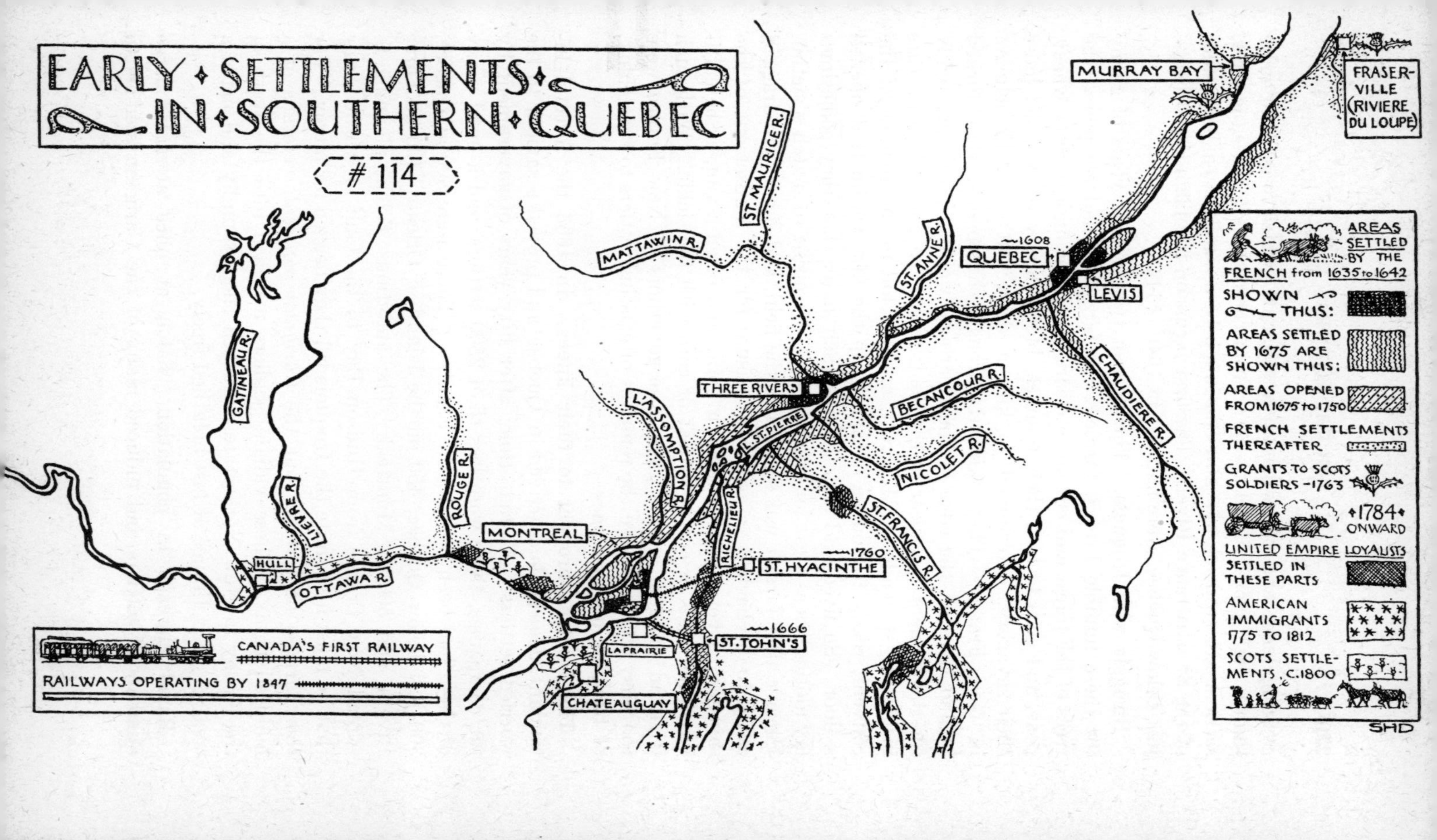
EARLY SETTLEMENTS IN SOUTHERN QUEBEC
114
MURRAY BAY
FRASERVILLE (RIVIERE DU LOUPE)
ST. MAURICE R.
MATTAWIN R.
ST. ANNE R.
1608
QUEBEC
LEVIS
GATINEAU R.
THREE RIVERS
BECANCOUR R.
CHAUDIERE R.
L'ASSOMPTION R.
L. ST. PIERRE
NICOLET R.
ROUGE R.
LIEVRE R.
RICHELIEU R.
ST. FRANCIS R.
MONTREAL
1760
ST. HYACINTHE
HULL
OTTAWA R.
1666
ST. JOHN'S
LA PRAIRIE
CHATEAUGUAY
CANADA'S FIRST RAILWAY
RAILWAYS OPERATING BY 1847
AREAS SETTLED BY THE FRENCH from 1635 to 1642 SHOWN THUS:
AREAS SETTLED BY 1675 ARE SHOWN THUS:
AREAS OPENED FROM 1675 to 1750
FRENCH SETTLEMENTS THEREAFTER
GRANTS TO SCOTS SOLDIERS -1763
1784 ONWARD
UNITED EMPIRE LOYALISTS SETTLED IN THESE PARTS
AMERICAN IMMIGRANTS 1775 TO 1812
SCOTS SETTLEMENTS C.1800
SHD

clearing land, and throwing up a stockade, which was little more than a high, close, picket fence. From these three forts, as centres, settlements crept slowly and continuously up and down both sides of the St. Lawrence River. The first farms on the Island of Orleans, just below Quebec, were laid out in 1640 (#114).

Struggle with Iroquois. Then began twenty long years of agony for these humble French farmers. Before they had conquered the trees of the forest and cut out a clearing large enough to grow wheat, peas, and corn to feed their families, blood-thirsty Iroquois let loose their savagery, determined to kill or drive out every Frenchman in New France. They skulked through the forests, they lurked behind trees on the edge of the clearings, waiting with rifles ready for the farmers to appear in the field; they even crept like wild cats inside the stockade of Montreal, and lay hid in the shadow of the convent, ready to pounce on and scalp the first victim that opened a door. So fixed and spiteful was their hate, that they sometimes lay motionless as a cat for two whole days, waiting to snatch a victim. Farmer after farmer was scalped in the field, and some were dragged away to be hacked to pieces at leisure for the amusement of the savages. Every settler carried a knife, and often a gun, as he cultivated his fields or cut his wheat or corn with the sickle. With such frightful dangers around them, the governors of New France were too busy trying to save the remnant of existing settlers to even think, of bringing out new ones.

The Iroquois brought to their knees. In 1663 the forceful intendant, *Jean Talon*, arrived in Quebec and proved a saviour to the worried settlers. At that time, after fifty years of government by fur companies, there were less than 2500 settlers, and a good many of these had been born in the country. There were 800 at Quebec, 300 just below Quebec, 450 on the Island of Orleans, 400 at Three Rivers, and 500 at Montreal. The ambitious Talon, planning to spread New France from Hudson Bay to the Gulf of Mexico, saw that he must first smite the Iroquois till they begged for mercy. For this purpose he brought out the *Carignan regiment*, veterans of a hundred fights. These valiant soldiers marched into the teeth of the enemy in New York state, thrashed them soundly and made a treaty. New France at last breathed freely.

Rapid increase in population. Talon at once attracted large numbers of settlers and induced many of the Carignan soldiers to swing the axe instead of the sword and to direct the plough instead of the rifle bullet. Between 1665 and 1673 he brought out several thousand settlers, who were placed along both sides of the St. Lawrence between Three Rivers, and Quebec, and on the south shore far beyond Levis. He also started settlements up the Richelieu as a defence along the line of march of the Iroquois. As there were twice as many men as women in the colony, he had shipload after shipload of carefully selected French girls brought out for wives. One hundred came the first year, two hundred the next, and still larger numbers in succeeding years. As soon as they stepped off the boat, zealous bachelors were waiting. The girls were eagerly wooed, the courtship was short, and they were almost immediately married, sometimes in batches of thirty couples. Each pair marched off to their new home in the forest with the king's bounty of an ox, a cow, two pigs, two fowl, two barrels of salt pork, and a small purse of gold. Further to spur early marriage, every father was fined for each bachelor son over twenty and each spinster daughter over sixteen. To encourage the rearing of large families, the king gave a bounty for every family of ten, with a steeply-graded increase of bounty for eleven or twelve children.

Early Quebec industries. Talon, the business-like intendant, not only poured people into the country but also founded useful industries by which they could earn a living, and arranged markets in France and the West Indies for the sale of their products. He built tanneries at Quebec and Montreal so that they could convert the skins of the farmer's cattle and sheep into leather, and make their own boots; he got good prices in France and the West Indies for canned salmon, salted eels, fish-oil, timber, and flour. He trained them to build oak ships and tried to break them off the habit of drinking imported French wines, by building a brewery in Quebec to convert their barley into beer. So sturdily did he build that the old stone building is still brewing beer. He established the first foundry and blast-furnace on the continent near Three Rivers (p. 244) where the stoves, axes, knives, ploughs, horseshoes and many other iron products of the inhabitants were made for two hundred years afterward. Indeed, this foundry was one of the busiest places in Canada, as many as one hundred and fifty mechanics working among the blazing furnaces, which made the night sky as red as blood.

The seignior and the curé. These French peasants, chiefly from Normandy and the west of France, had not owned a square foot of land when in France, but were serfs, tied to one piece of land all their lives and compelled to pay heavy taxes to their overlord, called the *seignior*. Along the St. Lawrence, Richelieu, and to a less extent the Chaudiere rivers, the land was divided into large estates, called *seigniories*, and these were bestowed upon French gentlemen and soldiers. For example, many officers of the Carignan Regiment were selected for this honour. It was to the interest of the seigniors to obtain settlers for their seigniories, though they did not always do so. The settler owned the land, could sell it or leave it to his heirs, but he had to work a few days each year for the seignior, had to give him certain presents, had to take him a small share of his crops, and, when he sold the land, had to give the seignior a small fraction of the selling price. The seignior would build a grist mill and a sawmill, at which the farmers were compelled to have their grain ground and their logs cut into lumber; the seignior kept a small portion for the work. As this method of holding land was somewhat similar to the method in France, but more favourable for the settler, it suited the French habitant very well. The seignior was usually a leader in the community, lived in a handsome manor-house, saw that the roads were kept up; and the humble peasants were pleased to bow before him, and look to him as an example. The real leader and guide of the people, however, was the *curé*, or Roman Catholic priest, who had a tremendous influence over the morals and devotions of these deeply religious people.

French defeats. Alas! the prosperity of these hard-working farmers was short-lived. Talon was recalled in 1665. France plunged into a series of costly wars, in which the British under the Duke of Marlborough humbled her pride in victory after victory, first *Ramillies*, then *Blenheim*, and finally bloody *Malplaquet*, where, were slaughtered greater numbers of the finest young men of France than the total number of French settlers brought to Canada. The climax to France's shame came later in the Seven Year's War, in which Wolfe wrested New France from Old France forever (p. 182).

The habitant spreads out. In this succession of wars which followed the recall of Talon, the interests of Canada were almost forgotten. No more settlers were brought out; many already there were used to fight her wars; some were taken from Quebec to settle Detroit, Louisiana, and other districts; the industries that Talon made to thrive were left to languish and many of them to die. But the French Canadians went on steadily clearing the forest, working bigger and bigger farms, and raising large families. As a result of natural increase, when Quebec was conquered by the British, the 7000 settlers, who lined the St. Lawrence in 1672 at the end of Talon's bright reign as Intendant, had become 65,000 people. As the people multiplied, the sons took strips of land alongside the parents, or the parents divided the land with the sons. As a result the farms became narrower, the houses along the river closer and closer together. As new generations arose, land farther and farther back from the rivers was occupied. This steady expansion has continued to the present day and is still at work, as new farming regions are being opened. #114 shows clearly the steady spread along the rivers, and, during the last seventy years, along the railways.

First British settlers. Almost immediately after the conquest of 1763, eager Yankee traders and British merchants to the number of perhaps two hundred, rushed to Montreal and Quebec, anxious to obtain the profits from whatever trade was available. Besides these, a large number of Wolfe's veterans, especially the Highlanders, who had captured Quebec, decided to settle in the new land. Some were given seigniories. For example, Captains Nairne and Fraser obtained the seigniory of *Murray Bay* (#114) and settled in that district, where many of their descendants, with Scottish names and French speech, still live. Another group settled at *Fraserville*, named after Alexander Fraser, who became its seignior. In 1919 this name was changed to *Riviere du Loup*. However, most of these soldiers remained in Montreal and Quebec, where they and their descendants have ever since played an important part in public life.

United Empire Loyalists. The next inflow of English speaking people began about 1784, when United Empire Loyalists came pouring up the Richelieu and over the border to the St. Lawrence. The most of these were settled in Ontario along the St. Lawrence (p. 322) but some were given land at *Sorel*, *Lachine*, *Fort Chambly*, *St. Johns*, and about 450 at *Gaspé* on the Bay of Chaleur. This last group has held its own ever since, and there is still an English speaking island among the sea of French-Canadians that occupy the rest of Gaspé.

Settlement of Eastern Townships. But the rich soil, and the dense forests of giant maple, ash, beech, and birch trees (#102) were so tempting that even before the American war of Independence was won, great numbers of settlers from New Hampshire and Vermont streamed over the border with all their belongings, settled down in the bush of the Eastern Townships, and suffered the same privations as the settlers of Ontario (p. 316). Soon they had formed thriving, progressive communities, which were thoroughly Canadian and loyal in their outlook (#114). The war of 1812 suddenly dammed back this flow of immigrants over the border. Then began the peaceful struggle between the quiet, tenacious, unchanging French-speaking people of Quebec along the St. Lawrence and the forceful, progressive, English-speaking settlers along the United States border. They came toward each other, soon met, began to diffuse, but never mixed. Many parts of the Eastern Townships, that fifty years ago were English, are now largely French.

About 1800, many Scots, evicted from their poor farms in Scotland, crossed the ocean and settled in Canada. A large number were attracted by the Scottish loyalists in Glengarry (p. 325) and settled south of the St. Lawrence near them. A settlement in the county of *Chateauguay* in the south-west corner of the province was started about 1800 (#114), and a goodly number of Canada's leading Scotsmen sprang from this small settlement. Others settled north of the Ottawa in what is now the county of *Argenteuil* (#114). A good many from the United States also settled along the Ottawa, attracted by the fine timber and the profits of the lumbering trade. The most outstanding of these was *Philemon Wright* from Massachusetts. His family and four other families, with 25 men, seven teams of horses, four yokes of oxen, a cow or two, sleighs, and provisions, set out overland from Massachusetts through the deep snow of this roadless country for the Ottawa River in the middle of winter. They reached the Chaudiere Falls in 33 days, founded *Hull*, and at once began cutting down the trees, to the surprise of the Indians, busy making maple-sugar. Soon Wright's rafts of square timber were floating down the Ottawa, the settlement rapidly grew, but no more quickly than Wright's fortune, for he soon became the first of many men about Ottawa who turned timber into gold.

Canada's first railway. Quebec was a pioneer in the building of railways. Montreal, always keen that none of her traffic should be switched to American ports, built the first railway from *Laprairie* to *St. Johns* on the Richelieu River, in order that the abundant produce of the progressive Eastern Townships should not go up to Lake Champlain and down the Hudson River to New York City, but to her port (#114). The rapids at *Chambly* on the Richelieu River had not at that time been overcome by a canal. The track of this first railway was made of wooden planks to which were spiked two strips of iron; in 1834 when it was first opened, horses were used to draw the cars, but by 1836 a strange-looking wood-burning locomotive puffed over the route at ten or twelve miles an hour. By 1847 steam railways from Montreal to Lachine and from Montreal to St. Hyacinthe were running, and during the next ten years a network throughout the region south of the St. Lawrence took form, and the meshes have been growing smaller and smaller ever since.

Manufacturing

Power everywhere. In no part of Canada are natural conditions more favourable for factories than in Quebec. Abundant, cheap hydro-electric power is at their very doors (#104). Immense developments on the St. Maurice River spread their electrical arteries west, south, east, and north-east to all the great centres, *Three Rivers, Montreal, Quebec, Sherbrooke,* and lesser towns. The still greater quantity of electrical energy, springing from fall after fall along the Saguenay, though too far away to be conducted to the great centres of the Province, is so cheap and convenient that it has drawn to itself during the last twenty years the greatest array of pulp and paper mills to be found anywhere on the continent except on the St. Maurice; and town after town has arisen, full-grown, along its shores. The falls of the *River Chaudiere,* tumbling down the Appalachian ridges, gives a whole series of hydro-electric power stations to run the pulp and paper mills and the textile factories of the progressive towns of the Eastern Townships. The mighty hydro-electric power generators along the Ottawa and its tributaries, the *Gatineau* and *Liévre* (#114), give energy to the sawmills, pulp and paper mills, and other factories of Hull and a dozen lesser towns of the district. All of these sources of power have been only well tapped and have reserves enough to make Quebec's cities and towns as busy as a great collection of beehives.

Good transportation, better workmen. Besides every town having such quantities of power at its finger tips, the deep channel of the St. Lawrence brings it to the front door of all the markets of the world and by the cheapest kind of transport, the water route. But to have the power to run the machines and the means of getting the raw materials whether cotton, wool, rubber, wood, or metal, to the factory, and to have the means of getting the finished article to every part of the world, is not enough. There must be proper workmen and workwomen to run the machinery. Fortunately for Quebec there is no more willing workman than the French-Canadian. He toils steadily and good humouredly, does as he is told, does not quarrel with his foreman, and is not likely to go on strike unless provoked by unfair treatment. French-Canadian women are eager to work in factories, and labour as patiently and efficiently as the men.

Capital for factories. Even these favoured features are not enough. It requires large amounts of money to build and to run great factories, and though Quebec has not the wealth of Ontario, capitalists are attracted to such a favourable field from Great Britain, the United States, and other parts of Canada.

Chief goods made. Pulp, paper, and lumber leave all other manufactures far behind (#104). Cotton spinning and weaving, the making of boots and shoes, the manufacture of all kinds of men's and women's garments and knitted goods, all of which use large amounts of female labour, are outstanding in Quebec (#113). Immense quantities of cheap electrical energy have also attracted the making of phosphorus for matches and phosphates for fertilizers to *Buckingham*, near Ottawa; the extraction of aluminium from clay to *Shawinigan Falls*, and *Arvida* near the Saguenay; and the making of a large array of chemical products to *Shawinigan Falls*. #113 shows the diversity of goods produced, and how they are distributed among the towns and cities.

Cities

Montreal. *Montreal*, the great commercial city of Canada, is situated on the south-east of the island of Montreal in the St. Lawrence River (#113). It grew back in a series of terraces from the river until it reached Mount Royal, which deflected growth to the sides and now it has surrounded this noble mountain. If the spectator stands on the peak of this park-like mountain, as did Cartier four hundred years ago, he will see the key to Montreal's greatness. The silvery St. Lawrence will be seen widening to the east, alive with great ocean liners and tramp ships (#103) laden with wheat, pulp, paper, lumber, flour, cheese, apples, and many other products of the farm, forest, mine, and factory, on their long journey to Liverpool, Southampton, France, Germany, Italy, South Africa, Australia, New Zealand, and the West Indies. Other ships, moving inward, laden with rubber, cotton, coal, and manufactured goods from Europe for the whole of Eastern Canada, will be seen steaming up to twenty miles of docks along the smoky waterfront. As the spectator looks west, he sees the river continued toward the Great Lakes, but the white foaming rapids of *Lachine,* just above the city, tell why all the inward ships are unloading, instead of going farther up to the Great Lakes. However, a continuous procession of smaller ships will be seen steaming down the St. Lawrence from the Great Lakes to be unloaded at Montreal's immense grain elevators. The rapids are overcome by the *St. Lawrence canals.* Nine out of ten of these ships are laden with wheat. Two other threads of water may be seen from this mountain. To the north-west stretches the muddy waters of the Ottawa, whose turbulent current has brought down so many million dollars worth of logs and lumber. Small steamers still run regular trips as far as Ottawa. Away to the east is the Richelieu, running straight north across the fertile plains. This connects Montreal by navigable water with the surge of traffic to and from New York. These four streams of water, crossed like the arms of an X, have brought their wealth for three hundred years to the great city at the point where the two arms cross. Looking northward, southward, eastward and westward from this mount the spectator may see a level plain, bursting with fertility (#103), carpeted with well-tilled fields, and settled by a thrifty, contented people. Montreal devours their milk, cream, vegetables, fruits, meats, in fact everything their half-million farms will produce, and in return sends them the textiles, clothing, boots and shoes, rubber, and a thousand other articles, which they require and which are made in the thousands of factories scattered through the central city as well as through *Lachine, Outrement, East Montreal, Maisonneuve,* and *Verdun,* which Montreal has gradually surrounded.

Montreal is the headquarters of the two great railways, the Canadian Pacific and the Canadian National, and their main lines feed the city with traffic from both east and west. It is the chief financial centre for Canada, many banks and insurance companies having their head offices there. *McGill,* the great English speaking University, and the *University of Montreal,* for the French, are both located in this city. The city is connected with the south shore of the St. Lawrence by three splendid bridges; and a tunnel, drilled right through Mount Royal, gives the Canadian National Railway access to the north side of the island.

This notable city is decorated with many splendid monuments, enriched by some of the most beautiful churches on the continent, such as St. Peter's and Notre Dame, and has interesting narrow old streets, bordered by buildings hoary with age. It has also fine wide boulevards, and splendid modern skyscrapers.

Quebec. *Quebec* city, the capital of the province, sits at the key position of Canada, where the broad mouth of the river St. Lawrence first narrows to a throat less than half a mile wide (#115). At first it was the centre of commerce, and long had its harbour crowded with ships laden with square timber and lumber. In due course the wooden ship, of which Quebec built hundreds, was replaced by the much larger iron and steel ship. Then with the deepening of the channel to Montreal, Quebec was passed by and ships pushed as far up stream as possible with their loads. Since then Quebec has become well supplied with railways; and her harbour has been expanded and

#115

AERIAL VIEW OF OLD QUEBEC

improved. A large grain elevator, well equipped warehouses, and the longest drydock in the world has been completed; so that her harbour has again become a hive of business. Since ships have increased very greatly in size, the largest, afraid of the channel above Quebec, make her harbour their stopping-place. With electric cables bringing power from *Shawinigan* on the west and *Montmorency Falls* on the east, she has turned her hand to manufacturing on a large scale. No less than forty shoe factories and tanneries, and fourteen fur shops turn out their valuable products.

To the tourist, Quebec is the most interesting spot in America. It is like a bit of old France of the Middle Ages dropped down on the beautiful bank of the St. Lawrence, where it is joined by the St. Charles (#115). The Lower Town, between the frowning, almost vertical, rock of Cape Diamond and the St. Charles River on the one side and the St. Lawrence on the other, is a charming disorder of narrow, old crooked streets leading nowhere. They are lined by grim, crumbling, old thick-walled stone buildings. The Upper Town is perched several hundred feet above on a rocky cliff that rises above the St. Lawrence. Here are found the most beautiful buildings, including the Basilica and the Parliament Buildings. Here are straighter, wider streets, and more modern homes. On the highest point stands the stone citadel, studded with great guns, now rusting away. A great wall with three gates surrounds the old part of the city. Farther west are the Plains of Abraham.

Quebec, the strongest fortress on the continent, has experienced attack after attack. Sir David Kirke took it for the British in 1629. but it was returned to France four years later. Sir William Phipps, with a New England fleet, beat against it in vain in 1690. Then General Wolfe finally captured it in 1759 for the British. It met one more attack during the war of American Independence when Generals Montgomery and Arnold marched up the Chaudiere River in winter, endeavouring to make Canada a fourteenth state. But the British and French beat them back from the cliff; Montgomery fell; the Americans retreated; and Quebec settled down to 150 years of peace, during which the walls of the fortress have crumbled, and the great cannons rusted.

Three Rivers (Trois Rivieres), half way between Quebec and Montreal, where the river St. Maurice empties into the St. Lawrence, is even older than Montreal. The St. Maurice, with its unrivalled electrical energy, has poured the life blood of industry into this city, so that during the last twenty years it has grown like a mushroom. It is the very heart of the greatest output of pulp and paper of any district in the world. Sawmills, a great cotton factory, and mills for making thread, gloves, shoes, and coffins add to its ever increasing prosperity (#113). Its two miles of docks and harbour, with fifty feet of water, are the greatest help in the export of many mill products. The export of grain, which Montreal so long looked on as her special business, is branching out. *Sorel* has already had for some time a great elevator where grain is transferred from lake steamers to ocean ships, and the alert business men of Three Rivers have recently followed Sorel's example and are obtaining a good share of this profitable trade.

Sherbrooke, the Queen of the Eastern Townships, is in the centre of a prosperous farming region, and has a network of railways that reach out in every direction. It is on the St. Francis River, from which, as well as from Shawinigan, it draws cheap electrical power to run a greater variety of industries than is found in any other Quebec town except Montreal. Cotton, woollens, silk gloves, hosiery, jewellery, and carriages are a few of a very long list (#113). There is situated Bishop's College, an Anglican institution.

CHAPTER XIV

PROVINCE OF ONTARIO

Another distinguished province. Ontario holds an enviable position. Though Quebec is half as large again (#101) and was settled nearly two hundred years earlier, Ontario's population is greater. She stands easily first among the provinces in total production, in products of the farm, the mine, the factory, and in the use of electrical power. Quebec surpasses her in the output of the forest, but she surpasses British Columbia, the lumbering province. Her agricultural output is almost equal to that of the three Prairie Provinces combined. She also leads in education, and has a smaller proportion of people who can neither read nor write than any other province.

Central position. It will be worth while to inquire what geographical advantages have given Ontario this remarkable leadership. She lies near the centre of the Dominion, with Quebec on the east, Manitoba on the west, the salt water of Hudson Bay and James Bay on the north, and the beneficial embrace of the fresh-water Great Lakes on the south. She extends much farther south than any other part of the Dominion. *Point Pelee* is as far south as the northern part of the state of California, and Toronto is in about the latitude of such genial winter resorts as *Nice* and *Monaco* in the south of sunny France. Though near the heart of the continent, she has a salt-water coast of nearly 700 miles, and one seaport, *Moosonee,* which is connected with the rest of the province by railway, but her numerous ports on the Great Lakes are of vastly more importance.

The Canadian Shield. The surface of Ontario spreads into three of the chief divisions of Canada; the whole northern part, including nine-tenths of the province, is in the Canadian Shield, whose origin and surface features will be later described (p. 341). This division sends a spur, called the *Frontenac axis* (#116), south to the *Adirondack Mountains* of New York. It crosses the St. Lawrence at the *Thousand Islands,* whose hard, red granite rocks have so well resisted the erosion of the St. Lawrence that they still stick up as a cluster of emeralds, bordered by garnet, and set in the sapphire blue of the majestic river.

St. Lawrence Lowlands. The second division, called the St. Lawrence Lowlands (#116), consists of the rich garden belt of Ontario and Quebec, bordering both sides of the St. Lawrence River. It is

ONTARIO
PHYSICAL FEATURES, EXPLORATION ROUTES, INDIANS, VEGETATION.
#116
SEVERN R.
TROUT L.
TROUT R.
WINISK R.
ENGLISH R.
LAC SEUL
LAKE OF THE WOODS
RAINY RIVER
FT. ST. PIERRE
ATTAWAPISKAT R.
ALBANY R.
L. NIPIGON
ALBANY R.
FT. KAMINISTIQUIA
LAKE SUPERIOR
MOOSE R.
MATTAGAMI R.
MISSINAIBI R.
ABITIBI RIVER
KAPUSKASING R.
LAKE ABITIBI
FT. SAULT STE. MARIE
LAKE HURON
L. NIPISSING
L. SIMCOE
THAMES R.
GRAND R.
FT. ROUILLE
TRENT R.
OTTAWA R.
LAKE ERIE
L. ONTARIO
FT. FRONTENAC
ST. LAWRENCE RIVER
PERCH AND PICKEREL
LAKE TROUT
WHITEFISH
STURGEON
PIKE
HERRING
WHITE WHALES
SEALS
OJIBWAY, CHIPPEWA, OR SAULTEAUX INDIANS
WOOD CREES
OTTAWA INDIANS
BEAVER INDIANS
HURON INDIANS
TOBACCO INDIANS
IROQUOIS INDIANS
NEUTRAL INDIANS
FUR BEARING ANIMALS
HERE HENRY HUDSON AND HIS MEN SPENT THE WINTER: 1610-1611
HUDSON'S VOYAGE
JAMES' VOYAGE, 1631-1632
DOLLIER AND GALINÉE 1669-70
HUDSON BAY LOWLANDS
CLAY DEPOSITS FROM ANCIENT GLACIAL LA
NUMBER OF FEET ABOVE SEA LEVEL
ST. LAWRENCE LOWLANDS.
CANADIAN SHIELD
SEA LEVEL
BORDERS OF THE FRONTENAC AXIS
THE NIAGAR ESCARPM
NORTHERN TREE LIMIT
NORTH OF THIS LINE THE TREES THIN OUT AND BECOME STUNTED
NORTHERN LIMIT OF SOUTHERN FOREST
N. LIMIT OF IMPORTANT HARDWOOD
RADISSON'S AND GROSEILLIERS' EXPEDITION IN 1660(?)
MARQUETTE'S AND JOLIET'S RO TO THE MISSISSIPPI R., 1673
LA SALLE, 1682
BRULÉ, TO LAKE SUPERIOR IN 1622
CHAMPLAIN - 1615
NICOLET IN 1634

hemmed in between the Canadian Shield on the north and the Appalachian Highlands on the east and south. The part of this region in Ontario is split into a smaller eastern and a larger western section by the intrusion of the Frontenac axis southward (#116). The St. Lawrence Lowland supports the chief part of the population and produces the greater part of the wealth in both Ontario and Quebec. It is a chequer-board of well cultivated fields, containing beautiful orchards, and gay gardens; grows luscious pastures dotted with sleek stock; and is closely marked from end to end with substantial farm houses and bulging barns. Its surface is thickly meshed with hard roads and neat railroads, whose intersections are knotted with quiet villages, and hustling towns and cities, which hum with factories of every description.

Hudson Bay Plain. The third region, called *Hudson Bay Plain* (#116), borders the south and west coast of James and Hudson bays. It is the smoothest, lowest area in Canada, sinking steadily about two feet per mile toward, and finally under, Hudson Bay. So flat is this plain that there is not enough slope to drain the water to the rivers, which run their slow and muddy courses directly to the bay, with scarcely a feeder from right or left. Every slight hollow has retained water and become a quaking *muskeg* of *sphagnum moss*, many feet thick. This flat area, like the St. Lawrence lowlands, is underlaid by horizontal limestones and sandstone. Though little is yet known about its resources, its farming possibilities are believed to be small on account of low temperatures and bad drainage. Its forest covering is thin, as trees refuse to prosper with their roots in water-soaked sphagnum moss; and its mineral wealth is largely unknown.

Lake Ojibway. The Canadian Shield was once crushed beneath ice-caps thousands of feet thick, which gradually pressed their way outward in every direction (p. 342). The ice scraped clean the thick mantle of soil, which formerly covered this area and deposited much of its valuable material in the St. Lawrence lowlands and farther south. But after all the ice south of the height of land had finally melted and a ridge of ice farther north still flung its mass across the courses of all the rivers emptying into Hudson and James bays, a great lake, called *Ojibway*, formed between the ice at the north and the height of land at the south. As in the case of Lake Agassiz, where the deep, black soils of the Red River prairie were laid down, so here a thick, flat layer of rich sediment was spread on top of the barren, granite rocks of the Canadian Shield. When the ice ridge, which dammed back the water melted, and the rivers took their present courses into Hudson and James bays, there was uncovered an immense area of flat, clay, fertile soil, much of it without a stone or pebble, which soon became covered with a dense forest. This is the origin of the famous *Clay Belt* (#116), which extends for many miles along both sides of the most northern line of the Canadian National Railway from north of Lake Nipigon far into Quebec on the east. A branch of Ojibway, called *Lake Barlow*, extended south as far as Lake Timiskaming, which is perhaps a remnant of it. Consequently, Ontario possesses the remarkable asset of a magnificent farming tract, as large as all southern Ontario, right in the midst of the bleak barrenness of the Canadian Shield. Lake Agassiz sent an eastern arm over part of the Canadian Shield into western Ontario, (#116) and the farm lands west of Rainy Lake and around Dryden are fertile sediment from this immense body of water.

Lake beaches. During the glacial period the ice-cap formed a dam across the outlet of the Great Lakes more than once, their size was much larger than at present, and much sediment was deposited beyond their present margins (#116). Wherever, in their vicinity, broad, perfectly flat areas of rich soil are found, in all probability these are deposits at the bottom of the enlarged lakes. The fertile farm lands around Sault Ste. Marie and Fort William are such accumulations, and many of the best farming townships of southern Ontario were the bottoms of these glacial lakes.

Surface moulded by ice. Generally the surface features of the St. Lawrence lowlands were moulded under the ice-cap, or at its edge as it retreated. The soil, dragged and pushed from the north under the ice, and the rock flour formed by stones frozen in the bottom of the ice grinding against the rocks beneath, became such a load that much of it was left under the ice when it melted, but some was sorted out into *sand beds* and *gravel pits* by the streams running out from the fast-melting rim of ice. These have given this region its irregular surface of ridges, domes, hollows, or its gently rolling character which is seen everywhere.

Niagara escarpment. There is one feature, older than the ice-cap, which was so bold that even the mountains of rasping ice slid over it without destroying it. It is the *Niagara escarpment* (#116), the most prominent surface feature of the south-west peninsula, which divides the part of the St. Lawrence plain west of the Frontenac axis into two platforms, the western being the higher. #116 shows the irregular course of the dividing escarpment, which reminds one of the cliffs separating the prairie steppes (p. 343). At Niagara River it forms *Queenston Heights*, over 250 feet high; at Hamilton, where it forms the mountain, it frowns over the city 180 feet below; in Bruce Peninsula, of which it is the backbone, it is nearly 300 feet high as it stands like a sentinel at *Cabot head;* and on the north shore of *Manitoulin Island* it rises sheer out of the water for 500 feet, showing most beautiful water sculpture at its base. As the slope from the crest of the escarpment is towards Lakes Huron and Erie, few rivers tumble over its crest, except the Niagara, which has cut a gorge back seven miles from the crest at Queenston Heights.

Mining

Sudbury mines. The ice-cap that stripped the Canadian Shield of its thick mantle of precious soil, made it a desert for farming, but a paradise for the mineral prospector; and what was considered a barren, rocky waste has become the richest group of mines on the continent. It exposed the rich mineral veins everywhere. For many years its harsh, forbidding surface and dense covering of trees made its development difficult; but as soon as railways were built across its face, it began to give up its rich mineral wealth. When in 1883 the main line of the Canadian Pacific was being put through the wilderness where the thriving city of Sudbury now stands, a workman observed a reddish patch on the right of way. He dug down and struck copper. Soon prospectors scoured dense forests, waded to their hips through slimy swamps, and scrambled over steep rocks, looking for reddish patches; and they found them. The great nickel-copper mines were all discovered in a few years, and *Sudbury* became the centre of the world's nickel industry (#117). The uses of nickel are now so varied that it is found in almost every article of metal from the armour plate of battleships to five-cent pieces. So vital is it to implements of all kinds that if nickel was extracted from an automobile, or a man-of-war, the whole structure would crumble to pieces at the slightest shock. For many years the nickel mines have been worked day and night, furiously trying to satisfy the hungry demand of the war-lords. But even if the world were at peace forever, it is certain that the hunger for nickel in the factory, office, and home, would take long to satisfy. For many years most

MAP OF ONTARIO,
SHOWING THE PRINCIPAL MINES, PULP AND PAPER MILLS, WATER POWER DEVELOPMENTS, RAILWAYS, STEAMSHIP ROUTES, SETTLED AREAS, PARKS, FOREST RESERVES, AND SUMMER PLAYGROUNDS.
117
WINNIPEG
KENORA
DRYDEN
RED LAKE
SIOUX LOOKOUT
FORT FRANCES
PORT ARTHUR
FORT WILLIAM
SAULT STE. MARIE
SUDBURY-1884
COPPER CLIFF
NORTH BAY
NEW LISKEARD
HAILEYBURY
COBALT-1903
TIMMINS
COCHRANE
IROQUOIS FALLS
KIRKLAND LAKE, 1911
TOWNS AND CITIES OF NORTHERN ONTARIO
KENORA
OTHER ONTARIO TOWNS AND CITIES
STEAMSHIP ROUTES
PROVINCIAL PARKS
FOREST RESERVES
SUMMER PLAYGROUNDS
SETTLED AREAS (AGRICULTURE)
UNSETTLED AREAS
MINERALS
GOLD
RADIUM REFINERY
SILVER
SILVER AND COBALT
CHROMIUM
COPPER
NICKEL
SALT
NATURAL GAS
INDUSTRIES
PULP AND PAPER OR PAPER MILLS
WATER POWER DEVELOPMENTS
SMELTERS, OR REFINERIES
GRAIN ELEVATORS
RAILWAYS
C.P.R.
C.N.R.
T.&N.O.R.
ALGOMA CENTRAL
AMERICAN RAILWAYS

of the nickel came from *Creighton Mine;* but the *Frood,* discovered in 1884, after lying idle and unknown for nearly fifty years, has now blazed forth as a star of the first magnitude. Besides its great output of nickel, its copper production is the largest in Canada and has raised our country to rank as one of the chief copper producers. This wonderful ore also contains such quantities of *platinum,* that Canada now leads in the output of this beautiful metal, more valuable than gold. With platinum are important quantities of three other

#118

PROSPECTORS SEARCHING THE CANADIAN SHIELD FOR GOLD

What is the hammer for? What lies to the left of the hammer?

rare platinum-like metals, *iridium, rhodium,* and *palladium,* all of whose uses in alloys are increasing.

The nickel saucer. The most valuable feature of the nickel-copper industry at Sudbury is the immense reserves of the ore. Imagine a deep elliptical saucer, 36 miles long and 16 miles wide, embedded in rocks, and with its hard edges sticking up above the surface. This thick saucer is nickel-copper ore, and up to the present miners have only nibbled the edges of the saucer at a few places.

Gold, gold, gold. But nickel-copper is as nothing compared with the gold that the Canadian Shield is pouring forth at a dozen places, scattered over northern Ontario (#117). New valuable gold mines are springing up in widely scattered sections of this invaluable Canadian Shield. Shortly after the *Temiskaming and Northern Ontario*

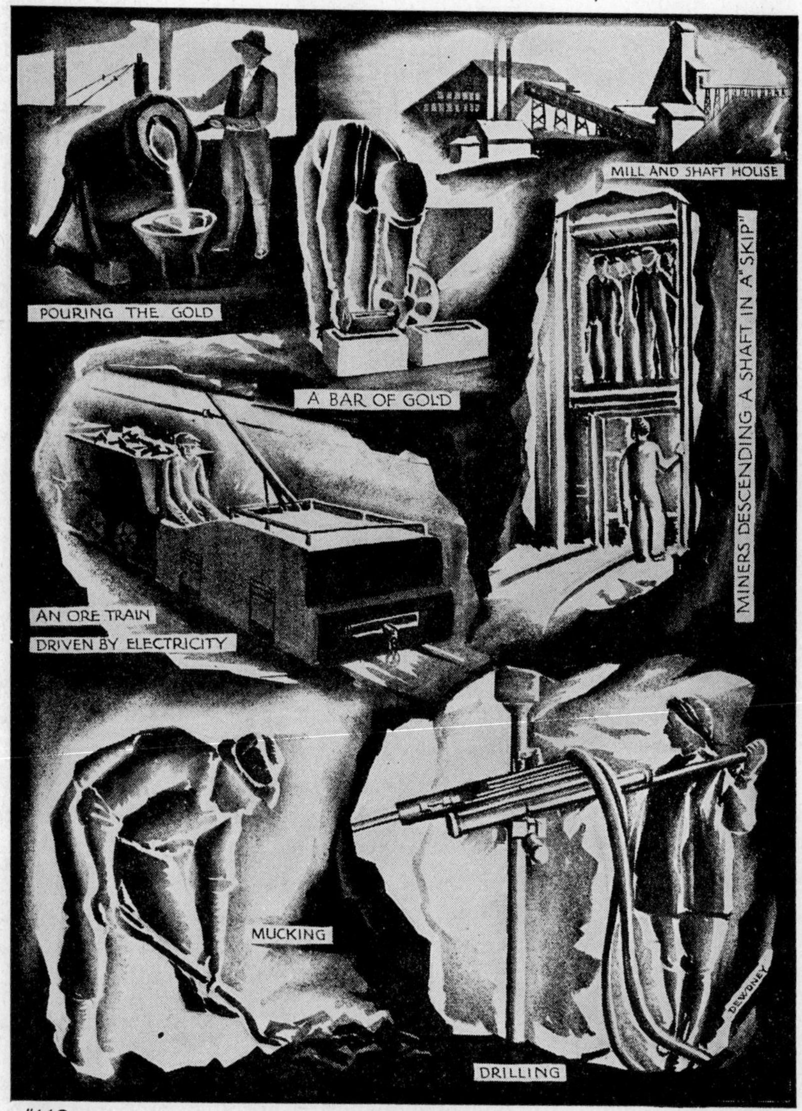

#119

EXTRACTING GOLD AT TIMMINS

Railway was completed, hardy prospectors spread out to right and left in their search for gold (#118), and soon two of the most important gold mining fields were discovered, *Porcupine* in 1909, and *Kirkland Lake* in 1911. *Porcupine* with *Hollinger, McIntyre,* and *Dome* mines led for many years, but Kirkland Lake is now forging ahead. New mines, which may soon rival these leaders, are producing gold north of Lake Superior and far in the north-west at Red Lake and Crow River (#117). Ontario's output of gold, which is steadily increasing, keeps Canada battling with Russia and the United States for second place as a producer.

Cobalt. When in August 1903, two timber rangers, McKinley and Darragh, nervously picked up pieces of shining metal from the shore of Long Lake (afterwards Cobalt Lake), excitedly tested it with their teeth, and found that they could dint it; their thoughts were more upon staking claims for a mine than on estimating the value of the timber. The dint of their teeth started one of the wildest stampedes in mining history, as thousands rushed north to stake claims in the new silver field. *Cobalt* soon became the most noted silver-mine in the world (#117). Some massive nuggets were almost pure silver; and veins, a foot wide, of almost pure metal, were exposed. In thirty years nearly four hundred million ounces of silver were extracted from this one field. Later, smaller fields at *Gowganda* and *South Lorraine* were worked; but they have gone the way of all mines; most are deserted, though in a few, miners are still picking the bones of mines that formerly were fat with meat. The town of Cobalt, which once was as busy as a beehive, filled with financiers, throbbing with hope and speculation, and crowded with industry, has now settled down to a quiet life, showing signs of the mildew of decay. Not only is silver extracted from Cobalt ore, but also cobalt and arsenic are extracted at the refinery at *Deloro* near Peterborough.

Chromium. Everybody is now familiar with the beautiful chromium finish that never tarnishes. This metal is also used as a valuable component of many alloys and has a great future. Ontario has now joined the company of chromium producers. A mine at *Obonga* (#117), west of Lake Nipigon, in 1935 raised nearly ten thousand dollars worth — the metal being extracted at Sault Ste. Marie.

Non-metals. Graphite, talc, feldspar, and mica are produced chiefly in the Frontenac axis (#116).

The St. Lawrence lowlands with their horizontal, layered rocks also produce many non-metallic minerals such as salt, petroleum, natural gas, and gypsum, the chief mining regions of which are marked in (#116). They are located mostly in south-west Ontario.

Drainage

Spangled with lakes. Providence has been lavish to Canada in the supply of that greatest of blessings, fresh-water. The Canadian Shield

#120

LOVELY LAKES OF CANADIAN SHIELD

A Forestry aeroplane on the look out for fires finishes its day's work.

is a close network of lovely lakes and interesting rivers (#120, #121). These lakes densely dot every part of this region in Ontario, as stars spangle the sky; and the thread-like rivers tie these lakes together into a pattern as complex as the shadow under a leafy tree. When the ice-gap gouged out the soil from every hollow, it left a stony framework as billowy as a storm-tossed sea. Then as the rain fell on this bare, uneven, rocky plain, the hollows gradually filled to make lakes. Each spilled over its lowest edge and swerved back and forth among rocks, now rushing swiftly, then tumbling down over a slope until it stumbled into another lake. Gradually at lower levels these became woven into a main stream, which, with its greater force and volume of water, flowed more directly to the sea.

Rivers of the plains. The rivers in St. Lawrence lowlands and Hudson Bay plains are very different. The layered rocks they flow over are flatter and softer than the granites of the Canadian Shield, and are covered with a deep mantle of soil through which the rivers have made decided channels. The lips of most of the lake-like

#121

INTERESTING RIVERS OF CANADIAN SHIELD

Tourists approaching the foot of steep rapids, past which they must portage.

hollows in the soft soil and rocks have been worn down and the lakes drained. Therefore in these regions the lakes are far fewer, the rivers more direct, with few rapids and waterfalls. The paths of the rivers flowing into James Bay and Hudson Bay show the difference: the upper part in the Canadian Shield is a maze, the lower part a steady, decided stream with no windings, few branches, and only occasional lake expansions.

Rivers of the north. The *Moose*, the *Albany*, and the *Severn* (#116) are great rivers, rising far back in the highest regions of the

Canadian Shield not far from the Great Lakes, and draining thousands of rocky lakes, each as charming as the emerald gem-like waters of Switzerland or Italy, but when they reach the Hudson Bay plains (#116), they flow forward steadily for hundreds of miles through the gloom of spruce forests with few rapids or waterfalls. For two hundred and fifty years they have been highways every spring for flotillas of canoes, directed by brawny arms of Indians and packed with furs, as they wended their journey over dangerous rapids, across the mirrored surfaces of sapphire lakes, and around waterfalls, to the trading-posts of the Hudson's Bay Company on the margin of the bay. They returned in the summer bearing winter supplies, now to battle against swift currents, tumbling rapids, and roaring cataracts. This is still their chief use; but the prospector can now be seen everywhere, his paddle replaced by the "kicker," or outboard motor, scanning every rock exposure for the glint of gold.

Playground of Ontario. The lakes and streams of the more accessible parts of the Canadian Shield, in *Muskoka*, *Haliburton*, *Algonquin Park*, and the *Rideau* region, are lined with summer cottages (#117). This district, with its waters well stocked with *bass*, *trout*, and *maskinonge*, with its abundant supplies of *deer*, *moose*, *waterfowl*, *grouse* and other game birds is a paradise for sportmen. With its long stretches of stream for motor-boats, its threads of water in such endless and entangling mazes one can, with a canoe, commence a trip at any point and emerge almost at any destination dedesired. With its safe bathing in clean, warm water, its network of trails through thousands of square miles of dense forests of stately spruce and pine, and shining, rustling oak, birch, and aspen, its sunshiny summer days and quiet, cool nights — this district is becoming a most delightful summer playground for the whole continent.

The Grand, Thames, and Trent rivers. The rivers of the St. Lawrence lowlands, all emptying into the Ottawa, the St. Lawrence, and the Great Lakes, are not so full of whims as those of the Canadian Shield; they run their steady course with few lakes, rapids, or waterfalls. They played an important role in the early history of Ontario, as they formed the safest and easiest routes for settlers in a country without roads and clothed with dense forest. The *Grand*, which rises on the western edge of the Niagara escarpment near *Dundalk* (1,705 feet high), the highest point of South Ontario (#116), has cut back a grand gorge to *Elora*, below which it emerges to flow tamely to Lake Erie. The *Thames* rises not so far back in the highland and runs its monotonous course, through such important cities as Woodstock, Stratford, London, and Chatham, to pour its dull and sluggish waters into the muddy shallows of Lake St. Clair. The *Trent* is of a different breed; it is just on the border of the Canadian Shield, is the mere remnant of a larger river that once drained the giant Lake Algonquin, and partakes of both types. Its lower course is normal and navigable; then it takes a series of zigzags, some of which show their former greatness by widening out into lake expansions of various shapes. Its northern tributaries open up a world of deep lakes, snuggled among the rocky domes of the Canadian Shield.

Lake girdle of Canadian Shield. While the whole surface of the Canadian Shield is spangled with beautiful little lakes, at the outer margin, where it meets both the St. Lawrence and the Great Plain, the lakes enlarge and blend to form a girdle of the mightiest masses of fresh-water on earth. The *Lake of the Woods* is the pivot of this marvellous chain which stretches eastward through a string of lesser lakes, through the *Great Lakes*, and the *St. Lawrence River* to Labrador, and which sweeps northward through *Winnipeg*, *Athabaska*, *Great Slave*, *Great Bear* and a thousand smaller lakes to the Arctic.

The Great Lakes. What the Yangtze is to China and the Nile to Egypt, the St. Lawrence Basin (with the Great Lakes) is to Ontario. It brings the sea to the door-step of a province near the middle of a continent. In no other part of the world are there such magnificent masses of deep, blue, fresh-water. The following table gives the names and some data regarding these mighty lakes:

	SUPERIOR	MICHIGAN	HURON	ST. CLAIR	ERIE	ONTARIO
Length in miles	383	320	247	26	241	180
Width in miles	160	118	101	24	57	53
Greatest depth in feet	1,180	870	750	23	210	738
Elevation above sea level	602	581	581	576	573	246
Area (sq. mile)	31,810	22,400	23,010	460	9,940	7,540

A procession of giant ships. The leading part this chain of lakes played in exploration and early settlement will soon be described (p. 300), but to-day it is as a means of transport that its value is overwhelming (#122). No other inland seas have such fleets of stately passenger ships, whose furnishings equal those of the most stylish

#122

SHIPPING ON GREAT LAKES

1. Great Lakes Freighter. 2. Ready to load grain at Port Arthur elevators. 3. A passenger steamer. 4. Loading coal at Lake Erie ports. 5. Coming through a lock on Canal. 6. A tug. 7. Unloading wheat from grain boat at Midland. 8. A Passenger Steamer.

ocean liners; but the vast throng of immense freighters is its chief boast. In normal times more freight passes through the canals at Sault Ste. Marie than through the Suez or the Panama canal. One may stand on the bank of the St. Clair or the Detroit river and, night or day, never be out of sight of one of these fresh-water giants, and often one can see a dozen at once, and a procession of two dozen may pass in a single hour.

The St. Lawrence waterway. The Great Lakes are deep throughout their courses, and it is only in the connecting rivers and in Lake St. Clair that channels have had to be improved. By deepening these with blasting and dredging, and overcoming rapids and waterfalls by canals, a channel, nearly 20 feet deep, has been completed from *Duluth*, at the head of Lake Superior, to *Prescott*, just above the first rapids in the St. Lawrence. With trifling cost the depth could be increased to 25 feet, which would be sufficient for ocean freighters; and the locks in the canals are already capable of passing boats nearly 1000 feet long. From Prescott to Montreal, about 90 miles, canals are only 14 feet deep and the locks will pass boats not more than 252 feet long. From Montreal, a safe channel, 30 feet deep, leads to the ocean. Thus in the 1900 miles of the magnificent course of the St. Lawrence from Belle Isle to Duluth, at the heart of the continent, only ninety miles, between Prescott and Montreal, hinders the path of the ocean ships. By the improvement of this short channel, grain boats from Port Arthur and Fort William could sail directly to Liverpool, and Nova Scotia coal could be transferred direct to all parts of Ontario. The improving of this ninety miles is what is called the *deepening of the St. Lawrence waterway*, and its completion is now being arranged between Canada and the United States.

A stream of iron ore. The chief freight down this waterway is iron ore and grain. Long, ugly freighters, which are little more than big steel tanks shaped like boats, (#122), drop in a steady procession under the docks at *Duluth* and *Superior;* chutes are opened; and in an hour a roaring flood of dusty, red iron ore weighs the boat almost to the water's edge. In a few minutes more they are on the journey south to be unloaded at *Hamilton*, *Toledo*, *Cleveland*, *Erie*, *Astabula*, *Buffalo*, *Sault Ste. Marie*, or other lake ports. As soon as each reaches the wharf, a pair of great metal clam-shells bury themselves in the ore, close their terrible jaws, and lift up 10 tons in a single bite. In a few hours the ship is steaming out of the harbour laden with coal for Lake Superior.

The flood of grain. *Port Arthur* and *Fort William* for Canada, *Duluth, Chicago,* and *Milwaukee* for the United States, are the feeders for the grain trade on the lakes. A tanker 600 feet long (#122,1) moves alongside an elevator, two dozen metal spouts at once begin to deliver streams of wheat, oats, corn, or flax to the hold of the boat, and she is off in an hour or two to southern ports. Most ships go to the big elevators at *Port Colborne* and *Buffalo,* but more and more of them are now going right through to *Prescott* and *Kingston.* Elevators at these terminal points are not so much for storing grain as for transferring it from these long tankers to small freight boats, less than 250 feet long, which can go through the lower St. Lawrence canals. There is a continuous procession of these smaller ships, which, night and day, throughout the autumn, keep the lock-masters busy. The grain is again unloaded at *Montreal* or *Sorel,* finally to be transferred to ocean steamers for *Liverpool, London,* and ports on the continent of Europe. Much grain is also shipped to Georgian Bay points, *Midland, Port McNicoll, Tiffin, Collingwood,* and *Owen Sound,* and to *Goderich* on Lake Huron. This wheat is either milled into flour in these towns (#133, #134), shipped to *Halifax, St. John,* or *Montreal* for export; or distributed for local use in eastern Canada.

The chief freight up the Great Lakes is Pennsylvania coal, which leaves ports on Lake Erie, already mentioned, for all Canadian and United States ports on the Upper lakes. *Fort William* is the greatest coal handling centre in Canada.

Besides these chief commodities, lumber, pulp and paper, stone, cement, and a thousand other commodities, are transferred anywhere from a mile to a thousand miles up and down the lakes.

Fisheries

Fish of the Great Lakes. The Great Lakes yield a bountiful harvest of fine, fresh fish, which are caught throughout the year, except during winter months (#116); even then some fishing is done through holes in the ice. Every part of the Great Lakes contributes its share, and #116 shows the chief kinds obtained in each district. Fish are caught chiefly in gill-nets, but trap, or pond, nets are used, especially on Lake Erie. The fish are usually packed in ice and shipped fresh to the markets of Canadian and United States cities. *Whitefish* and *trout* are of most value; the humble *perch,* dear to the heart of every small-boy fisherman, is rapidly becoming a market fish of value, no less than one quarter of a million dollars worth being marketed in 1935. Lake Erie supplies the largest quantities of these, which some outstanding experts consider the most tasty of all fresh-water fish. They now surpass in value such important fish as, pickerel, blue pickerel, herring, and tullibee.

Fish of Hudson Bay. Ontario has the most direct route by the Temiskaming and Northern Ontario railway from Hudson and James bays to the dense populations of fish eaters in Canada and the United States. The fish supplies of this immense inland sea, fifteen times as large as Lake Superior, have never even been tested, let alone developed. Where such immense eaters as whales, dolphins, (white whales) (#116), and walruses find sufficient food, and where seals, which are fish eaters, are plentiful, there must be abundance of fish. Whitefish, trout, capelin, and lake herring are known to swim in its cold waters, and only exploration is needed to find in what abundance.

Hydro-electric Power

The water-wheel. Wind blowing against the vanes turns a windmill, which can run a pump or other machine. If it is properly connected to an electrical machine, it produces a current of electricity. A stream of water, because it is more powerful, turns a water-wheel much more vigorously. Water, falling on the vanes or paddles of a suitable wheel, causes it to turn so vigorously that its shaft, when connected properly, will run very powerful machines. The farther water falls, the greater its speed, and the more rapidly and vigorously it turns the water-wheel.

The town used to come to the waterfall. While swift currents, rapids, and cataracts have injured streams of Ontario for navigation, they have doubly repaid the province by giving cheap power to turn the giant machinery of our factories; they help us to laugh at the lack of coal. At first the factory had to be built alongside the wheel, which was turned by falling water, as the shaft of one had to connect directly with the shaft of the other. Therefore in early settlements the mill and town had to come to the waterfall; and hundreds of villages and towns of Ontario crystallized around waterfalls and rapids, which ran grist or sawmills. But waterfalls usually are placed in very inconvenient places, quite unsuitable for towns, where the surface is rocky, the soil scarce and stingy, transportation bad, or raw materials for manufacture far away.

The waterfall now comes to the town. To-day we make the water power come to the mill. The water-wheel now turns an electric generator whose current can be sent through wires to wherever it is needed (#123, #124). This electricity is the life-blood of industry in Ontario. The heart of the system is the Central Generating Station at Niagara Falls, on the Abitibi, or at a hundred other localities, some far away and difficult to reach; the electric cables are the main arteries which lead to every town and village in Ontario and stretch

#123 HYDRO-ELECTRIC GENERATORS AT QUEENSTON PLANT, NIAGARA

along rural highways as well; and the wires leading into factories and houses are the capillaries. These run the mill and factory machines, light our houses, wash our clothes, cook our food, toast our bread, curl our hair, work our radios and gramophones. On the farm, electrical machines now milk the cows, separate the cream, churn the butter, chop the fodder, as well as light the house, barn, and stables.

The right arm of the mine and the forest. Hydro-electric power has been the strong right arm of two great industries in Ontario:

mining, and pulp and paper manufacture. Prospectors have the bad habit of finding valuable ore away on the rough and distant top of the Canadian Shield, where there are neither roads nor navigable streams. At a mine, power is required to raise the rock from the shaft, grind the ore, pump immense quantities of water, and for a thousand and one other operations. If coal had to be carried to these out-of-the-way mines, the expense would turn profit into loss, and the mine would lie dead. But rapids and waterfalls are to the

#124
DAM AND POWER-HOUSE

Canadian Shield what waves are to the sea. As soon as a valuable body of ore is discovered, a nearby cataract is sought, an electrical generator built, a cable line laid, and soon every machine, powered by electricity, throbs ceaselessly at its work. The southern edge of the forests of Ontario is moving farther and farther back on the Canadian Shield. Because logs are more bulky to handle than neat rolls of pulp or paper, it is becoming more profitable to send mills back to the logs than to bring logs down to the mills; consequently pulp and paper mills are being pushed farther and farther back wherever convenient hydro-electric power can be obtained (#117).

Exploration

Brulé discovers Ontario. *Etienne Brulé,* the French servant of Champlain, was the first white man to explore the land and water routes of Ontario. Champlain in 1610 loaned him to the Hurons in exchange for one of their young men, whom he took back to France for the winter. The bold young Brulé plunged fearlessly into the western forest wilderness to live among cruel, treacherous Huron Indians, of whom he knew nothing. He did not understand a word they spoke, knew nothing of their mode of living, and was quite in the dark as to how he would be treated by these uncertain savages, whose friendship might quickly turn to hate; and who with all the instincts of a cat, delighted to torture a victim with the most extreme and lingering cruelty. He probably lived in one of their villages on the Ottawa River or south of Georgian Bay, and for five years wandered through the wilderness with his dusky companions, as they hunted in the forest for fur and food. He was probably the first white man to explore the upper Ottawa, Lake Nipissing, Georgian Bay, Lake Simcoe, Lake Ontario, the Niagara River, and Lakes Huron and Superior (#116). Almost every year Brulé would travel by canoe down the Ottawa and St. Lawrence Rivers. His companions on those trips would likely be Indians who wished to trade with the white men. What stories would be unfolded and poured into the eager ears of Champlain on those occasions! They would be stories telling of the boundless forests, filled with game and furs; the rivers and lakes, rich with fish, and, above all, of the great boundless fresh-water lakes, larger than seas.

An alarming rumour and an alluring lie. In the meantime alarming news worried the French, and Champlain in particular; vague stories reached Paris that *Henry Hudson,* an Englishman, by a northern route had sailed into that Western Sea, or Northern Sea, which for two hundred years fascinated every adventurer who turned his course toward America (#116). Further, to spur the great *Samuel de Champlain,* founder of Quebec, to push westward to find this mythical inland sea by which he would reach China, a young Frenchman, *Nicholas Vignau,* whom he had sent to spend a year with the Algonquins on the Ottawa River, told how he had travelled north and in the short period of seventeen days had reached a great sea. Though this young man proved to be a more expert story-teller than discoverer of northern seas, his false story fitted in so well with French rumours of Henry Hudson, that Champlain in the spring of 1613 determined to travel west along the Ottawa, in spite of the protests of the Algonquins, who wished to keep the French east of Montreal, and themselves to bring the western Indians down to trade.

Champlain battles against the Ottawa. In the Spring of 1613 Samuel de Champlain, Governor of New France, with three other Frenchmen, one Indian guide, and the uneasy Vignau, launched forth from Montreal up the untrodden path of the Ottawa in two canoes toward that will-o'-the-wisp Northern Sea (#116). The Ottawa, as it forces its way across the rugged rockiness of the Canadian Shield, is a succession of swift currents, foaming rapids, and dangerous waterfalls. Champlain had to surmount these barriers, one by one, with ceaseless toil; and it is no wonder that he writes that "it makes one sweat." These adventurers made long portages over rough trails, with heavy packs on their backs; often they had to plunge into cold waters and, by ropes, drag their canoes among rasping rocks. Swarms of greedy mosquitoes and blood-sucking flies ceased not their teasing attacks, day or night. The anxiety of mind was equal to the pain and weariness of body, for they did not know when a hidden band of stealthy Iroquois savages, with blood-curdling war-whoop, would pounce upon their helplessly small band. But the determined Champlain pushed forward, ever allured by the Northern Sea. As they passed the first roaring waterfall, tumbling over a ledge or rock to swirl in an eddy below, Champlain felt how apt was the Indian name of "boiling pot," which name, in French, "*Chaudiere*," it yet retains. He must have noticed that grand picture where two splendid rivers, the *Gatineau* from the north, and the *Rideau* from the south meet the main stream, sternly decorated with the great fall above; but little did he dream that he gazed on the future capital of the great country, which he loved and which he did so much to found. At last the rushing currents, fierce rapids, and cruel rocks of the Ottawa were too much even for those dogged explorers; they hid their canoes and stumbled along through dense forest and over fallen trees, until they suddenly came to an Algonquin village, near the present Muskrat Lake, with corn growing in a clearing.

Vignau breaks down. They pushed on a little farther until they came back to the Ottawa River at *Allumette Lake*, near the present town of *Pembroke*. The mind of Vignau was troubled; it was here he had quietly spent the winter, when he was supposed to be pushing forward to the great Northern Sea, which he had so vividly described. The Indians of the village all knew him well. When Champlain urged him to lead them to the northern sea, the extravagant story-teller broke down and admitted there was no such body of water. Champlain's vision of a northern sea disappeared like a pricked bubble; the urge to go forward was cruelly destroyed, and he was soon back in Montreal. But Vignau's lie was not wholly bad, since it had stimulated the first exploration of the province of Ontario.

Trip to Georgian Bay. The *Hurcn tribes*, though kindred to the Five Nations, often called the *Iroquois*, were their deadly enemies (#116). They lived in a number of walled villages south of Georgian Bay, where they cultivated hemp, corn, squash, and sunflowers. A band of Hurons appeared at Quebec in 1615 to beg Champlain to assist them in an attack on the Iroquois in New York state. He foolishly consented and started for the Huron country on the safest route, because farthest from the Iroquois. He again fought the rapids and cataracts of the Ottawa, turned west up the *Mattawa River*, and was soon skimming, with the daring young Brulé and a number of French and Indians, over the beautiful waters of Lake Nipissing (#116). He shot down the swift currents of *French River*, and at last his startled eye gazed at a sea so wide that its farthest edge was beyond the horizon. Was it the Western Sea he sought so eagerly? Alas, its water was fresh! He was on the shore of *Georgian Bay*. Then in the middle of summer they paddled south through that fairyland of islands, until at last they reached the rich forest land between the Bay and Lake Simcoe. A missionary, *Father Le Caron*, had preceded them by a few days on the same trip.

Champlain crosses Ontario. But Champlain's chief business was war on the Iroquois; soon he and the Huron warriors set out for their enemies in New York, which led this French explorer over an entirely new field of discovery in Ontario.

No wonder, as Champlain paddled through the placid waters of the Kawartha Lakes (#116), like a string of pearls in a setting of green, that he never ceased admiring the quiet beauty of the glorious landscape. The canoes crossed Lake Simcoe, followed *Balsam* and *Cameron Lakes*, traversed the still waters of *Sturgeon*, *Pigeon*, and *Buckhorn Lakes*, followed up the *Otonabee River* to *Rice Lake*, and along the *Trent* till they finally paddled eastward on the peaceful waters of the *Bay of Quinte;* at last Champlain gazed on the broad waters of the second of the Great Lakes, as he skirted the shore of *Lake Ontario*. He was soon down near *Syracuse*, fighting an unsuccessful battle with well trained and fearless Iroquois. Champlain returned to a place near Kingston to recover from a wound. Here along *Cataraqui Creek* and *Loughborough Lake* he spent the autumn pleasantly hunting and exploring. Game was abundant; the party killed 120 deer, landed great numbers of trout and pike of immense size, and shot wild duck and other wild fowl to their hearts' content. Champlain must for the first time have felt the wild pleasure of living for weeks among the blaze of colour, as the autumn leaves took on their startling splendour of golden yellows, brilliant oranges, livid purples, and flaming crimsons. When the lakes froze, he, with his Huron companions, wandered back overland to Huronia, the land of the Hurons. Here he and Father Le Caron were the first white men, with the exception of Brulé, to spend a winter on Georgian Bay. Champlain not only visited all their villages, lived in their unclean houses, ate their loathsome food, endured continual noise, foul talk, and still fouler conduct; but he explored much of the counties of Grey and Bruce (#116), where a kindred tribe, called the Tobacco Indians, lived (#116).

The tragic life and death of Brulé. Because Champlain was the chief man in New France, his explorations have been emphasized by writers, but those of his servant, Brulé, though they shine out with daring and danger, are often scarcely mentioned. This daring young man had already spent five years among these Indian tribes and was more learned in their language, more skilful in their ways, and had a wider knowledge of their country than any other white man. When Champlain started east to attack the Iroquois, Brulé set out to arouse another enemy of the Iroquois in the Susquehanna valley south of Lake Ontario to join the coming attack with Champlain. His route bristled with danger, as it passed directly through enemy country, and he had only a dozen fighting men. He opened a new route to Lake Ontario (#116), he paddled south on Lake Simcoe, entered the *Holland River*, portaged from its upper waters to those of the *Humber River*, whose beautiful valley he followed past the site of Toronto to Lake Ontario. This later became known as the *Humber route* and was long considered one of the best between the upper lakes and St. Lawrence River (#131). The rest of his route is uncertain. He may have skirted the north shore of Lake Ontario to Burlington Bay, crossed to the Grand River, followed the Thames to its mouth, then plunging into the dense woods south of Lake Erie, followed the Alleghany, and finally reached the Susquehanna. What seems more likely, he may have followed Lake Ontario to the Niagara River and forced his way through the dense woods of New York, for he had to avoid the Iroquois-haunted trails. When he could not rouse the Indians in time to assist Champlain in the attack, he decided to spend the winter with the tribe. He followed the Susquehanna River (#73) to the sea through unfriendly tribes of warlike Indians and dared to return across the dangerous country of the savage Iroquois. He was captured, pegged to the ground and tortured by having his nails torn off and his whiskers pulled out. A furious thunderstorm saved his life. The superstitious Iroquois, in great fear, bound up his wounds, nursed him back to health, and, when he was able, sent him by Lake Ontario and the Humber route to the Huron country. He was at Three Rivers in 1618 to report to Champlain. If he had wished to cross the ocean to Paris he could have told a tale so thrilling that he would have become the idol of the great city. But his heart was in the wilderness of Ontario, which he did so much to explore. He was soon up in the Huron country again, where he remained for some years. How far he travelled nobody knows, but from a map, published by Champlain a few years later, many details (which only Brulé could have supplied), show that this marvellous man must have travelled south-western Ontario more thoroughly than most of the people living in it to-day. He brought back from one trip to Sault Ste. Marie and beyond (#116), a big slab of copper, which must have come from Lake Superior, which he was the first white man to see. Later, Champlain was greatly shocked to find him with the British under Kirke at the capture of Quebec in 1629 (p. 167). He was soon again with the Hurons, whom he had lived among for twenty years. His death was terrible. In a dispute with his treacherous friends, their cruelty conquered every kind instinct, they clubbed him to death and cooked and ate his body. Though no great monument is erected to this strange man, no man was more fearless, none suffered more brutal treatment, none met a more terrible death, and no explorer trod more Ontario trails and paddled along more Ontario streams and lakes for the first time than this humble and obscure "runner of the woods." A modest tablet at Sault Ste. Marie, which he first discovered, is a suitable monument to this modest adventurer.

Exploration from Huronia. Soon a large band of fearless Jesuit missionaries were zealously working among the friendly Hurons on Georgian Bay, and the trip to Quebec by the Ottawa was frequently made by both French and Indians, but never without danger from an Iroquois attack. Undoubtedly the missionaries made many trips through western Ontario to visit the Tobacco Indians in Bruce and the Neutral Indians along Lake Erie (#116). One of the Jesuits, *Father de la Roche d'Aillon* spent five months in 1626 exploring the region to the south of Huronia and probably reached as far as the Grand River. *Jean Nicolet,* a French adventurer, who had been drawn from civilized life by the lure of Huronia, the charm of the forest, and the excitement of the hunt, was sent as a peace envoy to an Indian tribe to the west of Lake Michigan; perhaps he reached the Mississippi (#116). Soon Huronia was utterly destroyed by the Iroquois, the villages were burned, the missionaries were tortured and murdered, and it ceased forever to be a centre for exploration. The frightened remnant of Hurons gathered at *Sault Ste. Marie* and at *Michilimackinac,* with *Father Marquette* as their missionary; the Ottawas fled to Manitoulin Island.

Master among the Indians. *Pierre Esprit Radisson* was the most daring and the most interesting and forceful of all the young French-Canadians who preferred the ever present dangers and gnawing hardships of life among Indians in the wilds of the forest, to the tame life on farms or the more settled life in the towns of Quebec. While a mere boy of seventeen years, he was pounced on outside Three Rivers by the panther-like Iroquois, tortured in their country, escaped after treacherously tomahawking his captors, and was soon back with the French attacking their pitiless enemy. He was the first white man to explore the west, north-west, and north, of Ontario, and his best work was done before he was thirty years old. He understood the Indians, had masterful control over tribes on his first meeting with them, and, what delighted the merchants most of all was his ability to bring down to Three Rivers and Quebec hundreds of canoes, heavy with bales of furs and paddled by dusky Indians from every part of the far west and north-west.

Radisson explores north-west Ontario. He was usually accompanied by his brother-in-law, *Chowart des Groseilliers,* whose name was so hard to pronounce that the British usually called him Mr. Gooseberry. In 1658 these young men passed up the Ottawa (#116), fighting Iroquois every inch of the way, crossed by Lake Huron to *Michilimackinac* and then across *Lake Michigan* to *Green Bay,* where they spent the winter. Even here they were not safe from their fierce foes, until Radisson led a party and killed the last brave in an Iroquois band, lurking around the fort. After remarkable explorations far beyond the Mississippi and Missouri in 1659 these young adventurers returned in the autumn to the north of Lake Superior perhaps as far as Sault Ste. Marie, where Groseilliers remained to attend to trading, while the restless Radisson pushed through the frozen woods, and probably explored as far west as Manitoba. These routes from Port Arthur to the Lake of the Woods, which later were to witness a constant stream of traders bringing furs from as far west as British Columbia, were blazed for the first time by Radisson. This event took place before he was yet twenty-six years old and was through a rough and unbroken wilderness. Laden with furs, and accompanied by hundreds of canoes, they prepared to return to Quebec. But the most ugly rumours, that great bands of stealthy Iroquois were lying in wait, so depressed the redskins that it took all the eloquence of Radisson to spur them to the trip. Before they reached the Ottawa, the Iroquois hordes were so battered and frightened by a fight with seventeen heroic Frenchmen under *Dollard* at the *Long Sault*, that Radisson's band reached Quebec in peace.

Starving west of Lake Superior. Radisson's restless spirit could not remain quiet long. The *Cree Indians* west of Lake Superior (#116) had made him curious by telling of a salt sea in the north where they went to hunt deer every spring; he determined to find this northern sea. In 1661 he was again fighting, foot by foot, his way up the Ottawa, against Iroquois, doubly enraged by their inglorious fight of the year before with Dollard. Though his companions were slaughtered at his side, Radisson seemed to have a charmed life. By November he was skirting the south shore of Lake Superior through snowstorms and ice. Somewhere west of Duluth, perhaps near the Lake of the Woods, he and Groseilliers built a little triangular fort and settled down for the winter among hundreds, if not thousands, of Cree Indians, who might treacherously pounce on them at any moment. They strung a row of bells around the cabin to warn them of the approach of a savage beast or of a still more savage redskin. Hunting was bad; the Indians starved to death by hundreds, and the two Frenchmen were shrivelled to skin and bone. In their endeavours to secure nourishment they had even gone to the extent of eating leather. A stray dog in camp was pounced on by Radisson and soon boiled in a pot.

British trading-posts on Hudson Bay. The spring arrived and with it abundance of deer; the famine was over. Still weak and thin, Radisson eagerly started on his search for the Northern Sea. He reached Hudson Bay or James Bay, though the course of the journey is quite unknown. By 1663 he was back in Quebec. From this point his movements are as mysterious and uncertain as those of a jumping-jack. Soon we hear of him in London, England, a man at court, describing the wealth of furs to be obtained at the great inland sea. He was interviewed by shrewd British merchants; perhaps he met gallant Prince Rupert or even discussed his travels with England's merry monarch, Charles II. One thing is certain, the British were soon building trading-posts on Hudson Bay, and in five years the great *Hudson's Bay Company* was launched on its long and successful career. Radisson's later years were chequered: at one time he was on the side of the British, at another he was raiding British trading posts on Hudson Bay; finally, the fearless, reckless explorer, who had discovered half a continent, sank into obscurity and died in poverty.

The Iroquois block the way. In 1667 Georgian Bay was well known to the French; they had missions on Lake Superior; Lake Michigan and Green Bay were familiar ground; Hudson Bay had been visited by Radisson; the Mississippi had been crossed by him and perhaps by others; even the Lake of the Woods had been seen by French eyes. Lake Ontario was almost a sealed book, no French ear had heard or French eye had gazed at the majestic Niagara cataract; peaceful Lake Erie, inviting the traveller, was nearly an unknown body of water; even the St. Lawrence River above Lachine Rapids, right at their door, was less known to the French than the Mattawa River much farther away. The fierce Iroquois hung above this main entrance to the Great Lakes like a blood-red cloud. To reach these attractive bodies of water the French had to pass the Iroquois front door, whose villages lurked back in the state of New York, a few miles from the head of the St. Lawrence (#116). While the Ottawa at this time was as dangerous as dynamite, the upper St. Lawrence was sure death to the adventurous traveller. During the next twenty years, the Iroquois by their constant raids and attacks, had almost emptied south Ontario of inhabitants and had made a desert of the settlements. The Iroquois were masters of New France, Montreal was never free from danger, and even under the shadow of the walls of invincible Quebec they occasionally sniped off French peasants.

Iroquois conquered. The French at last were revenged, when De Tracy, with old world soldiers, marched into the centre of the Iroquois settlements and dealt their strongholds such a slashing blow that they never recovered their impudent boldness, were compelled to sue for peace, and no longer laid in wait to pounce on every French or Indian canoe that passed.

A complete circuit of old Ontario. At once exploration revived. A big slab of copper brought to Quebec from Sault Ste. Marie was an attractive sight to the eyes of the ever watchful trader. The active intendant, *Talon*, appointed a refined, scholarly, young giant, *Louis Jolliet*, to search for the mine from which it came, as he had visions of wealth, such as the Spaniards had obtained in Mexico and Peru, pouring into Quebec from these rich mines. For the first time in history, this explorer started on the route to Lake Superior that has been used ever since. In 1669, just two years after the peace with the Iroquois, he, with paddlers, started up the St. Lawrence from Montreal; they entered Lake Ontario (#116), paddled along its northern shore to the vicinity of Hamilton, portaged to Lake Erie; but after going a part of the distance, for some unknown reason turned back, hid their canoe at Port Stanley, and followed the Thames and Grand Rivers back to *Burlington*. What was their surprise to learn that another band of French adventurers was only a few miles away. Indeed, this expedition had set out not long after their own. It included the greatest of the early explorers, *La Salle*, and two Sulpician missionaries, *Dollier* and *Galinée*. Jolliet joined the missionaries, and the unsociable La Salle went off on the first of his marvellous journeys (#116). He followed the Ohio for hundreds of miles, even though deserted by his companions. Jolliet, Dollier, and Galinée crossed over from Dundas to the Grand River, which they descended to Lake Erie (#116). As the autumn was getting cold and stormy, they built a hut at Port Dover in the midst of abundant waterfowl and other wild game, and there they joyfully spent the winter. Next spring they proceeded on their way, and were the first to pass through *River Detroit*, *Lake St. Clair* and *River St. Clair*, and *Lake Huron*. They reached Sault Ste. Marie in safety but were received coolly by the Jesuit missionary, Marquette; before autumn, 1670, they were all back in Montreal, by the Ottawa route, with no knowledge of the origin of the piece of copper. They were the first explorers to make a complete circuit of Ontario by the Great Lakes.

La Salle on the Great Lakes. Fort Frontenac, now Kingston, was built in 1673 (#116), and La Salle was made its seigneur. He dreamed of a great French empire, which would stretch from Hudson Bay to the Gulf of Mexico, and determined to play his part in making the dream come true, by tracing the great river that emptied into the latter and securing it for France. The expedition started out in a grand manner from Fort Frontenac, when in 1678 the Indians were dazzled with wonder to see four splendid sailing vessels on the waters of Lake Ontario, which up till then had only known the frail, birch-bark canoe. These early sailors were men of metal. At the present day even large steamers hesitate to brave the stormy waters of Lake Ontario in late November, but these French-Canadians, nourished in danger, snatched from death dozens of times, did not hesitate on November 18, 1678, to launch forth a tiny ten-ton sailing ship on the freezing storm-tossed waters of this lake, with the prow headed for the Mississippi. They took refuge in the mouth of the Humber and finally reached Niagara River in December. Here a fort was built. The roar of Niagara could be heard, and *Father Hennepin*, one of the explorers, in graphic words has given the first description, and made the first sketch, of that wonder of the world. He paints it in rather vivid colours, as he describes the fierce currents above, sweeping the wild beasts off their feet and hurling them down six hundred feet. At the head of the river they built the *Griffon*, the first ship that ever sailed the lakes above Niagara Falls. In this grand craft they sailed to Michilimackinac (#116). La Salle started the Griffon on the return trip with a valuable cargo of furs; but reverse after reverse dogged the footsteps of this persistent explorer. The Griffon was never again heard of. After several trips back to Fort Frontenac, on one of which, 1682, he returned by the Humber route, La Salle finally got started on that long, last, immortal journey to the mouth of the Mississippi.

By 1685 the Great Lakes were all well known; three routes from the upper lakes to Montreal were regularly travelled — by the Ottawa River, by Lake Simcoe and the Humber, and by Lake Erie. The internal streams, the Grand, the Thames, and the Trent, with connecting trails, were highways for canoes.

Climate

Ontario has its feet in the same latitude as sunny Rome and northern California, while its head is on a level with the stern climates of northern Scotland and Moscow. With such a vast inland sea as Hudson Bay at its head, and the Great Lakes at its feet, it does not feel, like Manitoba or Saskatchewan, the full strength of the frigid winters and the boiling summers of a typical continental climate.

Climate of Hudson Bay coast. Winters along the fringe of the province on Hudson Bay and James Bay are somewhat subdued by the water, but not much. Summers, though short and cool, are long enough and warm enough to grow potatoes, oats, barley, root crops, and even to ripen tomatoes at *Moosonee.* July, August, and early September are summer months with warm days and cool nights. The all too short autumn, extending to October, is delightful. Ice closes all rivers in November, and a long, cold winter, with short days and long black nights, lit up frequently by northern lights, sets in. About five feet of snow falls during this season. Rivers begin to open the last of April, but the icy waters of Hudson Bay delay the full bloom of spring, so that it is June before trees are well covered with leaves.

Climate on the Canadian Shield. Up on the Canadian Shield, where the full sweep of north-west winds is not softened by any large bodies of water, and where the altitude is great, the climate is extreme. Winter is exceptionally severe, but the intense cold is not made unbearable by many storms. Summer, which is cool and quiet, however, is longer than at Hudson Bay; there are longer periods free from frosts, and the same crops that are grown on the prairies can be ripened. As the Clay Belt is in this section, the longer summers with much sunlight are of great value for growth of crops. Winter is also long, while spring and autumn are very short. The coldest area is on the height of land well away from both Lake Superior and Hudson Bay. *White River* has the record in Ontario for cold, the lowest being 60° F. below zero, but *Kapuskasing* is a close rival. Rainfall is usually sufficient and, fortunately, is most abundant in summer. But summer rainfall is not certain, and when it fails, not only do crops suffer, but forest fires cannot be controlled and leave their loathsome trail behind.

South Ontario. Southern Ontario is almost embraced by the benign waters of the Great Lakes, which take the cold sting of winter away, and put cool breezes into summer. The dampness they bring, however, often gives a rawness to moderate winter temperatures and a depressing humidity to hot summer days. They also assist rainfall, which is 10 inches more than in northern Ontario. Not only is there sufficient rain, but it is well spread over every month in the year. Except along the lower lakes snow is deep in winter. Along Lake Erie winter lasts only three months; on Lake Ontario, four months; and from the Ottawa valley to Georgian Bay, five months.

Forests and Lumber

Except for a bare, narrow strip along Hudson Bay and James Bay, the whole of Ontario was originally covered with a crowded forest of majestic trees (#116). In southern Ontario such hardwood as birch,

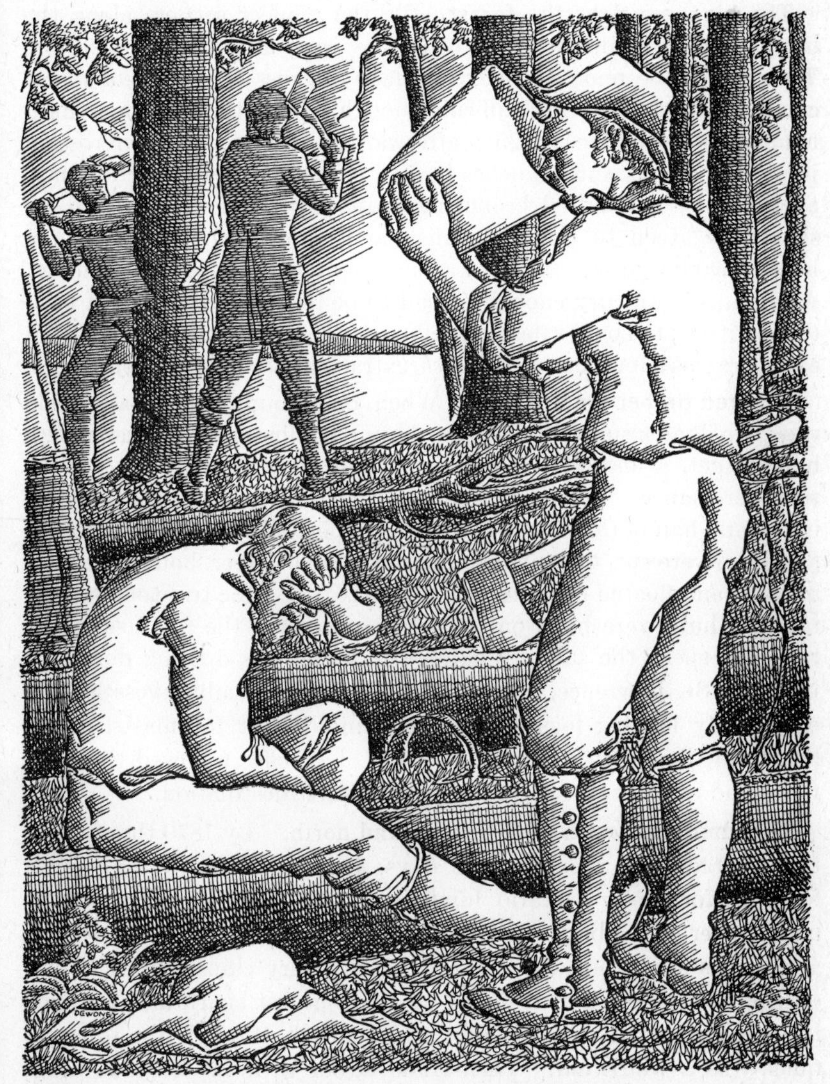

#125 PIONEERS CLEARING THE LAND IN OLD ONTARIO

oak, ash, elm, beech, and basswood clothed the fertile soil; in the cold and rocky north these were replaced by pines, cedars, spruces, tamaracks, and balsams; and there was a region between in which the gayer hardwoods with their broad, shiny, light green leaves were mixed with the more sombre evergreen softwoods (#116).

The pioneers fight the forest. To the pioneer settlers along the lakes these magnificent hardwood trees were a continual vexation. They had to be removed before a blade of grass or a stem of wheat could be grown. It was difficult enough to cut through their hard trunks with an axe (#125); afterwards the tangled, intercrossing jumble of trunks and branches had to be removed, and each stump, anchored by a hundred branching roots, grubbed out. All honour should be given to the brave men and women who faced such a heart-breaking task. There was no market anywhere for timber. As soon as it was dry enough, it had to be burned; which was a hard task in itself; the only saleable product of the many tons of wood from each acre was a few pounds of *potash*, extracted by water from the ashes.

Squared timber for England. When, on account of the Napoleonic wars, England's supplies of timber from Scandinavia and Russia were hard to get, a market was opened to Canada. She was not slow to seize her chance. All along the lower Ottawa River and its tributaries, in the first half of the nineteenth century splendid white pines, straight as masts, were cut, trimmed, hewed to squared timber, bound together in rafts, and floated down to Quebec (#126). As the trip took several months, huts were built on the rafts, and often the families of the raftsmen spent the summer happily in their huts drifting down the beautiful St. Lawrence. At Quebec hundreds of sailing vessels were continually loading logs and select round timber for masts. Soon logs from the Trent River and upper St. Lawrence followed the same route and often took two years to reach their destination.

The lumbering centre moves west and north. By 1870 the squared timber trade had seen its best days, but not before it had made fortunes for many Ontario lumbermen. It was very wasteful, as one quarter of the best wood was hewn away to make the log square; and it was dangerous, for the loose layer of dry chips and slash, left on the floor of the forest, licked up the slightest spark and whipped it instantly into a sheet of fire, which swept through the forest as quickly as a whirlwind. Then trade began with the United States, and from 1870 to 1880 the centre of the industry shifted to Georgian

Bay and spread along the north shore of Lake Huron (#127). White pine logs, poured down to the mouth of every river in the spring, were united into rafts, and towed down the Great Lakes to be cut into lumber in immense sawmills on the United States ports of Lake Huron and Lake Erie. When in 1898 Ontario prohibited the export of sawlogs, sawmills were quickly transferred from these American ports to the mouths of streams emptying into Lake Huron; *Parry Sound, French River, Spanish River, Blind River, Thessalon, Sault*

#126
A RAFT OF SQUARED TIMBER IN THE ST. LAWRENCE RIVER

Ste. Marie, Fort William, Fort Francis, and *Kenora* sprang into importance as lumbering towns (#117). The centre of the industry is being pushed further and further north, and now lies in the belt between the Canadian National railway and the Great Lakes and Lake Nipissing (#134).

At first only white pine and oak were cut; but the best of the former wood is gone, and spruce, cedar, hemlock, and birch, even of moderate or small size, are now eagerly sought.

Pulp and paper. About 1900, methods were discovered by which wood could be shredded by rubbing against rotating stones, or loosened into threads by chemicals. The felted fibres, when pressed into sheets, made paper good enough for newsprint (p. 261), or even for

#127

LOGGING IN ONTARIO AND QUEBEC

1. Lumber shanty. 2. Enjoying the evening meal. 3. A load of logs. 4. Whip-sawing. 5. Using the cant-hook. 6. Using the pike-pole. 7. A sawmill.

books, and so cheap that newspapers could afford to increase their size. As a result newspapers multiplied in number and expanded in the number of pages. Ontario and Quebec had many thousands of square miles of small spruce, balsam, and poplar trees, too small for

#128

A FOREST FIRE

lumber but suitable for pulp. These were found on areas that had been burnt over or had been logged for the larger trees. Pulp and paper mills, of immense size, run by hydro-electric power, sprang up like mushrooms across the Province from the Lake of the Woods to Lake Timiskaming (#134), and to-day the making of pulp and paper is the leading manufacture of Canada, and one of the most important in Ontario.

Agriculture

Gnawing away the forest. The work of the lowly and obscure pioneer farmers of Ontario, as they faced the closely packed forest of giant hardwood trees, was heroic (#125). Though to many Canadians these humble toilers are unknown, and though their unmarked graves in many places are overgrown with weeds, they laid the foundation of Ontario's great prosperity, which is their fitting monument. After a long winter's chopping, and after the spring burning (p. 312), they had perhaps ten acres of rough land, pock-marked with black stumps. As ploughing was impossible, the surface among the stumps was scratched with a crude harrow or chopped with a hoe and potatoes or wheat planted. Scarcely enough food for the family could be raised. Only after six or seven years could the stumps be removed and the land properly cultivated. The clearing of the Ontario forest can be compared to the eating of a big cheese by a swarm of insects; they nibble at it from various parts, chew their way in every direction beneath the crust, until it becomes so honeycombed that at last it crumbles like a house of cards and disappears. These farmers' clearings wormed their slow and tedious course into the pathless forest, the burrows became longer and wider; they crossed each other, until at last the whole forest fell away, and we have to-day in southern Ontario a many-coloured chequerboard of trim farms.

Early grain growing. There was little but potash, wheat, and hides that the earliest farmers could sell, and even these had to be taken great distances over trails so crooked, muddy, and rough, that even oxen could hardly make their way. The spade, the home-made plough, harrow, sickle, scythe, and flail, were the slow and crude implements which made the work long and hard, and the output small and coarse. The Napoleonic wars made the price of wheat so high after 1812, that Ontario directed her farmers to growing that grain. The invention of the reaping machine in 1814 put a power in the farmer's hand that greatly increased his output; and it was

SOUTHERN ONTARIO
AGRICULTURE LIVESTOCK
RAILWAYS
CANALS
A LIST OF SUCH CITIES, TOWNS, AND VILLAGES AS ARE MENTIONED IN THE TEXT
1 AMHERSTBURG
2 BLENHEIM
3 CHATHAM.
4 GODERICH.
5 LISTOWEL
6 LONDON
7 NORWICH
8 OWEN SOUND
9 PORT McNICOLL
10 PORT STANLEY
11 SANDWICH
12 SARNIA
13 SOUTHAMPTON
14 STRATFORD
15 ST. THOMAS
16 TILSONBURG
17 WALKERTON
18 WINDSOR
19 WOODSTOCK
20 BARRIE
21 BRADFORD
22 BRANTFORD
23 COLLINGWOOD
24 DELHI
25 ELORA
26 GALT
27 GUELPH
28 HAMILTON
29 KITCHENER
30 MEAFORD
31 MIDLAND
32 ORILLIA
33 OSHAWA
34 PARRY SOUND
35 PETERBOROUGH
36 PORT COLBORNE
37 PORT DOVER
38 PRESTON
39 SIMCOE
40 ST. CATHERINES
41 TORONTO
42 WATERLOO
43 ARNPRIOR
44 BEAMSVILLE
45 BELLEVILLE
46 BROCKVILLE
47 CORNWALL
48 KINGSTON
49 LANARK
50 OTTAWA
51 PEMBROKE
52 PRESCOTT
53 SMITHS FALL.
CITIES HAVING A POPN. OVER 100,000
-HORSES
-MILK CATTLE
-BEEF CATTLE
SHEEP-
HOGS-
POULTRY-
CHEESE-
-HAY
-OATS
-BUCKWHEAT
-RYE
FALL WHEAT-
SPRING WHEAT
BARLEY-
-CORN
-FLAX
-BEANS
-PEAS
POTATOES-
SUGAR BEETS-
TOBACCO
-APPLES AND PEARS
-GRAPES AND PEACHES
-SMALL FRUIT.
SHD
#129

only the first of a long series of inventions of farm implements, which have multiplied his production, lessened his labour, and improved the quality of his products.

Rise of dairying. Sluggish, plodding oxen did the farm work, for nervous horses were useless among stumps; there were few cattle; wolves pounced on sheep, and bears raided the pig-sty. After 1840 great numbers of skilled immigrant farmers from England, Ireland, and Scotland, settled in groups all over the province (#131) and brought with them their love of good stock. While wheat still held first place, cattle raising steadily increased in importance, but low-grade, scrub stock were common everywhere. Choice stock began to be brought in from New York state, and settlers from the British Isles, after visits home, brought back high-grade animals from their districts; thereafter the quality rapidly improved. As towns became larger, the demand for milk and butter increased; and as settlement became denser, the cheese factory became possible. The first one was built at *Norwich* in Oxford county in 1864. By 1880 there were 500, and by 1897 over 1100. The opening of the wheat fields of the Canadian prairies drove the Ontario farmer more and more to mixed farming and dairying, for in the growing of wheat he could not compete with the boundless virgin fertility of the western plains. Butter making was a later industry than cheese making, and in a broad way, to-day, creameries dominate the west and cheese factories the east of southern Ontario (#129).

Special kinds of farming. Ontario agriculture is still changing. Now it is becoming more and more specialized, each district raising those products for which it is best suited (#130). For example, along the shore of Lake Erie, from Essex to Norfolk, rich green fields of tobacco and neat curing-barns are seen everywhere; in the counties on Lake Huron gorgeous fields of flax in flower, look as blue as the skies; in Essex and Kent thousands of acres of sugar-beets are attended by industrious new Canadians from Belgium; the highways of Niagara peninsula between the escarpment (p. 284) and Lake Ontario are avenues of sweet-smelling, pink peach blossoms in early May, and in September in the same district many miles of trellises groan under clusters of luscious green, red, and, purple grapes. #130 shows the special kinds of farming for each district.

Leadership in farming. Ontario's leadership in farming is remarkable; she stands first in production in every department: field crops;

farm animals; wool; dairy produce; poultry and eggs; fruit and vegetables; fur farming; tobacco; clover and grass seed; and honey.

#130

DIVERSITY OF FARMING IN ONTARIO

1. A litter of pigs. 2. Picking small fruit. 3. Ploughing. 4. Thinning sugar-beets. 5. Tobacco Plant. 6. Disking with a tractor. 7. Holstein dairy cow. 8. Apple orchard. 9. A vineyard.

Among field crops she is first in the growth of peas, beans, corn, potatoes, and alfalfa, and while falling far behind the Prairie Provinces in output of wheat, oats, and barley, she surpasses all the other provinces. She is easily first in the number of cattle and swine and stands second in the number of sheep and horses. In dairy produce, she leads in production of all manufactured products, such as butter, cheese, and condensed milk products. Her fowl, ducks, and geese far surpass those of any other province, Saskatchewan leading her in the number of turkeys. In all fruits, except apples, she is first and sometimes leads in apples also; she supplies the whole Dominion with peaches and grapes. Nevertheless, the farming industry is not satisfactory; she is not increasing her output, but in most departments production is at a standstill, or, if anything, slipping back.

Of course Ontario has great advantages. Her many large, wealthy, manufacturing cities and towns supply the farmers with a good home market. She is endowed with a mild climate, large areas of fertile soil, and excellent transportation facilities. These, together with an intelligent and industrious Anglo-Saxon and French population, all combine to stimulate the growth of agricultural prosperity.

Transportation and Settlement

A naked province at the British conquest. When in 1763 Canada became British, the Province of Ontario was almost empty of inhabitants. The savage terror of the Iroquois had destroyed or driven out almost all Indians, and with the exception of a group of Iroquois that settled on the *Bay of Quinte*, no other tribes had taken their place. There had been a trading post and a Jesuit mission at *Sault Ste. Marie* since 1669, to which a few frightened fragments of the once mighty Hurons had fled in terror; there had been a mission for the Iroquois on the Bay of Quinte, and a fort near Kingston built in 1673 (*Fort Frontenac*). These were scarcely settlements, as they contained only a few traders and missionaries. *Detroit* was founded by *Cadillac* in 1701, and, on account of its key position for trading, soon attracted settlers. A remnant of the Huron tribe from Georgian Bay, calling themselves *Wyandottes*, were settled on Detroit River, and an Indian mission was established in 1728 in Ontario at Sandwich, called *L'Assomption*. Here along the Detroit River and the south border of Lake St. Clair was gradually established a French settlement, the first in Ontario; and there it remains to-day surrounded

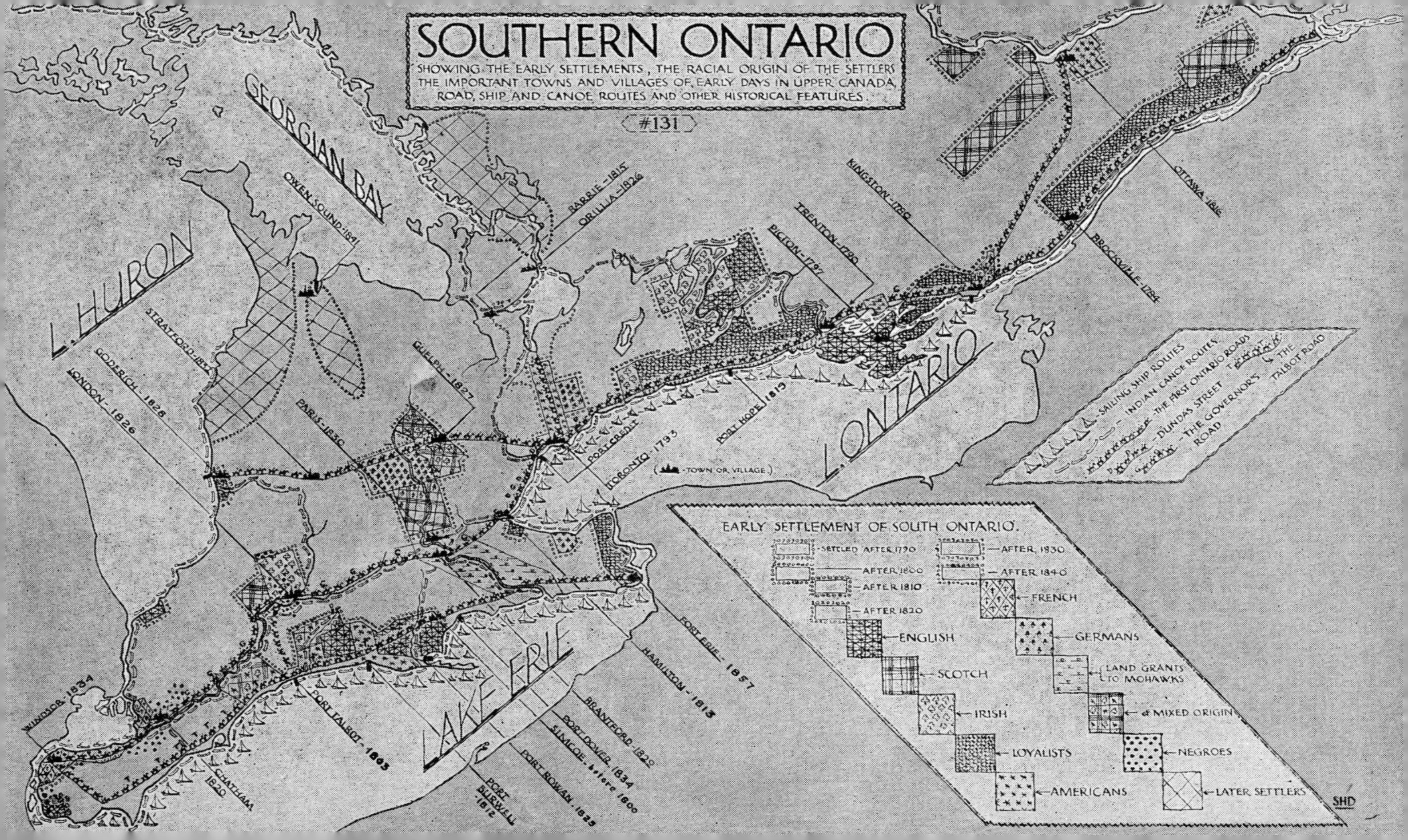
SOUTHERN ONTARIO
SHOWING THE EARLY SETTLEMENTS, THE RACIAL ORIGIN OF THE SETTLERS THE IMPORTANT TOWNS AND VILLAGES OF EARLY DAYS IN UPPER CANADA, ROAD, SHIP AND CANOE ROUTES AND OTHER HISTORICAL FEATURES.
#131
GEORGIAN BAY
L. HURON
L. ONTARIO
LAKE ERIE
OWEN SOUND-1841
BARRIE-1815
ORILLIA-1826
KINGSTON-1780
TRENTON-1790
PICTON-1797
OTTAWA-1816
BROCKVILLE-1784
STRATFORD-1832
GODERICH-1828
LONDON-1826
GUELPH-1827
PARIS-1830
PORT CREDIT
PORT HOPE 1815
TORONTO-1793
(-TOWN OR VILLAGE.)
FORT ERIE-1857
HAMILTON-1813
BRANTFORD-1820
PORT DOVER 1834
SIMCOE before 1800
PORT ROWAN 1825
PORT BURWELL 1812
PORT TALBOT-1803
CHATHAM 1820
WINDSOR 1834
SAILING SHIP ROUTES
INDIAN CANOE ROUTES
THE FIRST ONTARIO ROADS
DUNDAS STREET
THE GOVERNORS ROAD
THE TALBOT ROAD
EARLY SETTLEMENT OF SOUTH ONTARIO.
SETTLED AFTER 1790
AFTER 1800
AFTER 1810
AFTER 1820
AFTER 1830
AFTER 1840
FRENCH
ENGLISH
GERMANS
SCOTCH
LAND GRANTS TO MOHAWKS
IRISH
of MIXED ORIGIN
LOYALISTS
NEGROES
AMERICANS
LATER SETTLERS
SHD

by British, but nevertheless almost as French as the day it was formed. It contained nearly one thousand people at the conquest in 1763. For twenty years afterwards not another settler entered the fertile lands of Ontario.

Early trails. By canoe and afoot were the first methods of travel in Ontario. Trails became worn by the habitually stealthy steps of moccasined feet through the woods from one body of water to another, and these were the paths followed by the white men (#131). Trails were gradually trodden to bridle paths for horses; these, when trees had been cut and stumps removed and the ways smoothed, were suitable for carts and later for wagons; to-day these old trails have become the great trunk highways with their ceaseless throb of trucks and motor-cars. One of the longest and most travelled of Indian trails went from Hamilton to the source of the *Thames* near *Woodstock*, crossing the *Grand River* at the ford (*Brantford*) (#131); then by canoe, the broad and quiet *Thames* carried the dusky paddlers to *Lake St. Clair* and *Detroit.* It is possible that the ill-fated Brulé travelled this route (p. 300).

Branches went from this trail to Lake Erie at *Chatham* and *London,* and one from Brantford to *Port Dover.* The route to Lake Huron was up the north branch of the Thames and across by trail to the *Maitland River* and to *Goderich.* Another trail to Detroit was along the north shore of Lake Erie. There were two well beaten trail and canoe routes from Georgian Bay to Lake Ontario. One by the *Severn River, Lake Simcoe, Holland River,* a trail portage to *Humber River,* and down to *Toronto* (#131); the other, by the Trent waterway, was followed by Champlain in his unfortunate raid on the Iroquois in New York (p. 302).

United Empire Loyalists. The loss of the United States in 1783 helped Ontario. Many of the best and bravest citizens not only refused to join in rebellion against the mother country but rose in arms to support the British. When the American colonies succeeded and became independent, thousands of these loyalists preferred to sacrifice all they possessed to remain in the British Empire, and others were so cruelly treated that they were forced to leave. Ontario was the gainer. The British government offered them homes and farms in the forests of Ontario; and thousands of these cultured, well-to-do United Empire Loyalists plunged into the dense forest to hew out for themselves homes under the Union Jack. They were the first colonists, who came principally from New York state, and for the most part settled along the adjoining lakes and rivers. The first of these began clearing land and farming along the Niagara River even before the war was over (#131). Immediately after, beginning in 1784, thousands settled along the St. Lawrence between Brockville and Cornwall and along the Bay of Quinte. Many of these were Highland soldiers who settled in *Glengarry*, and a good many who settled in *Dundas* county were Germans. Another group clustered around old *Fort Malden*, now *Amherstburg*. The United Empire Loyalists were not all British and Germans; many of the Iroquois, under the leadership of Joseph Brant, supported the British in the unsuccessful struggle and, still desiring to live under the British flag, moved over the border in 1784. They were given a strip, six miles wide, on each side of the Grand River from the mouth to Elora, and the townships along this river to this day show plainly this tract, their names commemorating the Iroquois tribes. As this was more land than they required, some of it was soon sold in blocks to new settlers. A part of the Mohawk tribe was placed on the Bay of Quinte near *Deseronto*, where they still live (#131).

United States Settlers. As Governor Simcoe was anxious to have the province settled with people loyal to Britain, he offered free land to any in the United States who preferred British rule. Large numbers, more interested in cheap and fertile land than in the oath of allegiance to British institutions which they took, flocked in and settled along Yonge Street and on the shore of Lakes Erie and Ontario. Many were republicans at heart, used American school books, and scoffed at Canadian ways. Some proved traitors during the war of 1812.

A stream of Quakers. Many Quakers in New York and Pennsylvania, who had taken no part in the American war, preferred British government to the lack of law and order in the republic, and about 1800 settled several townships in North York, Prince Edward, and Norwich, near London. The number of these devout industrious people steadily increased up to 1825, and everywhere they became model, thrifty, prosperous Canadians.

German Mennonites. Not only Quakers but German Mennonites from Pennsylvania, many of whom originally came from Switzerland, were attracted by the law and order of the British flag, and the opportunity to secure good land at cheap rates. As early as 1794 a band settled in Markham township near Toronto, and their descendants to-day cultivate some of the finest farms in that banner township. This was the first settlement back from the Great Lakes. But the branches of Rouge Creek spread all over the township, and the main stream enters Lake Ontario. While at the present time during the summer there is a mere thread of stagnant water in a good part of the course of this stream, a hundred years ago, when the rain was held in the spongy soil of the shady forest and only gradually reached the streams, this creek was sufficient for the transport of their goods and themselves in canoes and small boats in summer, and on sleighs in winter. Another group settled on the south shore of Lake Ontario in the vicinity of Beamsville. Their descendants still farm the land in these districts; and their piety, industry, honesty, and intelligence has added much to the life and prosperity of the province.

Waterloo Mennonites. But the largest and most notable of German settlements was in the township of Waterloo, which was bought from the Iroquois Indians. From 1799 to 1835 large numbers of these stalwart immigrants drove in covered wagons into this roadless wilderness and made their clearings forty miles back from the Great Lakes; they often drove their cattle before them. From time to time they have received additional settlers from all parts of Germany. To-day the county of Waterloo is covered with prosperous farms, interspersed with neat villages, and thriving factory towns. It centred in *Kitchener*, one of the most hustling cities in Canada.

Settlements along the Trent River. After the war of 1812 American immigration ceased. Some treacherous settlers from the United States had aided the Americans, and even a member of Parliament conducted a raiding party in west Ontario. Several of such parties crossed the St. Clair River and penetrated far into the province, burning buildings, seizing stock and grain, leaving the farmers hungry and helpless, and their farms desolate. Such actions resulted in settlers from that country being no longer made welcome. Indeed, they were under such suspicion that the government made every effort to fill the province with loyal, British stock, especially those with military experience, who in case of another war could defend the country. Land along the rivers was most attractive. A considerable number of English officers and other educated settlers took farms along Rice Lake and the Otonabee River in Peterborough county (#131), and, in 1825 and succeeding years, *Peter Robinson* brought out thousands of settlers from the south of Ireland. These "wild Irishmen," as they were called, proved hard-working, peaceable, kindhearted farmers, who settled several townships in Peterborough and Victoria counties; their descendants are still the chief farmers of the districts. Peterborough was named after the founder. Well-to-do English and Scottish settlers followed back the string of lakes that are drained by the Trent River, the first five settlers on *Sturgeon Lake* being university graduates.

Scottish settlers along the Rideau Canal. The fear of American invasion hung like a nightmare for many years after their unprovoked attacks in 1812; and the St. Lawrence River with New York state bordering one shore, was considered a sore artery connecting the two provinces. To overcome this threat to transportation, the British government undertook to make a canal from Kingston to Ottawa using the necklace of beautiful lakes and the Rideau River. This *Rideau Canal* was completed in 1832, as was the canal on the Ottawa River to overcome the Long Sault Rapid, thus completing a new water route from Kingston to Montreal, by Ottawa. The government was anxious that, for better protection, loyal British settlers, with a fair mixture of soldiers, should be placed on the fertile lands along the course of this canal. So in 1817 a large group of Scottish soldiers, many of whom had helped to defeat Napoleon at Waterloo, were placed in several townships with the town of *Perth* as a centre (#131). Such favourable reports from these settlers were carried to their friends in the counties of Lanark and Renfrew, Scotland, that they petitioned the government to aid them in migrating to this good, new land. By 1821, more than 2000 of these thrifty, industrious Scots were clearing the land in a row of townships in what is now the county of Lanark, and the towns of *Lanark* and *Smith's Falls* soon began to crystallize amid the prosperous settlements.

The MacNab clan arrive in kilts. A very different motive led to the founding of Renfrew county. *Archibald MacNab* was a proud but poor highland chief, who slipped out the back door of his ancestral home while the sheriff's men entered the front to seize him or make him sign away his estate. He fled to Canada to redeem his estate by making a fortune. Between 1825 and 1834 he placed hundreds of his clansmen in the township of MacNab (#131), and built himself a fine log house overlooking the Ottawa River, determined to rule like a laird in highland state. As these Scots prospered in the free Canadian air they were less and less willing to serve as clansmen to a tyrannical chief, until finally MacNab was driven out of the settlement which he did so much to make. The town of *Arnprior* soon sprang up to supply the needs of the Canadian division of the MacNab clan.

Scots from New York state bought a township from the Iroquois on the Grand River and founded and settled around Galt in 1816. A group of Irish took possession of the district north of London in 1818, and various settlements spread along the Governor's Road east and west from London.

The early roads. Settlers soon had to go back from the front, and roads were required to meet the needs of the farmers. Simcoe, the first governor of Ontario (Upper Canada) was not slow to recognize their value not merely for the settlers but also for rushing troops to vital points if the United States should again attack the colony. As said before, the first roads largely followed the old Indian trails. One of the very earliest, the *Governor's Road,* was completed from Kingston to Sandwich by 1800. The western part of it was called the *Longwoods Road*; between London and Dundas, it was called the *Gravel Road*; between Dundas and Toronto, the *Dundas Road,* and between Toronto and Kingston, the *Danforth Road.* As the St. Lawrence handled traffic east of Kingston, it was not completed to Montreal until 1830. Yonge Street from Toronto to Lake Simcoe was one of the earliest built, being in use by 1796. Another early road followed the trail along Lake Erie from Amherstburg to Fort Erie and went through Blenheim, St. Thomas, Tillsonburg, Delhi, Simcoe, and Port Dover. Part of this was the famous *Talbot Road.* Before 1813 there were settlements along most of these roads.

These early roads, which were largely built by the settlers, were fearful traps. At first the stumps were not removed, and as no gravel was applied, they were rivers of mud in the spring and fall, in which wagons sank to the axles. Across swamps, logs were placed crosswise often with no earth and even with gaps between the logs. A wagon went over these *corduroy roads* in a succession of jerking bounds. Often these roads could only be used in the winter.

The Canada Company. A little later settlement in a big way was undertaken by the *Canada Company,* financed in England by *John Galt,* a soldier, writer, and philanthropist. An immense tract of land running from Lake Huron eastward to Guelph, as primitive as the day Columbus discovered America, was surveyed into townships and put up for sale. It had neither road, village, nor settler. With much display, in a downpour of rain, a distinguished company laid the foundation of the chief town of the tract in April 1827, and called it *Guelph* after the reigning king. Viscount Goderich, the English Chancellor of the Exchequer, who had assisted the company, thought the town should have been named after him. However, the company soon established a second town at the other end of the *Huron tract*, as it was called, and soothed the pride of the Chancellor by calling it *Goderich*. Soon a road was built from Goderich, through Stratford, to Guelph and continued to Toronto (#131), and large numbers of settlers from the British Isles began to spread over this vast tract of fertile land.

Talbot settlement. *Thomas Talbot* was one of the most distinguished of the early settlers. He was a dashing young Irish officer, friend of the Duke of Wellington, a close friend of the king, wealthy, and belonging to one of the oldest and most aristocratic families in England. While touring western Ontario with Governor Simcoe, the soft murmuring of the pines, the rustle of the oaks, the great stretches of forest, the blue face of Lake Erie, and the wild freedom of the uninhabited stretches of western Ontario wooed and won him. By 1803 he had left all the glamour of the camp and the glitter of the court behind him to hew down the forest on the high banks of Lake Erie near Port Stanley. Here he had his estate at *Port Talbot* for nearly fifty years. He devoted his whole life to building roads and selecting suitable settlers for the district. Anybody with an Irish name received special welcome; but a settler from the United States was viewed with suspicion. While perhaps he did not actually bring out settlers, he was the magnet that drew the immigrants to Elgin and Middlesex, and the power that could get roads built, schools established, and law and order enforced. His settlement extended almost the length of Lake Erie, though most compact in Elgin, Norfolk, and Middlesex. It was by far the most extensive of the early settlements and contained tens of thousands of people. Scottish Highlanders, especially from Argyleshire, were numerous, but Irish, English, and Quakers from Pennsylvania were attracted by fertile land, good roads, and law and order.

Negro settlements. The numerous negroes in Essex, Kent, and Elgin entered the province from the Detroit River, mostly before 1860. In 1830 some negroes of Cincinnati were driven out of the city and were given land by the Canada company at the village of *Lucan*, north of London. Many came in, between 1840 and 1860, by what was known as the "*underground railway.*" This was a secret organization among the haters of slavery in the United States. Runaway slaves were stealthily passed along regular routes in the darkness of night, through dense forests, across lonesome swamps, and over spongy marshes till they reached the Detroit River, which they crossed in leaky punts, on rafts, or even straddling logs. When they reached Amherstburg, often half dead and nearly naked, they, with tears in their eyes, kissed the earth of freedom, or as they sometimes put it "shook the lion's paw." Benevolent men in the United

#132

PIONEERING IN NORTHERN ONTARIO

States sometimes bought slaves their freedom, and established them in settlements in western Ontario (#133). The most notable of these was established by Rev. Wm. King at the village of *Buxton* in Kent County, where they still live modestly as a group of contented farmers. Most of the negroes that came in by the "underground railway" settled in the larger towns, *Windsor*, *Chatham*, and *London*. In recent years many of them have gone back to the United States.

The snort of the iron horse. By 1850 the second generation had boiled over and flowed out from the early settlements into the townships behind these. Then a new and powerful factor caused a rapid spread of settlement which is still pursuing its course. The first railway was built from Toronto to Bradford in 1853. With feverish haste the people offered ruinous bonuses to railways to pass through their towns and cities, and wherever the railways passed, a fringe of farms and a necklace of villages formed along its course. Indeed, the settlement of the whole of northern Ontario is along the railways, and the settlement in each case began with the completion of the railway (#117, #134).

Manufacturing

Manufactures in the farm home. The pioneers manufactured nearly everything they used. They hewed the logs into square timber, and cut them into boards with a whip-saw (#127). With these they built their houses and barns, shaped their crude tables, benches, and sleeping bunks, built up barrels and tubs, and carved their troughs and bowls. From the flax, which they grew, and the wool which they clipped from the sheeps' backs they spun the warp and weft, which they wove into good, warm serviceable homespun. This they made into both underclothes and outer garments for the whole family. They even pounded their grain into meal in a maple stump, if they were not able to buy a hand-mill. Of course they made their own butter and cheese, and cured their own meat, which was chiefly fat salt pork. They knit their stockings and caps from homespun wool, and made their moccasins from deerskin or boots from cowhide. The ashes from the immense quantities of wood which was burned to try to keep the loose log-houses warm during the winter, were soaked in water in the spring by the women, and the liquid then boiled in iron kettles with waste fat, lard, and drippings, to make soft soap, which was used freely in bathing the body, scrubbing the floors, and laundering the clothes. They had to buy salt, writing-paper, and a few articles of pottery and metal.

Village factories. As villages began to crystallize around cross-roads, grist mills, and sawmills, many tradesmen such as, shoe-makers, weavers, blacksmiths, tailors, coopers, carpenters, and plasterers, undertook the special manufacturing indicated by their trades. As the settlements became denser, towns larger, roads common, and railways more widespread, the grist mill became a flourmill; the small sawmill became a large one; the weaver's shop became a woollen mill; the blacksmith shop expanded to a foundry; the cooper shop turned into a stave mill; the carpenter shop was enlarged to a furniture factory; and the plasterer or mason started a limekiln or a brick yard; and Ontario was fairly launched on her prosperous career of manufacturing.

Electricity, the life-blood of manufacturing. To-day old Ontario has become the centre of the greatest diversity of factories to be found anywhere on the continent. She manufactures nearly half of the total output of Canada. Two events are landmarks in her rapid strides. First, the discovery that the energy of falling water could be cheaply changed into electricity, which would flow through wires to wherever it was needed, wiped out the steam engine and the tall factory chimney in a few years. *Sir Adam Beck* formed a provincial commission and took control of most of the hydro-development for the people of the province. To-day through this commission, abundant supplies of cheap electricity are furnished to run all kinds of machines, to extract or shape metals, to melt the most resistant materials, to blend mixtures of molten metals into alloys, to form the most curious and obstinate chemicals — yes, such cheap supplies are available that the Ontario manufacturer can laugh at the lack of coal.

From buttons to bridges. The other event was the cutting off from Canada of supplies of many kinds of manufactured goods during the Great War. At once the adaptable Canadian manufacturers undertook the making of so many new commodities that to-day there is scarcely an article in a great departmental store that is not made in some Ontario factory (#133, #134). Whether it is buttons or bridges, tacks or tractor-engines, locomotives or liver-pills, one can depend they are made somewhere in the province.

Distribution of factories. While factories are found in every town, they are most numerous in south-western Ontario (#133), between

KEY TO NUMBERS ON ILLUSTRATION FACING THIS PAGE

1. Acton
2. Amherstburg
3. Aylmer
4. Barrie
5. Bowmanville
6. Bracebridge
7. Brampton
8. Brantford
9. Burlington
10. Campbellford
11. Chatham
12. Clinton
13. Cobourg
14. Collingwood
15. Dundas
16. Dunnville
17. Galt
18. Goderich
19. Grimsby
20. Guelph
21. Hamilton
22. Hespeler
23. Huntsville
24. Ingersoll
25. Kincardine
26. Kingsville
27. Kitchener
28. Leamington
29. Lindsay
30. Listowel
31. London
32. Meaford
33. Merritton
34. Midland
35. Milton
36. Mitchell
37. Mount Forest
38. Newmarket
39. Niagara Falls
40. Oakville
41. Orillia
42. Oshawa
43. Owen Sound
44. Paris
45. Parry Sound
46. Penetanguishene
47. Peterborough
48. Petrolia
49. Port Colborne
50. Port Credit
51. Port Hope
52. Preston
53. St. Catharines
54. St. Thomas
55. Sarnia
56. Simcoe
57. Southampton
58. Stratford
59. Thorold
60. Tillsonburg
61. Toronto
62. Waubaushene
63. Walkerton
64. Wallaceburg
65. Waterloo
66. Welland
67. Weston
68. Whitby
69. Wingham
70. Windsor
71. Woodstock

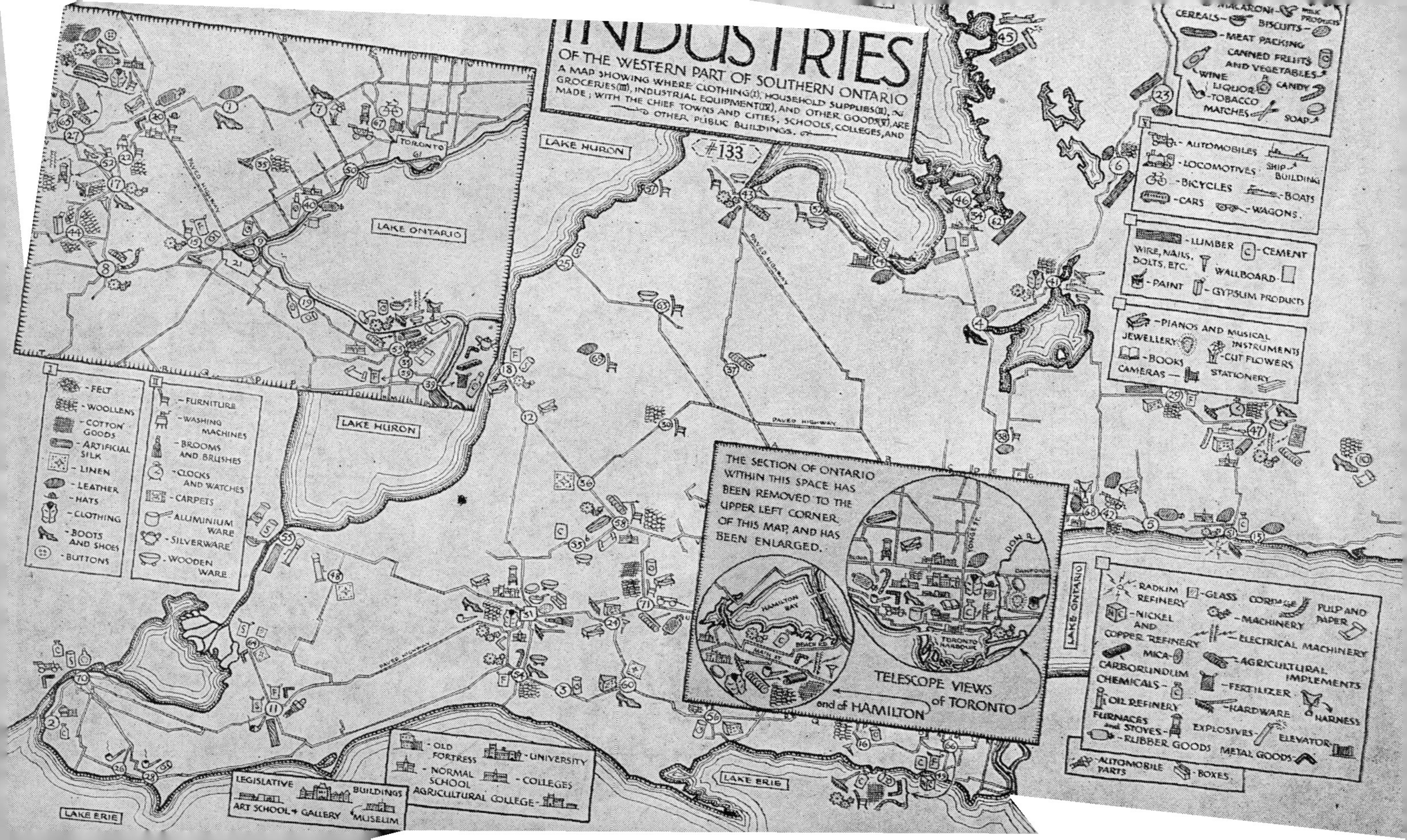
INDUSTRIES
OF THE WESTERN PART OF SOUTHERN ONTARIO
A MAP SHOWING WHERE CLOTHING(I), HOUSEHOLD SUPPLIES(II), GROCERIES(III), INDUSTRIAL EQUIPMENT(IV), AND OTHER GOODS(V) ARE MADE; WITH THE CHIEF TOWNS AND CITIES, SCHOOLS, COLLEGES, AND OTHER PUBLIC BUILDINGS.
#133
LAKE HURON
LAKE ONTARIO
LAKE ERIE
TORONTO
PAVED HIGHWAY
I
FELT
WOOLLENS
COTTON GOODS
ARTIFICIAL SILK
LINEN
LEATHER
HATS
CLOTHING
BOOTS AND SHOES
BUTTONS
II
FURNITURE
WASHING MACHINES
BROOMS AND BRUSHES
CLOCKS AND WATCHES
CARPETS
ALUMINIUM WARE
SILVERWARE
WOODEN WARE
CEREALS
BISCUITS
MILK PRODUCTS
MEAT PACKING
CANNED FRUITS AND VEGETABLES
WINE
LIQUOR
CANDY
TOBACCO
MATCHES
SOAP
AUTOMOBILES
LOCOMOTIVES
SHIP BUILDING
BICYCLES
BOATS
CARS
WAGONS
LUMBER
CEMENT
WIRE, NAILS, BOLTS, ETC.
WALLBOARD
PAINT
GYPSUM PRODUCTS
PIANOS AND MUSICAL INSTRUMENTS
JEWELLERY
CUT FLOWERS
BOOKS
CAMERAS
STATIONERY
THE SECTION OF ONTARIO WITHIN THIS SPACE HAS BEEN REMOVED TO THE UPPER LEFT CORNER OF THIS MAP, AND HAS BEEN ENLARGED.
TELESCOPE VIEWS of TORONTO and of HAMILTON
YONGE ST.
DON R.
TORONTO HARBOUR
HAMILTON BAY
MAIN ST.
RADKIM REFINERY
GLASS
CORDAGE
PULP AND PAPER
NICKEL AND COPPER REFINERY
MACHINERY
ELECTRICAL MACHINERY
MICA
CARBORUNDUM
AGRICULTURAL IMPLEMENTS
CHEMICALS
FERTILIZER
OIL REFINERY
HARDWARE
HARNESS
FURNACES and STOVES
EXPLOSIVES
ELEVATOR
RUBBER GOODS
METAL GOODS
AUTOMOBILE PARTS
BOXES
OLD FORTRESS
UNIVERSITY
NORMAL SCHOOL
COLLEGES
AGRICULTURAL COLLEGE
LEGISLATIVE BUILDINGS
ART SCHOOL & GALLERY
MUSEUM

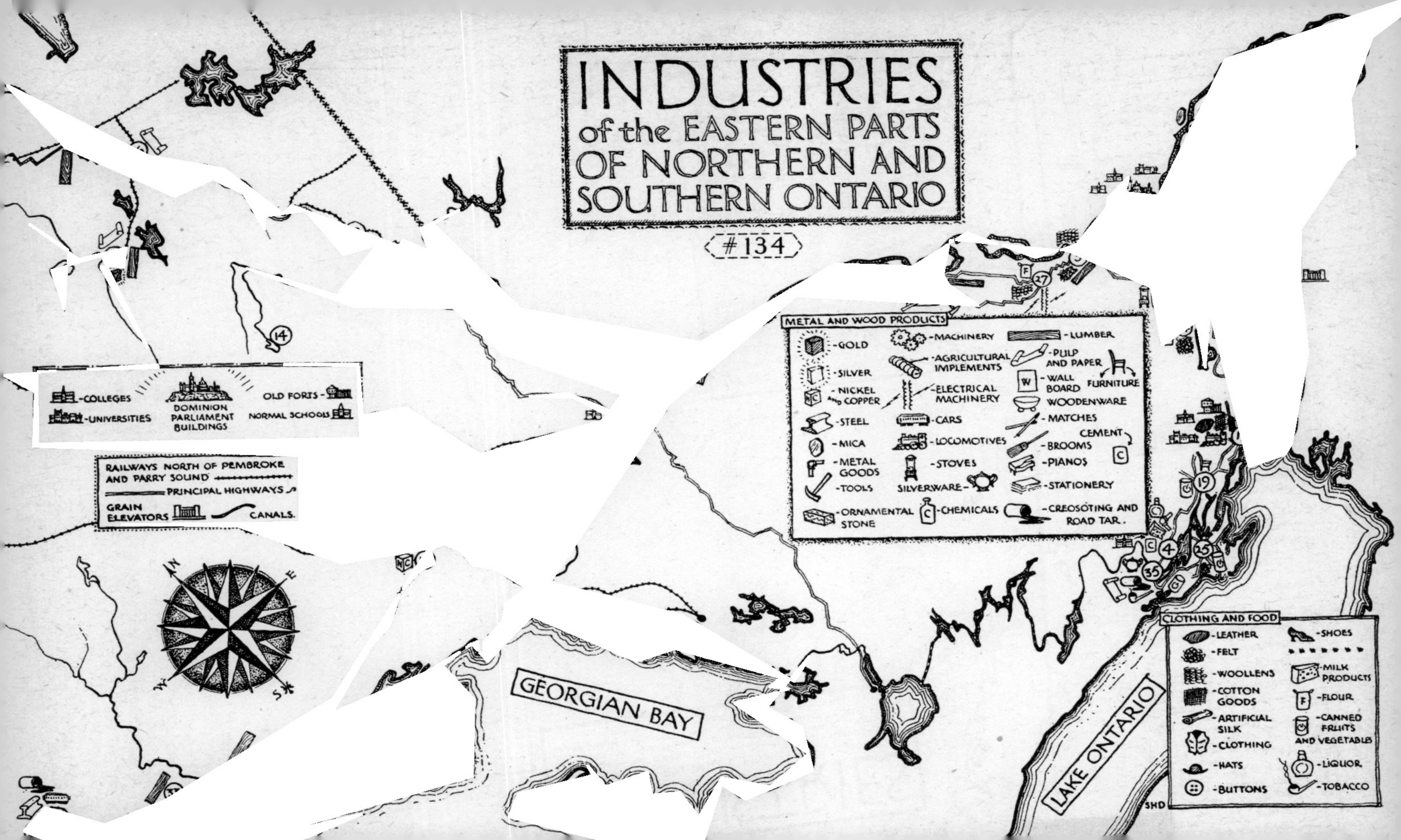

INDUSTRIES of the EASTERN PARTS OF NORTHERN AND SOUTHERN ONTARIO
#134
METAL AND WOOD PRODUCTS
-GOLD
-SILVER
-NICKEL AND COPPER
-STEEL
-MICA
-METAL GOODS
-TOOLS
-ORNAMENTAL STONE
-MACHINERY
-AGRICULTURAL IMPLEMENTS
-ELECTRICAL MACHINERY
-CARS
-LOCOMOTIVES
-STOVES
SILVERWARE-
-CHEMICALS
-LUMBER
-PULP AND PAPER
-WALL BOARD
FURNITURE
WOODENWARE
-MATCHES
-BROOMS
CEMENT
-PIANOS
-STATIONERY
-CREOSOTING AND ROAD TAR.
CLOTHING AND FOOD
-LEATHER
-FELT
-WOOLLENS
-COTTON GOODS
-ARTIFICIAL SILK
-CLOTHING
-HATS
-BUTTONS
-SHOES
-MILK PRODUCTS
-FLOUR
-CANNED FRUITS AND VEGETABLES
-LIQUOR
-TOBACCO
-COLLEGES
-UNIVERSITIES
DOMINION PARLIAMENT BUILDINGS
OLD FORTS-
NORMAL SCHOOLS
RAILWAYS NORTH OF PEMBROKE AND PARRY SOUND
PRINCIPAL HIGHWAYS
GRAIN ELEVATORS
CANALS.
GEORGIAN BAY
LAKE ONTARIO
N
E
S
W
SHD

Toronto and the Niagara Peninsula on the east, and Windsor and Sarnia on the west; but such towns as *Peterborough* and *Cornwall* (#134) in the east, which have hydro-electrical power at their back-doors, also have large and numerous factories. Pulp, paper, and lumber mills fringe the southern edge of the hardwood forest from *Pembroke*, through *Sturgeon Falls*, *Thessalon*, *Sault Ste. Marie*, *Nipigon* *Port Arthur*, *Fort Frances*, and *Kenora* or extend well into it as at *Dryden*, *Kapuskasing*, and *Iroquois Falls*. Flour mills are chiefly located on the line along which wheat moves from Winnipeg to Montreal, as *Kenora*, *Fort William*, *Owen Sound*, *Midland*, *Goderich*, *Sarnia*, *Port Colborne*, and *Toronto*. Furniture is made in the towns near the fine hardwood forests in the high land between Lake Simcoe and Lake Huron, such as *Walkerton*, *Waterloo*, *Preston*, *Kitchener*, *Goderich*, *Stratford*, *Listowel*, *Meaford*, *Southampton*, *Owen Sound*, etc. The flax mills are confined to the counties bordering Lake Huron, where flax is grown. As mixed farming and sheep raising are wide-spread, woollen mills and tanneries are scattered over the whole province (#133, #134). A shrewd man with abundant capital can do much to overcome unsuitable location, and often successfully launches a factory in his home town whether it is in the right position or not.

Many commodities are made almost exclusively in Ontario. Automobile manufacture is concentrated in *Windsor*, *Toronto*, and *Oshawa*, just as in the United States Detroit is pre-eminent; nine-tenths of the agricultural implements, four-fifths of the rubber goods and leather, seven-tenths of the furniture, fruit and vegetable canning, three-fourths of all electrical goods, more than half of steel, iron, castings and forgings, half of the meat packing and flour, and two-thirds of knitted goods, are made in Ontario. Its ten chief manufactures are electricity, flour and cereals, pulp and paper, meat packing, electrical apparatus, butter and cheese, rubber goods, gasoline, and knitted goods.

Cities and Towns

Ontario, being the leader in manufacturing, has many large manufacturing towns and cities, the most of which are in the south-west

---

KEY TO NUMBERS ON ILLUSTRATION FACING THIS PAGE

1. Alexandria
2. Almonte
3. Arnprior
4. Belleville
5. Blind River
6. Brockville
7. Carleton Place
8. Cobalt
9. Cochrane
10. Copper Cliff
11. Cornwall
12. Englehart
13. Gananoque
14. Gowganda
15. Haileybury
16. Iroquois Falls
17. Kingston
18. Mattawa
19. Napanee
20. New Liskeard
21. North Bay
22. Ottawa
23. Pembroke
24. Perth
25. Picton
26. Prescott
27. Renfrew
28. Rockland
29. Sault Ste. Marie
30. Smith's Falls
31. Sturgeon Falls
32. Sudbury
33. Thessalon
34. Timmins
35. Trenton

peninsula. Usually they manufacture a number of different commodities (#133,#134), and often the same commodity is made in different towns widely scattered over the province. This diffusion of manufacturing, results from the fact that most of the manufactured commodities are used in the home market, often not far from where they are made. #133 and #134 show what a great variety of outstanding products are made in the chief towns and cities and how well most of them are distributed. It also shows other facts of note concerning the towns. Many towns on the Great Lakes are chiefly commercial in importance, other towns like *Timmins*, *Kirkland Lake*, *Sudbury*, *Haileybury* are important due to extensive mining development (p. 284), and a few, like *North Bay*, *Cochrane*, *St. Thomas*, and *Smith's Falls* are notably important because they are railway centres.

Toronto. There was an important trading post, called *Fort Rouillé*, a couple of miles east of the mouth of the Humber River, before the British conquest in 1763 (#116). After ten years, it was deserted and burnt by the French and the present city was founded in 1793. It was at first called York; later, when it became the capital of the province, it was named Toronto. From the day of its second birth it has never looked back, is still growing steadily, and is now the second largest city in Canada and the greatest manufacturing centre. Its first great asset is its broad harbour, perfectly protected by islands, and recently, by deepening and by building a beautiful waterfront, made into as noble a piece of water as is found on the Great Lakes. Though miles of the harbour are lined with elevators, warehouses, coal docks, and wharves, and though lake traffic is rapidly increasing, its chief commerce is by means of the spoke-like railways and radiating highways throbbing with puffing trains, and grinding motor-trucks. It is the distributing centre through which pass most of the goods from the busy factories of south-western Ontario to the buyers on the prairies.

Though it does not compare with Montreal in commerce, it now surpasses it in manufacturing. The cheap, abundant electricity, supplied by the provincial Hydro-Electric Commission makes the machinery of the factories hum. Its manufactures are of every class; automobiles and parts, agricultural implements, machinery, furnaces, and metal goods of all kinds, chemicals and paints, musical instruments, radios, and electrical goods, leather and knitted goods, hats, books, shoes and garments of all kinds, glassware, candy and chocolate, aluminium ware, carpets, silks, and cottons, rubber goods, and meat packing are a few of the outstanding commodities. It is also the chief centre of book publication in the country, many of the chief publishers of Great Britain and the United States having branches there.

Toronto is a beautiful city, with lovely green ravines, numerous well-kept parks, an impressive landscaped water-front with a handsome drive, the most complete exhibition grounds on the continent; but better than all these are the hundreds of clean, well-paved, well-lighted streets, lined by comfortable brick houses, usually owned by the family living within.

As the capital, it is the political centre with dignified parliament buildings in Queen's Park: just across from the parliament buildings are the many stone and brick buildings of the University of Toronto, the key-stone to education for the whole province, by far the best equipped in Canada, and with the largest attendance of any in the British Empire. Its fine college and its spacious gallery make it the Mecca for art in Canada. Its technical schools and numerous private schools for both boys and girls are unequalled anywhere else in Canada.

Hamilton. Until 1850 it was uncertain whether Hamilton or Dundas was going to be the leading town at the head of Lake Ontario. Then Hamilton leaped ahead, has grown rapidly ever since, until now she is only surpassed in size by Toronto, and is the third largest manufacturing city in Canada. It is delightfully situated at the head of Burlington Bay at the base of the noble Niagara escarpment, with the rich garden of the Niagara peninsula behind it. It has unlimited electrical power from Decew Falls and Niagara right at hand; natural gas is near; lake boats are at her front door; and the main lines of all the railways have sidings to her factories. Hamilton has proven especially attractive to United States capital, and a number of American industries have important branches on her waterfront. Steel, coke, agricultural implements, wire-fences, electrical machinery, tobacco, silk, cotton, woollens, pressed brick, chemicals, clothing, hardware, knitted goods, machinery, soap, and tools, are important commodities made in this industrial city. Hamilton does not neglect education. Her public schools, collegiate institutes, and technical schools are all modern and efficient; and McMaster, the Baptist university, has a beautiful group of impressive buildings in a delightful setting.

Ottawa. *Ottawa*, while still a village, was wisely selected by Queen Victoria for the capital of Canada. It was partly chosen because it was midway between the two provinces of Ontario and Quebec, which then made the whole of Canada, but chiefly because the attack of the United States in 1812 so haunted men's minds that it was considered wise to have the capital well back from an unfriendly border. Though any threat from the south is long past, Ottawa's situation has, nevertheless, proved admirable, and with numerous railway services it is as convenient to both east and west as can be expected in a long, narrow country.

Its setting is superb, surrounded by maple and pine forests, on a high rock overlooking the swift waters of the Ottawa, amidst the roar and the mist of the tumbling *Chaudiere cataract;* the view is picturesquely rugged in every direction. The Gatineau comes down in torrents from the north, the Rideau Canal winds quietly among the rocks from the south, and the lordly Ottawa surges along both east and west. Ottawa is a city of splendid buildings. It is crowned by the magnificent new Parliament Buildings, which is bordered by numerous, elegant departmental structures. Many wealthy citizens have built spacious homes and set them in beautiful gardens. Ottawa contains the Royal Victoria Museum, the National Art Gallery, a Catholic university, a Provincial Normal School, as well as elementary and secondary schools. The wonderful Dominion Experimental Farm, where so many scientific discoveries have been made, is in the suburbs. Close at hand is found beautiful scenery of rocky lakes, erratic rivers, broad forests, and the Laurentian hills which attract tourists from all over the continent. No city in Canada has so many rich sources of electrical power staring it in the face from every direction. It was long the centre of the lumber trade and still has large sawmills, pulp and paper mills, and factories for making woodenware. It has a number of foundries; manufactures articles of mica; and has many manufacturing stationers.

London. Even before London was a village, its attractiveness so appealed to Governor Simcoe that it was his first selection for a provincial capital. Though his choice was rejected, the good qualities which he saw still remained and have made it one of the most home-like cities in Canada. Though it is on no large body of water — for the Thames River at that point is scarcely navigable—it is embowered near the heart of the garden of Ontario. If a circle of fifty miles radius is drawn around London, it is doubtful if you could, anywhere else on the continent, find a similar circle containing such well cultivated fields, sleek high grade beef and milk cattle, well-shaped bacon hogs, beautifully trimmed orchards, prosperous looking farm houses and barns, and such an intelligent, industrious group of farmers. London arose as a farmer's town, manufacturing those commodities which they required and those which their raw materials would make. Though now she has out-grown being a mere farmer's town, and become a busy manufacturing city, she still shows those sturdy qualities learned in her favourable youth.

The rising young University of Western Ontario has a fine group of buildings, picturesquely placed overlooking the Thames. London also has a theological college and a normal school among her educational buildings.

From #133 and #134, the characteristic output of the other towns of Ontario can be studied.

CHAPTER XV

PRAIRIE PROVINCES

Uninteresting boundaries. The three Prairie Provinces — Manitoba, Saskatchewan, and Alberta — form a rectangular block 1,000 miles long and 760 miles wide between Ontario and British Columbia. Their boundaries are uninteresting straight lines, except where the Rocky Mountains separate Alberta from British Columbia and where, for four hundred miles, Manitoba is pressed into by Hudson Bay (#135). As the straight lines forming the north and south borders are the forty-ninth and sixtieth *parallels of latitude*, which all run parallel to the equator, the length of each from north to south is the same, namely eleven degrees, and as the distance between each two parallels is about 69 miles, each province is about 760 miles long. The straight north and south lines forming the chief part of the boundaries between the provinces are *meridians*, which, since they run from the equator to the pole, gradually come toward each other. Consequently these provinces become narrower toward the north, except Manitoba, whose eastern boundary diverges to give her a fair share of the coast of Hudson Bay. These purely artificial boundaries are the result of the land being so flat and formless that no natural boundaries could easily be found.

Ranges and townships. These convenient lines are not the only straight ones used as boundaries in the prairies. Indeed, a detailed map of one of the provinces looks like a drab network of minute square blocks. But what appears at first sight so uninteresting is, when understood, really very surprising, efficient, and useful. When in 1870 Canada had bought the whole of the North West Territories from the Hudson's Bay Company, it was necessary at once to mark it out in lots, so that the farms selected by settlers could be properly located and registered in their names. The Dominion government sent out a band of surveyors, who did their work so well and so quickly that in about three years the whole of the available land was divided and named. This close-grained network was the result. Starting at a meridian they divided the whole province into narrow north-south strips, six miles wide, called *ranges*, which they numbered from one to thirty, beginning at the east (#136). Each vertical strip, or range, was marked into blocks, six miles square, which

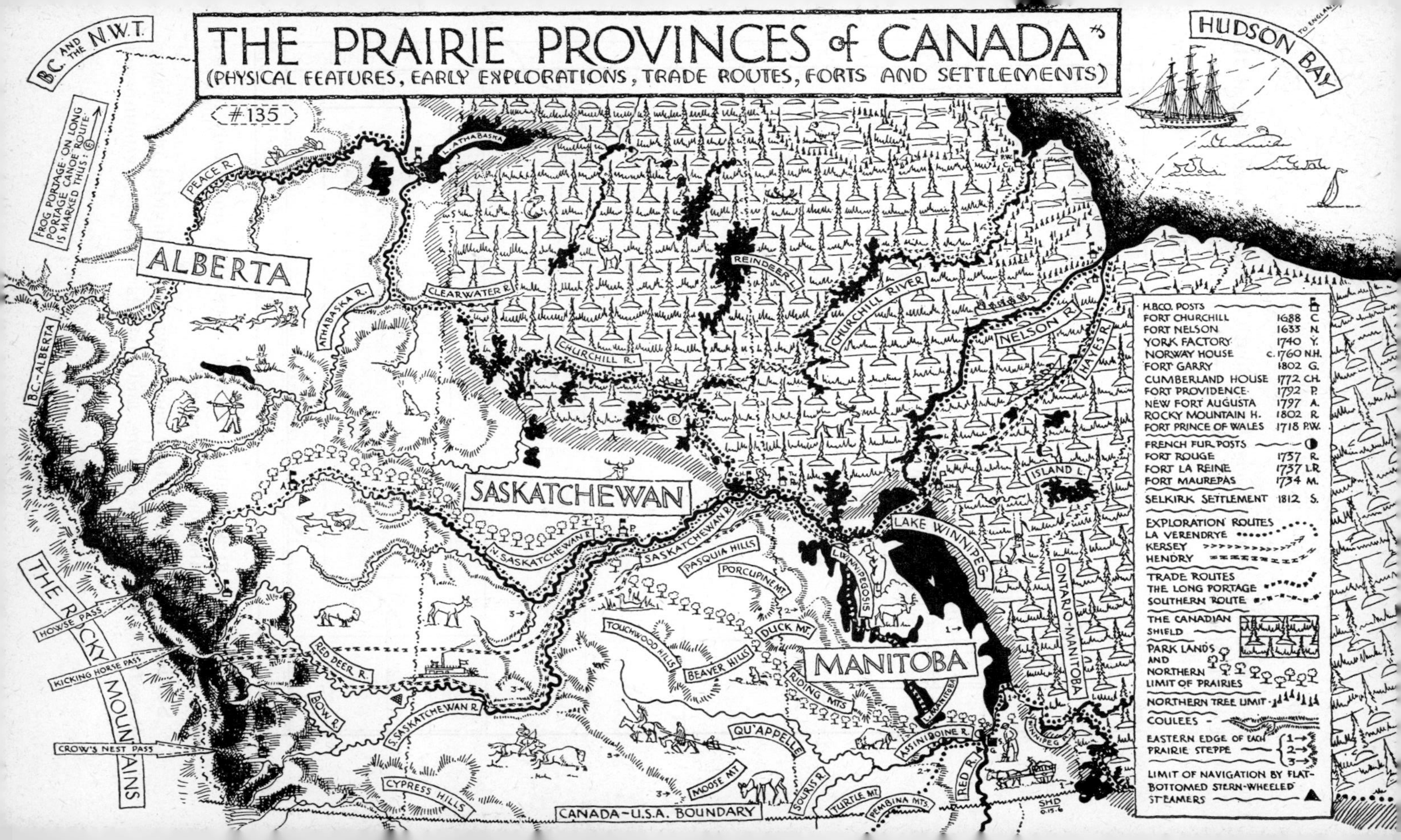
THE PRAIRIE PROVINCES of CANADA
(PHYSICAL FEATURES, EARLY EXPLORATIONS, TRADE ROUTES, FORTS AND SETTLEMENTS)
#135
B.C. AND THE N.W.T.
FROG PORTAGE ON LONG PORTAGE CANOE ROUTE IS MARKED THUS: (F)
HUDSON BAY
TO ENGLAND
ALBERTA
SASKATCHEWAN
MANITOBA
B.C.-ALBERTA
THE ROCKY MOUNTAINS
HOWSE PASS
KICKING HORSE PASS
CROW'S NEST PASS
PEACE R.
L. ATHABASKA
ATHABASKA R.
CLEARWATER R.
CHURCHILL R.
REINDEER L.
CHURCHILL RIVER
NELSON R.
HAYES R.
ISLAND L.
LAKE WINNIPEG
L. WINNIPEGOSIS
L. MANITOBA
ONTARIO-MANITOBA
WINNIPEG R.
ASSINIBOINE R.
RED R.
N. SASKATCHEWAN R.
SASKATCHEWAN R.
S. SASKATCHEWAN R.
RED DEER R.
BOW R.
PASQUIA HILLS
PORCUPINE MT.
DUCK MT.
RIDING MTS.
TOUCHWOOD HILLS
BEAVER HILLS
QU'APPELLE
SOURIS R.
TURTLE MT.
PEMBINA MTS.
MOOSE MT.
CYPRESS HILLS
CANADA-U.S.A. BOUNDARY
H.B.CO. POSTS
FORT CHURCHILL 1688 C.
FORT NELSON 1633 N.
YORK FACTORY 1740 Y.
NORWAY HOUSE c.1760 N.H.
FORT GARRY 1802 G.
CUMBERLAND HOUSE 1772 CH.
FORT PROVIDENCE 1792 P.
NEW FORT AUGUSTA 1797 A.
ROCKY MOUNTAIN H. 1802 R.
FORT PRINCE OF WALES 1718 P.W.
FRENCH FUR POSTS
FORT ROUGE 1737 R.
FORT LA REINE 1737 L.R.
FORT MAUREPAS 1734 M.
SELKIRK SETTLEMENT 1812 S.
EXPLORATION ROUTES
LA VERENDRYE
KERSEY
HENDRY
TRADE ROUTES
THE LONG PORTAGE
SOUTHERN ROUTE
THE CANADIAN SHIELD
PARK LANDS AND NORTHERN LIMIT OF PRAIRIES
NORTHERN TREE LIMIT
COULEES
EASTERN EDGE OF EACH PRAIRIE STEPPE
LIMIT OF NAVIGATION BY FLAT-BOTTOMED STERN-WHEELED STEAMERS
SHD

were numbered successively from south to north, beginning at the United States border. These square blocks were called *townships*. Hence if a farm is on meridian, second; range, 21; township, 25;

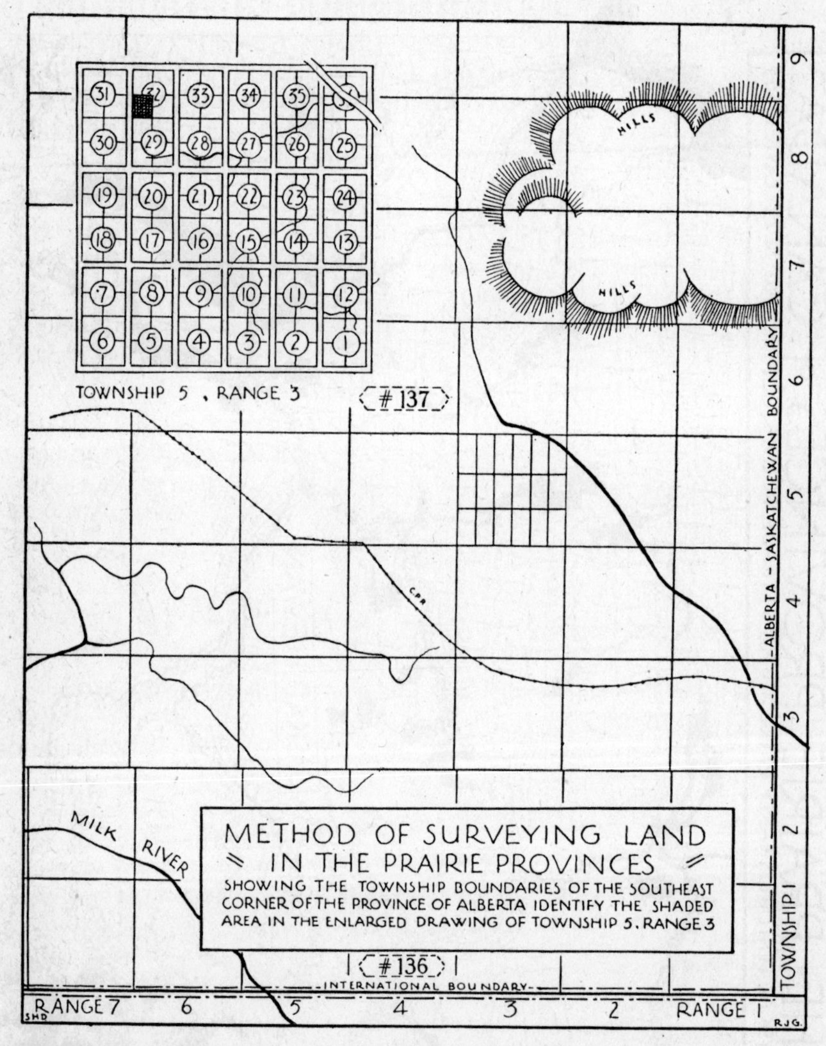

METHOD OF SURVEYING LAND IN THE PRAIRIE PROVINCES
SHOWING THE TOWNSHIP BOUNDARIES OF THE SOUTHEAST CORNER OF THE PROVINCE OF ALBERTA IDENTIFY THE SHADED AREA IN THE ENLARGED DRAWING OF TOWNSHIP 5, RANGE 3

its position can in an instant be more accurately and easily located than a house number on a city street.

Sections. #137 shows how each township is divided into thirty-six *sections*, each a mile square, and how the sections are numbered.

Moreover, the double lines show the land set aside for roads, 60 feet wide. Each section is itself sub-divided into four quarter-sections of 160 acres. For many years a settler was given free a quarter-section of good farm land. You can now readily see that if some kind friend gave you a present of the south-west quarter; section, 6; township, 27; range, 27; west of the fourth meridian; you could not only locate its exact position on a map, but in a trip with your motor-car you could find it with ease. Nowhere in the world is land divided more scientifically.

Joys and sorrows of surveyors. It was because the land was so level, offered so few hindrances to the surveyor, and had scarcely any settlers, that this wonderfully regular and simple division was possible. But the surveyors had their troubles. French half-breeds, already settled near the forks of the River Saskatchewan and having their land marked out in long strips running back from the river, became alarmed when strange men appeared with mysterious-looking instruments and began to mark the land in square blocks. They thought their farms were going to be taken from them, and sometimes drove away the surveyors, and finally broke out into rebellion in 1885, which caused considerable loss of life.

Divisions. These three provinces are composed of two very distinct regions, separated by a line shown clearly in #135. The eastern and northern part is in the Canadian Shield, the southern and western part in the Great Plain region.

More than two-thirds of Manitoba, more than one-third of Saskatchewan and a small part of Alberta north and south of Lake Athabaska, are in the Canadian Shield. It forms a great shield-shaped mass of granite rocks with Hudson Bay at the centre and the St. Lawrence River and a string of Great Lakes from Huron and Superior to Great Bear on the circumference (p. 293).

The Canadian Shield. This Canadian Shield, which is the oldest and most deeply worn part of the world, lays bare the very foundation of the crust of the earth. In the far distant past awful volcanic outbursts convulsed all this vast area. Great fiery seas of molten rock, pressed from below, bulged up the layers of rock above them in hugh bubbles, hundreds of miles in diameter (#138). Then the lava beneath the bubbles cooled and left dome-shaped mountain masses many thousands of feet high, with a great sea to the west and south where the Great Plain now stands (#138). For countless ages air, water, and ice eroded the high domed mountain masses, until they have been worn to their very bases and everywhere exposed the volcanic roots beneath (#138). This region has been left com-

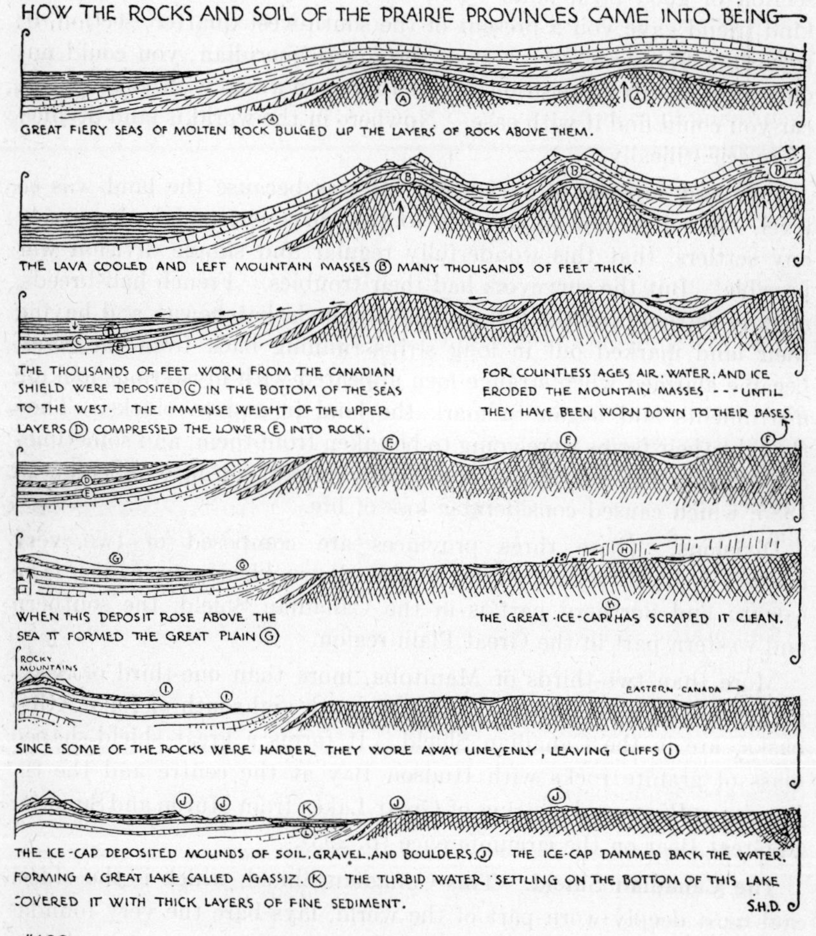

#138

FORMATION OF PRAIRIE PROVINCES

paratively low and flat; but as the harder rocks project, the surface is broken by irregular hills, mounds, valleys, and ravines. At one time this area was covered deeply with soil, but the great ice-cap, already mentioned, pushing out from this region in every direction has scraped it clean, and only along quiet parts in river valleys has new soil been laid down since. It is in the depths of the earth where heat is intense and pressure enormous that valuable minerals are collected in cracks to make *veins* (p. 28). Nowhere in the world have these veins been exposed more completely by the wearing away of the covering rock than in the Canadian Shield. The foundations exposed should be creased with mineral veins almost as much as a skirt is creased with pleats. It is not surprising that this region is proving as rich in valuable metals as the prairies further west are in fertile soils.

The Great Plains are formed. The thousands of feet of surface worn from the Canadian Shield were deposited in nearly horizontal layers of sand, clay, and lime, thousands of feet thick, in the bottom of the seas to the west, where the Great Plains now stand (#138). The immense weight of the upper layers compressed the lower into rocks of sandstone, shale, and limestone. When this area rose above the sea, to form the Great Plain, the layers were nearly flat and level, as they are to this day. Since, however, some of the rocks were harder than others, they have worn away unevenly, leaving layers of hard rock projecting as cliffs. The surface has been further roughened by the ice-cap, which has deposited great mounds and lines of a mixture of soil, sand, gravel, and boulders, which, on such a flat surface as the prairies, are called hills and even mountains (#135).

The prairie steppes. The surface of the Great Plain in the Prairie Provinces rises slowly and steadily from the Canadian Shield to the Rocky Mountains, but the steady rise is broken by two sudden steps, which form cliffs running obliquely across the provinces (#135). Viewed from the east they look like three broad platforms, one rising above and beyond the other. These are called the *three prairie steppes*, and the margins of the second and third form ranges of hills which are marked in #135.

The lowest steppe lies between the Canadian Shield and the first cliff, or escarpment, which is marked by the *Pembina*, *Riding*, *Duck* and *Porcupine mountains* and the *Pasquia Hills* (#135). It is the flattest and lowest of the three steppes, especially in the Red River plains of Manitoba. From east to west, first the woods vanish, rough places become smooth, the surface soon becomes as level as a bowling green, lakes and rivers lose their vigour, become flat, slow, and muddy, and the horizon widens. The soil is as rich as its surface is flat.

The Red River plain that once was a lake. This whole region, which is now drained by the Nelson River into Hudson Bay, was formerly a great lake, called *Agassiz* (#138K). An ice-cap, thrown around Hudson Bay, at that time dammed back the water so that it had to drain south in a swollen stream through the Red River into the Mississippi River. As the Saskatchewan River was blocked, turbulent currents of muddy water at the same time streamed through the South Saskatchewan, Qu'Appelle, and Assiniboine, which were one huge river, whose deep coulee, with shrunken streams, can still be traced through the flat prairie. The mud from the turbid water of this enlarged Assiniboine, settling year after year on the bottom of Lake Agassiz, levelled its hollows and covered it with thick layers of fine sediment. As the ice-cap across the Nelson River melted, the lake level sank, and the higher land was laid bare; to-day *Winnipeg*, *Manitoba*, *Winnipegosis*, and many lesser lakes, are fragments of the deeper parts of the giant that once was Lake Agassiz, a lake larger than all the Great Lakes combined. The parts of Agassiz that were not so deep are now the many marshes that surround these lakes, and the shallowest part of all has become the fertile prairie plain of Manitoba, which in some places is so level and smooth that, after rain, water may stand on the ground in sheets. Hence the marshes and dry land were once all lake; the dry land of to-day was the marsh of yesterday, and the lake of the day before.

The second and third prairie steppes. The second prairie steppe, averaging 1,600 feet in height, is more rolling than the plain of the Red River, its sameness being broken into gentle waves and broad, gradual slopes. The third prairie steppe, averaging 3,000 feet, rises from 2,000 feet at its eastern edge to 4,000 feet at the foot of the Rockies (#135). It is not only higher, but drier and barer than the other steppes and changes from east to west like a calm sea passing into a stormy one, until the confused foothills of the Rockies sweep away all likeness to a plain.

Mining

The minerals of Canadian Shield. For countless ages nature gathered together in veins in the bowels of the Canadian Shield a huge assortment of gold, silver, copper, zinc, nickel, platinum, and other metals. Then for countless centuries she wore away the great plateaus of rock that covered the veins and exposed them

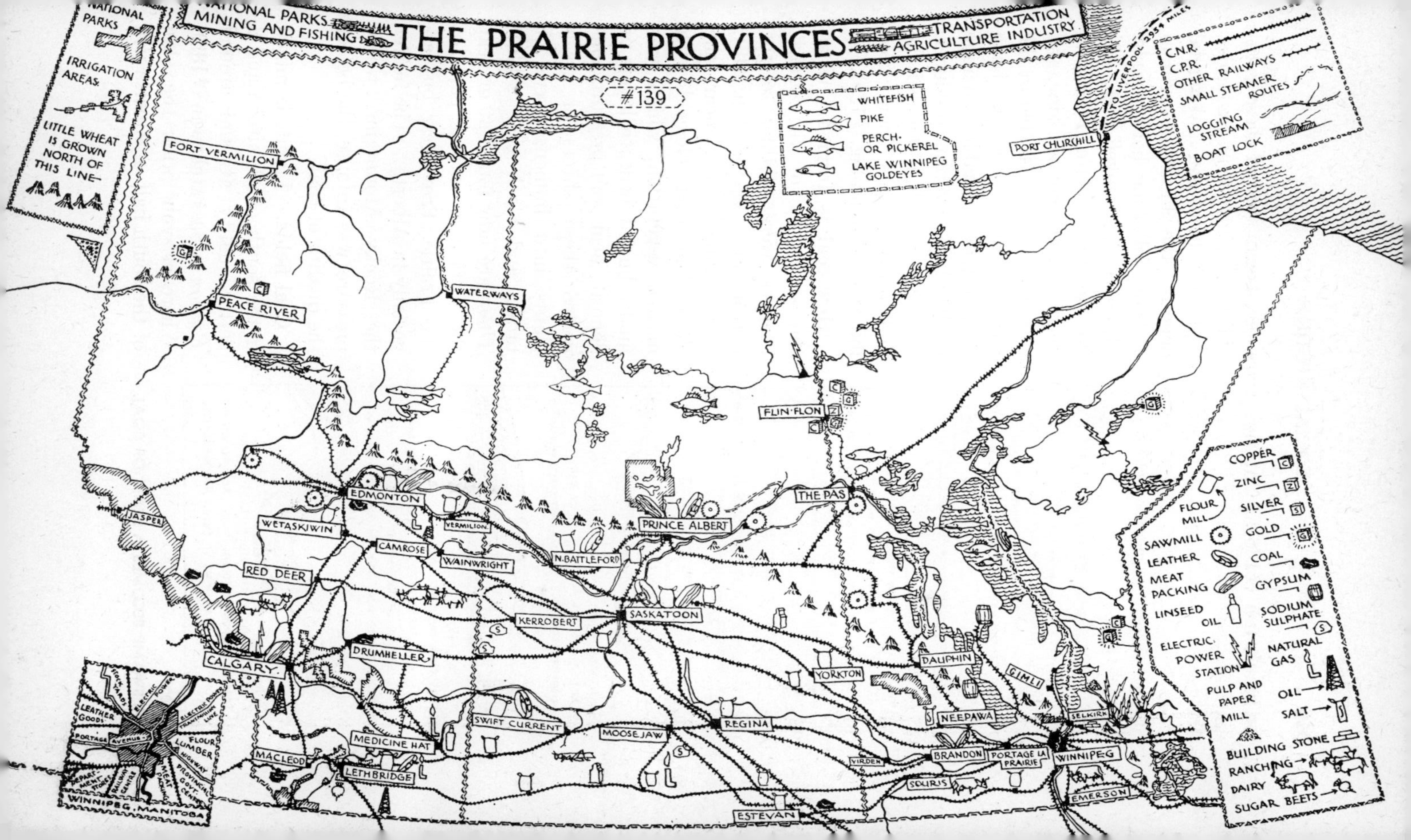
THE PRAIRIE PROVINCES
NATIONAL PARKS
MINING AND FISHING
TRANSPORTATION
AGRICULTURE INDUSTRY
#139
NATIONAL PARKS
IRRIGATION AREAS
LITTLE WHEAT IS GROWN NORTH OF THIS LINE
WHITEFISH
PIKE
PERCH- OR PICKEREL
LAKE WINNIPEG GOLDEYES
C.N.R.
C.P.R.
OTHER RAILWAYS
SMALL STEAMER ROUTES
LOGGING STREAM
BOAT LOCK
TO LIVERPOOL 2954 MILES
FORT VERMILION
PEACE RIVER
WATERWAYS
PORT CHURCHILL
FLIN FLON
THE PAS
EDMONTON
JASPER
WETASKIWIN
VERMILION
CAMROSE
WAINWRIGHT
PRINCE ALBERT
N. BATTLEFORD
RED DEER
SASKATOON
KERROBERT
CALGARY
DRUMHELLER
YORKTON
DAUPHIN
GIMLI
SWIFT CURRENT
MOOSE JAW
REGINA
NEEPAWA
SELKIRK
MEDICINE HAT
MACLEOD
LETHBRIDGE
VIRDEN
BRANDON
PORTAGE LA PRAIRIE
WINNIPEG
SOURIS
EMERSON
ESTEVAN
COPPER
ZINC
SILVER
GOLD
COAL
GYPSUM
SODIUM SULPHATE
NATURAL GAS
OIL
SALT
FLOUR MILL
SAWMILL
LEATHER
MEAT PACKING
LINSEED OIL
ELECTRIC POWER STATION
PULP AND PAPER MILL
BUILDING STONE
RANCHING
DAIRY
SUGAR BEETS
STOCK YARDS
ELECTRIC POWER
ELECTRIC POWER TRANSMISSION LINE
LEATHER GOODS
PORTAGE AVENUE
FLOUR
LUMBER
HIGHWAY
PROVINCIAL GOVT. CENTRE
MEAT PACKING
RAILWAY CENTRE
DEPARTMENT STORES
WINNIPEG, MANITOBA

for man's use (#138). Ontario, years ago, began to find these valuable seams, in most places hidden under a film of soil and glacial deposit; the Prairie Provinces are now following her example. Manitoba has moved first. In the north, at *Flin Flon* and *Sherridon*, are copper, zinc, and gold, which are extracted at the former place. East of Lake Winnipeg one valuable gold mine is already a vigorous producer, and further north, at God's Lake and other points, new gold mines are developing (#139).

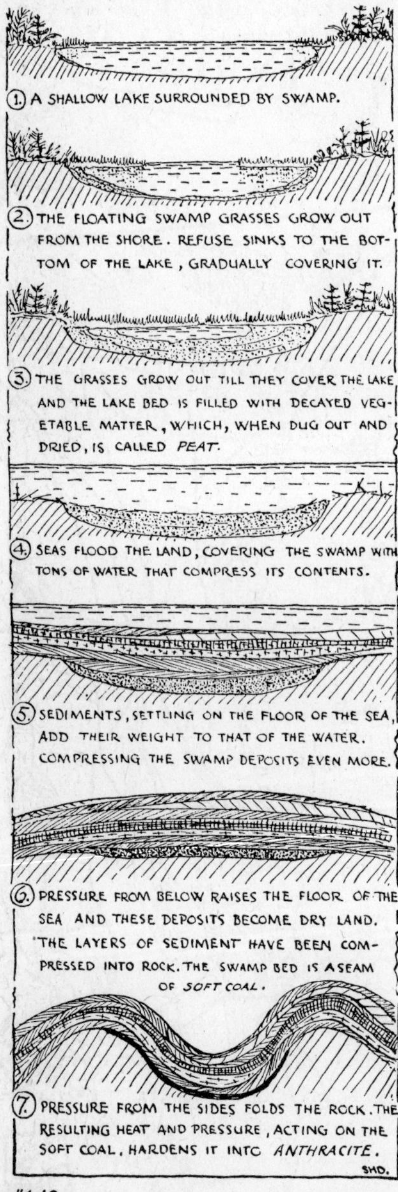

#140
A SWAMP BECOMES A SEAM OF COAL

How coal is formed. When the sediments of the Prairie Steppes were settling to the bottom of a vast sea (p. 343), the shallow water near the shore was often clogged with matted vegetation, which was finally buried beneath layers of deposit; heat and pressure converted these sheets of green plants into immense beds of brown and black coal (#140). These are now found at or near the surface in various parts of the Prairie Provinces, but especially in Alberta. The nearer the Rocky Mountains, where pressure was intense, the better the quality of coal.

Coal fields. Lignite is mined in the south of Manitoba and a better quality at and around *Estevan* in Saskatchewan (#139). Wherever you prick the skin of the south half of Alberta you are almost sure to find coal, and the seams in many places even protrude along the steep banks of rivers. *Fernie, Lethbridge* and *Drumheller* in the south, and *Mountain Park* and *Edmonton* in the north contain important collieries.

Turner Valley. Alberta is also Canada's chief producer of petroleum, though the product of the chief wells at *Turner Valley,* south of Calgary, is not really petroleum but a mixture of liquid naphtha

#141

FIREWORKS IN TURNER VALLEY

and an inflammable gas. After condensing as much of the gas as possible into liquid naphtha, the rest is led away and wilfully wasted. In order that it may not poison the air, it is allowed to burn in great flaming torches, which at night make the whole district as light as day (#141). This valuable gas is wasted because at present there is no large surrounding population to use it. Besides this more valuable product, ordinary heavy petroleum is obtained in moderate quantities and refined at *Wainwright.* There are many other signs that petroleum is wide spread in Alberta. Small quantities are mined on the south border, and far in the north, near *McMurray,* are immense areas of tar sand, which is probably the residue left from pools of petroleum, and which is valuable paving material for roads. Natural gas from *Viking* is piped to *Edmonton*, and *Medicine Hat* also has abundant supplies.

Non-metallic minerals. Gypsum, which as "gyproc" now rivals lumber as a building material, is extensively quarried in Manitoba at *Gypsumville*, and a limestone at *Tyndall*, north-east of Winnipeg, is so ornamental that it was transported all the way to Toronto to be used in the building of the T. Eaton Company's beautiful departmental store. The rainfall is so small and the summer evaporation so intense in some parts of the prairies that white salts gather on the edges and bottoms of lakes. One of these salts, *sodium sulphate*, which is largely used in nickel extraction, is now shipped from several towns in Saskatchewan (#139) to Copper Cliff smelters in Ontario.

A second story for the Prairie Provinces. The Prairie Provinces have trusted too much to the growing of wheat and oats for their prosperity and are now looking eagerly to the north to develop their mineral wealth, so that they will no longer have all their eggs in one basket. In the north they have water-power to run mining machinery, and the bountiful bosom of the Canadian Shield begging them to seek its riches. Prospectors already are swarming over the field; mineral veins around Lake Athabaska are at the present time being tested, and undoubtedly in the next few years these provinces, through the development of their northern mines, will build up a second story.

Drainage

The Churchill River, a cluster of stars. The rivers flowing through the Canadian Shield are decidedly different from those in the Great Plain. The *Churchill* is a good example of the former, as almost the whole of its fantastic course is in this region (#135). It is hardly a river, but is like a cluster of stars of all sizes scattered over the face of the Canadian Shield. Each star is a crystal lake nestling among the hills of granite and dotted with a swarm of beautiful islands. These lakes are joined by threads of water running in every direction like the fibres of a piece of lace. Streams of sparkling water flow from the higher to the lower lakes in foaming rapids and tumbling cataracts, marking a disorderly and complicated embroidery of currents, that move in a easterly direction, to eventually makes up the Churchill River, which empties into Hudson Bay. The *Nelson,* draining Lake Winnipeg, and the *Hayes* are very similar to the Churchill.

Rivers of the plain. The rivers that writhe and wriggle away from the Rocky Mountains through the plain are entirely different. They are much older, have eroded deep valleys through the soft, horizontal layers of rock, and stagger back and forth in their valleys like drunken men. Their deep valleys (#135), called *coulees,* are the most startling breaks in the monotonous level of the Prairie Steppes. As the traveller, moving over the flat, dry, treeless prairie, comes without warning upon one of these chasms, generally a mile wide and several hundred feet deep, he is filled with astonishment. Far below he will see a thin, muddy stream swaying back and forth in its channel like a thread unravelled from the toe of a woollen sock. It looks a mere shadow of what it ought to be, to have worn such a wide, deep valley. The contrast of lusty growth of sedges, shrubs, and trees along its moist banks with that of sallow stretches of yellow grass, sage-bush, wormwood, and cactus at his feet adds to the surprise. Streams like the *Athabaska, Saskatchewan,* and *Bow,* which rise in the Rocky Mountains have dug deeper and broader coulees than the *Qu'Appelle, Souris,* and *Assiniboine,* which begin apparently nowhere in the plain.

Rivers and transport. The Saskatchewan-Nelson system is the longest and drains a very large area, for it includes the Assiniboine, which empties into the Red River at Winnipeg, and the Red itself which rises far to the south in the United States. These rivers of the plain are for the most part navigable by flat-bottomed boats with big paddle-wheels at the stern (#135). Before railways were built, steamers used to sail from Selkirk on the Red River, through Lake Winnipeg, to Prince Albert, Battleford, and even to Edmonton, and, on the South Saskatchewan, to Saskatoon and even to Medicine Hat. The chief transportation system of the north is still the steamers on the Peace River, Athabaska River and Lake Athabaska. Lake Winnipeg, and to a less extent Manitoba and Winnipegosis, are used for transport of fish, lumber, and minerals. Nevertheless the rivers of the Prairie Provinces are not well suited for shipping. They are frozen for nearly half the year, are the longest rather than the shortest distance between two points, are shallow, blocked with sandbars, swift in places, subject to sudden rise or fall, and are constantly shifting their channels on account of the great quantities of mud they carry. When they cross the Canadian Shield they are broken by many rapids and falls. When one sets beside these hindrances the ease with which one can go on foot, on horseback, or by road in any direction over a level and for the most part treeless plain, the prospect of any increase in river navigation is small.

A lock, to overcome the rapids at Selkirk, allows Lake Winnipeg traffic to go to Winnipeg (#139). There are no steamers on the Red River above this city, though at one time the competition between steamers to supply Winnipeg with goods from the United States was keen and selfish.

Fisheries. Many large lakes, whose cold waters are well stocked with fine fish, are scattered over the Prairie Provinces, but most of these are so far north that the supplies are only available to the Indians and trappers for their personal use. The large lakes of southern Manitoba, however, are even larger than the Canadian sections of Lakes Superior, Huron, and Erie, from which Ontario gets her chief supplies (#139). Pickerel, whitefish, goldeyes, trout, and pike are obtained in large quantities and for the most part shipped frozen in carload lots to the United States and Ontario. Smoked and coloured goldeyes from Manitoba are one of the most attractive products on Ontario fish markets. In Alberta, Saskatchewan, and northern Manitoba fishing is a winter occupation, since the product often has to be carted hundreds of miles to bring it to market. Nets are set through holes in the ice, and the whitefish and pickerel are frozen by nature in these rigorous climates as soon as they are taken from the water. Formerly, they were then brought south on horse-drawn sleighs, but in recent years aeroplanes, equipped with landing skis, have carried out most of the catch. The largest northern lakes, such as Reindeer, Wollaston, and Athabaska, have not yet been touched.

Climate

Clear, cold winters. By the time the westerly winds, which bring bountiful rains, cool summers, and mild winters to the Pacific coast, have hurdled over range after range of British Columbia mountains, they are too exhausted to have much influence on the Prairie Provinces. This region has the typical climate of the interior of a continent, hot in summer and very cold in winter. These conditions are most pronounced in Saskatchewan, which has not the balmy *chinook winds* descending from the mountains to soften the intensity of the clear cold of winter, as has Alberta; nor the moderating influence of Hudson Bay and the Great Lakes, as has eastern Manitoba. Winter begins in earnest in November, when rivers and lakes freeze firmly, snow covers the ground, the days become short, and the long nights have sparkling clear skies, often blazing or glowing with the northern lights. Because the air is so clear and dry, and winds are few and weak, this season is not so severe as one is likely to think it might be, with temperatures ranging for weeks at a time well below zero. It is a stimulant tonic to the vigorous, children play out of doors, men go on with their work, and the cattle chew tranquilly the cud about the sunny farm-yards; nevertheless, it is trying for the aged and sickly. Spring comes in so suddenly that by April the ground is clear of snow and dry enough for seeding to get well under way; spring comes almost as early in the *Peace River country* as in the south. The nights are still cold and frosts threatening even in June.

Delightful summers. Summer is delightful. The days are hot, bright, with few strong winds or gales. Fifteen hours of sunshine daily stimulates the most vigorous growth in vegetation. The nights are pleasantly cool and short, with clear skies. By the end of August the nights are colder, and killing frosts may do great damage. September and October, though cool, are pleasant; before October is past ice appears on lakes and rivers, and a warning that the stern arm of winter is about to embrace the country.

The Chinook wind. Winter in western Alberta, under the shadow of the Rocky Mountains, is more variable than anywhere else in the world. Sometimes the rigours of winter do not relax for months. Again a south-west chinook may come down from the mountains (#149), and for days, or even weeks, these balmy breezes may be well above freezing (p. 379). A change of sixty degrees may occur in a few hours.

Rain, hail and snow. While the long sunshiny days of summer so suitable for agriculture, can be depended on year by year, the rainfall, equally vital to growth, unfortunately is not so reliable; nowhere is it in abundance, and in most places there is such a narrow margin of safety, that if it is even a little below the average, grain crops, which are the life blood of the prairies, suffer. The rainfall is from ten to eighteen inches, and in a general way the quantity increases from west to east and from south to north. Hence the greatest danger spot from drought is southern Alberta and Saskatchewan. Fortunately most of the rainfall is in the growing season, from May to July. The snowfall everywhere is light and increases from western Alberta, where there is little snow on the ground during the winter, to eastern Manitoba, where there may be fifty inches. Most of the summer rains fall in thunderstorms, and these may be accompanied by hail, which, where it strikes, may suddenly convert a waving wheat field of beautiful greenery into a pitted desolation of black mud.

Manitoba blizzards. In the Prairie Provinces strong winds are uncommon and gales rare; but during late winter, especially in the eastern part, they may be visited by fierce north and north-west gales, called *blizzards*. These furious storms may lash the country for many hours. They lift the fine snow from the ground, drive it in fury through the air like a raging snowstorm, and deposit it in immense drifts. The deadly below-zero wind, the fine snow crystals cutting the face as though shot by a gun, and the masking of the view, play havoc with men and beasts. Woe betide those who may be overtaken on the open prairie; they may find it impossible to move against the blustering blizzard, soon lose their way, and sink down exhausted to die in the fearful cold. These cases are fortunately rare, and are unknown in the western parts.

Exploration and Fur Trade

Discovery of Hudson Bay. The exploration of the Prairie Provinces and the whole North-West Territories is closely bound up with the discovery of Hudson Bay. In 1600, the only passage known from the Atlantic to the Pacific Ocean was the distant, storm-tossed mazes among the islands of Tierra del Fuego, discovered by Magellan. Jacques Cartier and Samuel de Champlain had sought in vain for a water passage through the continent to the Pacific (p. 251), and now efforts were directed to find a sea path to the riches of China and India by a *north-west passage* around the continent. *Martin Frobisher* had already skirted the bleak shores of Hudson Strait, but his men were so busy digging boat loads of glittering ore, which proved to be fool's gold, that he never reached that great inland sea, Hudson Bay. *Henry Hudson*, after exploring the New England coast, sailed up the Hudson River past the site of New York City. Still led by the lure of the north-west passage, he sailed further north in 1610, passed through the 400 miles of the ice-infested waters of Hudson Strait finally to be rewarded by the sight of a great inland sea, which might be the north-west passage, that he and other explorers had been seeking. He sailed its icy and uncharted waters to the extreme south of James Bay (#116), where he was overtaken by a winter that so crippled his crew with sickness and hardship that they were driven to enact one of the north's most pathetic tragedies. They cast their heroic leader and his young son adrift in a small boat amidst the ice of James Bay in June, 1611. The two were never heard of afterwards. But this daring explorer was even more powerful in his death than in life, for when his pathetic story reached Europe, explorer after explorer searched this inland sea. One can imagine with what eagerness they scanned every bend in its western coast for that alluring passage into the Pacific Ocean.

Hudson's Bay Company. Shrewd business men in England, more intent on making money than on reaching China, soon realized the value of this bay for trade in furs. Here was a great sea in the centre of a continent, easily reached in summer through Hudson Strait. It was a giant hub, from which radiated far back in every direction numerous rivers, along which Indians could bring their furs. The *Hudson's Bay Company* was soon established, and at the mouth of every important river erected a fort, or trading-post. One of the very earliest of the traders selected *Fort Nelson*, at the mouth of the Nelson, as the chief port. One marvels at his wisdom, for he could not have had any exact knowledge that this was a focus from which the best canoe routes spread out through the Churchill and Nelson to the Saskatchewan, Athabaska, Peace, and Mackenzie Rivers. This fort, changed over later to *York Factory* at the mouth of Hayes River, held its position as the hub of Hudson Bay for more than two hundred and fifty years, until Churchill, with its railway terminus, displaced it in the last few years and became the leading port in the bay.

The struggle for Hudson Bay. It is not surprising that fur traders at Montreal became alarmed. To reach that eastern port the Indians had to travel in canoes many hundreds of miles, through dangerous rivers, make at least fifty portages, and finally pass through a bottleneck on the Ottawa River, infested with cruel Iroquois. The hated English had worked in behind them on Hudson Bay, to which every river outside of the narrow basin of the Great Lakes led, and held a clever position to deflect the fur trade. It is no wonder that the French attacked Hudson Bay posts by land and sea, that many bloody battles were fought, fortresses raised only to be destroyed, and trading posts passed back and forth from French to English according to their strength. But the British were final masters.

The Prairie Provinces and the North-West Territories remain the oldest British part of Canada (p. 170), and the Hudson's Bay Company still is master of the fur trade on the bay (#51).

La Vérendrye on the Saskatchewan. When they recognized that the Bay was lost, the French pushed their traders farther back into the western wilderness to get behind the British on Hudson Bay and deflect trade. *Sieur Pierre de La Vérendrye,* a native-born French-Canadian, was the leader in promoting French trade west of Lake Superior. He was, like so many others, searching for the Western Sea. In 1731 he came up the Ottawa, across Lake Huron, past Sault Ste. Marie with fifty French-Canadian voyageurs, in birch-bark canoes of great size and packed with goods to trade to the Indians for furs (#116). They were accompanied by soldiers, decked in gorgeous uniforms. Fierce winter was upon him by the time he reached *Fort Kaministikwia,* which is now Fort William (#116). He had his three young sons with him. From beginning to end of his expedition he was dogged by disaster, and threatened with starvation because his financial backers in Montreal were hungrier to receive furs than to send forward sorely-needed food and powder and shot. Nearly half the precious time of this eager explorer was wasted going back and forth, on the long tiresome route to Montreal, trying to satisfy the greed of the Montreal fur traders. But in spite of poverty, the unfriendliness of Indians, the massacre of one son, and the death from privation of another, he pushed tenaciously forward to Lake Winnipeg, whose waters he was the first white man to see. There he built Fort Maurepas (#135). He ascended the brown waters of the Red River, built Fort Rouge near its juncture with the Assiniboine River, ascended the latter, established a fort at Portage la Prairie, called *La Reine* (#135), marched overland to the Missouri, explored Lake Manitoba and Lake Winnipegosis, and the Saskatchewan as far as its forks (#135), and perhaps saw the Rocky Mountains.

British explorers on the prairies. At first the Hudson's Bay Company did not encourage exploration, since their policy was to let the Indians bring the furs to the trading posts at the mouths of the rivers on Hudson Bay. But as early as 1690, *Kersey,* a mere boy, went westward on the Churchill River (#135), and perhaps reached the Saskatchewan. He saw musk-oxen and the American buffalo,which he graphically described. In 1754, *Anthony Hendry* left York Factory, which had, by this time, replaced Fort Nelson (p. 353), ascended the mazes of the Hayes River, paddled along the swift Saskatchewan and Carrot Rivers, where he cached his canoes, took to the prairie, crossed the south branch of the Saskatchewan, near Saskatoon, and went as far as Calgary. There he made the acquaintance of the Blackfoot Indians, the fiercest and most intelligent tribe of the plains, who chased the buffalo on horseback. He returned along the South Saskatchewan and the Hayes Rivers to York Factory. While the heart-breaking mountains and the deadly rapids and waterfalls, met by the explorers of the Pacific, were absent, he faced many shoals to be portaged; he faced drenching rains, hot midsummer weather, and a trying country infested with clouds of thirsty blood-sucking mosquitoes. The explorations of the *Frobishers, Samuel Hearne, Peter Pond,* and *Peter Fidler* added to our knowledge of the Prairie Provinces and northern Canada. It required courage, patience, and doggedness to pass into these unknown regions among Indians whose attitude was sometimes unfriendly and always uncertain.

Canoe routes to the west. Until about 1800, the chief canoe routes of both explorers and fur traders were from York Factory, or Fort Nelson, up the Hayes or the Nelson River to Cumberland House on the Saskatchewan River (#135), through a necklace of lakes by Frog Portage (#135F) to the Churchill River, which was followed through *Churchill* and *Peter Pond Lake, Methye Portage* to the *Clearwater* and *Athabaska Rivers.* This opened up the route along the Peace and the Mackenzie Rivers, a route that Mackenzie and Fraser followed to the Pacific (p. 382). About 1800, this, called the *Long Portage,* began to give way to the more southern route from Cumberland House through the North Saskatchewan to Edmonton. This was the route followed by Thompson.

Vegetation

The prairies, "boundless and beautiful." Only the southern part of the Prairie Provinces can be classified as grasslands. These areas of tall grass and brilliant flowers, stretching away in every direction to the far distant horizon like a green sea, made billowy with every gust of wind, are a solemn and impressive sight; it was still more so, eighty years ago, when thundering herds of thousands of buffalo roamed in every direction, with not a fence to hinder nor a house to alarm. Nowhere is the prairie entirely treeless, since along the moist banks of streams, deep down in the coulees, the rustle and shimmer of poplar leaves greets the ear and eye. With increase in dryness toward the west, the vegetation changes; the continuous carpet of tall grass becomes broken into tufts of *bunch* and *buffalo grass.* Thirsty *sagebrush* and *spiny cactuses*, those symbols of the desert, begin to show their sallow faces. It is usually stated that the absence of trees is due to the niggard rainfall, but the streets of the prairie cities like Regina, Brandon, and Winnipeg, overspread by the cool shade of lines of trees, indicate that trees can be grown, and many a farm-yard decked with trees tells the same story. Probably the fires, set by the Indians every autumn to improve the hunting, and which swept like a whirlwind over the prairies, consuming every living thing that was not able to flee before it, has much to do with its nakedness.

The northern woods. As one goes northward, the increasing moisture, due to less evaporation, gradually changes the prairie into forest. First scattered clumps of trees appear, like green islands on a brown sea, in what are called *parklands* (#135); then the clumps of trees expand in size and become more numerous until at last in the northern parts of all the Prairie Provinces there are dense, softwood forests. Spruce is by far the most valuable for lumber and pulpwood, the jack-pine is not to be despised for lumber and makes satisfactory railway ties and fuel. The paper birch, the only important hardwood, is eagerly sought in the wood-hungry prairies for furniture. Small sawmills are widely distributed in the north to cut the logs floated down the rivers (#139), and one of the most important mills is located at *The Pas* in Manitoba. *Prince Albert* is the centre of lumbering in Saskatchewan. In Alberta extensive lumbering is confined to the slopes of the Rocky Mountains, and the chief saw mills are at *Edmonton* and *Calgary* (#139).

Agriculture

The home of Manitoba, number one, hard. Little, red, glassy kernels of Marquis wheat, almost as hard and brittle as a rock, rule life in the Prairie Province with a tyrant's rod; these same small grains stand as dictators over the prosperity of all Canada more than all the gold from the mines, the fish from the sea, or the timber from the forest. The soil and climate of the prairies are just to the liking of this aristocrat among the grains. For thousands of years the glacial soils, which are spread unevenly over the country, have been deeply buried under rank grasses. Each autumn these plants wither and decay, to add the dead matter of leaves, stems, and flowers that enrich the soil, until the accumulation from long centuries has made it rich with nitrogen, phosphorus and mineral salts. The cool, wet weather of spring and early summer pleases the dainty wheat seedling, which is excited by the coolness to send out many branches, or *tillers*, and coaxed by the moisture to lengthen each branch into a long, strong, straight stem with a well-filled head of fat grains on the top. Then the warm, dry sunshine of late July and August converts the swollen milky grains into dry red grains, bursting with gluten, and as hard as rock, which grain inspectors christen with the noblest name in the wheat world, *Manitoba, Number 1, hard*; the wheat which bakers all over the world insist on having to make the highest grade of bread.

Sowing the seed. The land may have been ploughed the summer before to kill the weeds and preserve the all too scanty moisture, or it may merely be necessary to run a disk plough through the stubble of last year's wheat field. Whatever the preparation, it must be done with feverish speed as soon as the melting snow lays bare the soil, for the period before the first autumn frost is all too short. On the average, a farmer plants nearly one hundred acres with wheat; and planting must be completed in May. Then, as the crop looks after itself until harvesting, he can plant oats, flax, alfalfa, corn, sunflowers, vegetables; harvest his hay, and begin summer-fallowing for next year.

Harvesting the grain. Before the end of August four-horse binders of eight-foot cut, flash their reels over the wide fields of golden grain, and from dawn to dark move round the field in smaller and smaller circles; there dare be no pause, as cold autumn, with perhaps snow, is not far away. The sheaves are not stacked or stowed in a barn as in eastern Canada, but drawn at once to the thresher. The grain is either stored in big movable metal bins in the field or drawn by truck or team to a country elevator. Some farmers, who have not bins, even pile the grain on the ground. Because of the scarcity of labourers and the need for speed on account of the threat of winter, an increasing number of farmers are using what are called *combines* (#142). These wonderful machines, drawn by a tractor, cut off the heads in a swath fifteen feet wide, thresh the grain, and store it in a bin on a truck as the combine moves through the field.

A struggle for life. The wheat grower of the prairies is more the creature of wind and weather than any other farmer. If June is dry and hot, dust storms may not only destroy his whole crop, spray every article in his house and barn with dust and dirt, but remove completely the layer of soil, and leave his farm a barren desert. This is one of the most threatening dangers of the prairies. Hail,

#142

A COMBINE

This cuts off the heads of wheat, threshes them, and runs the grain in the small cart to the right.

in a few minutes, may batter a smiling field of swaying wheat into a mangled mess of mud and straw. If late July or early August is wet and hot, rust fungus creeps like a snake through the tissues of the wheat, and leaves it a lifeless, blighted, crumbling mass of dust and dirt. Occasionally if spring is late, growth too slow, or autumn early, the standing grain may be shrivelled by frost. The wheat farmer of the prairies wages a constant battle with noxious weeds, and in some cases the weeds win and actually drive the farmer off his land.

Scientists are striving to overcome all these evils. Sir Charles E. Saunders, at the experimental farm at Ottawa many years ago produced a new kind of wheat, called *Marquis*, which besides having the best qualities for making good flour, was strong of stem to resist wind and hail, a heavy producer, and ripened earlier than the Red Fife wheat, which it replaced. Special rust-resistant wheats, such as Thatcher and Apex have been developed; and it has been found that these are also specially adapted for certain types of soils in the drier years. The government, in co-operation with the provincial universities, has carried out extensive soil surveys with a view to

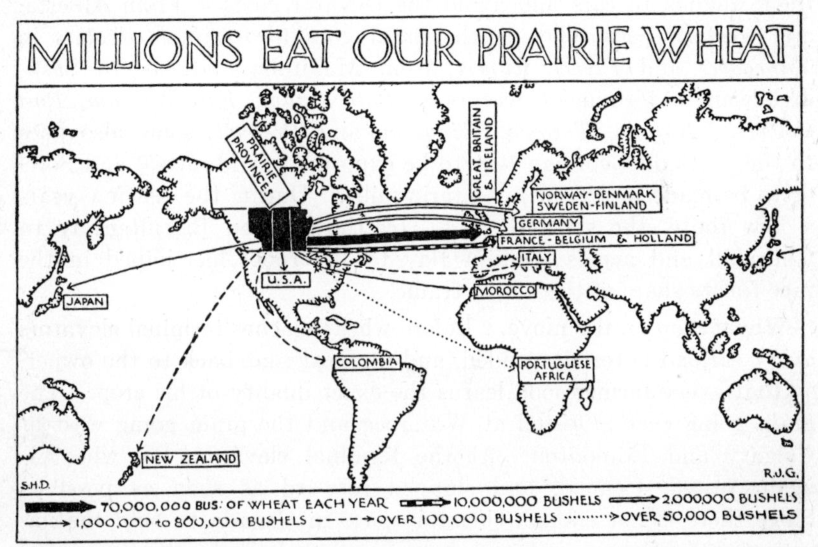

#143

helping farmers make the most effective use of their land. Drought problems are being studied, farmers are being encouraged to conserve spring moisture in *dugouts*, and by small irrigation dams. Wheats that ripen still earlier than Marquis are allowing the wheat fields to be extended farther and farther north until at present the Peace River district, two hundred miles north of Edmonton in Alberta, raises the choicest wheat in the world. Undoubtedly the northern range of wheat in the Prairie Provinces will be greatly extended (#139).

The stream of wheat. The most striking sight in the Prairie Provinces is one, or a row of tall, sentinel-like grain elevators (#142), along every railway siding. These are made of wood, covered with sheet iron, and contain a number of vertical bins, whose contents can be rapidly unloaded into box-cars by means of pipes. An endless chain with buckets, which runs on pulleys, can lift grain quickly from the receiving hopper to the top of the elevator, and by means of a horizontal, concave belt can transfer it into any bin. The farmers usually have no large granaries, and as soon as they can get it threshed, bring their grain in wagons and trucks to the receiving platform, weigh it, then dump it into the receiving hopper, and it is carried to one of the bins, which is the beginning of its long journey to every corner of the earth. Some farmers shovel their grain from their wagons to cars and avoid the elevator costs. From Alberta, and western Saskatchewan the grain goes to *terminal elevators* in *Vancouver* and *Prince Rupert*; from Manitoba and eastern Saskatchewan to terminal elevators at *Port Arthur*, *Fort William*, *Port Colborne*, *Buffalo*, *Toronto*, *Kingston*, and *Prescott*; some also goes to the ports on Georgian Bay to be exported via Montreal and Sorel or to be made into flour in Ontario mills. During the last few years a new route, the shortest of all, by the Hudson Bay Railway to Churchill and across Hudson Bay to Liverpool has joined in the race for its share of the wheat traffic.

Wheat crop on the move. Before wheat reaches terminal elevators every carload is tested, graded, and a report sent back to the owner, so that every farmer soon learns the exact quality of his crop. The grain going east is tested at Winnipeg and the grain going west at Calgary and Edmonton. At the terminal elevators the wheat is again tested, cleaned, and shipped forward as soon as possible. Since there is not enough elevator space in Canada to hold the immense crop of three or four hundred million bushels, there is a desperate struggle in the autumn to get as much as possible down to Montreal and Vancouver and away to Europe before the lakes are frozen. The elevators at *Fort William* and *Port Arthur*, Port Colborne, Prescott and the ports on Georgian Bay are packed to the top, and the immense grain boats on the lakes become storage bins for grain during the winter, since every one is tied up in port, lying low in the water with a full load. The Dominion Government also has large storage elevators at Winnipeg, Moose Jaw, Saskatoon, Calgary, Edmonton, Medicine Hat, and Lethbridge to assist in the struggle to find sufficient winter storage for grain.

Durum wheat. Manitoba, in the search for a wheat, less subject to rust than Marquis, has found *durum wheat* satisfactory, and nearly half of her crop is of this hard variety, which is largely used to make macaroni.

Oats, barley, and flax. While wheat is the main crop in the Prairie Provinces, *oats* and *barley*, of the very highest quality, are grown in large quantities. The cool, wet weather of northern Saskatchewan and Alberta is most suitable to grow abundant crops of oats. Barley is more important in Manitoba. *Flax* plants for seed thrive on deep, rich soil newly broken in, with a climate such

#144

COWBOY ON ALBERTA RANCH

as the south of the Canadian prairies. Saskatchewan and Alberta have large crops.

Stock raising. It would be a great mistake to think that all the Prairie farms are just broad, unfenced fields of wheat and oats with squatty stables, and no barns; with the fodder being stored in ugly stacks, from which it is brought in to feed the few work-horses which are necessary to run the farm implements. More and more, as circumstances permit, the prairie farmer is reaching out to new kinds of production, increasing the number and variety of his stock, and erecting suitable buildings to house them. But it must not be forgotten that, as lumber is scarce and dear on the prairies and as there is no other cheap material to take its place, that the building of splendid barns and houses is a very expensive work. Near the cities, where the farmer has a good market for milk products, dairy farming is advancing rapidly, and butter factories, whose products are unequalled in quality in other parts of Canada, are springing up rapidly in all the provinces. As farms are far apart and fresh milk would have to be gathered daily from a wide area to supply cheese-factories, these have not made much headway. With abundance of oats and hay, both wild and cultivated, the Prairie Provinces lead in the rearing of horses, great numbers of which are required for local farm work.

Ranching. The drier parts of southern Alberta contain great ranches, of which the one at *High River*, owned by the Duke of Windsor, is most noted. From about 1870 to 1900 great herds of cattle, attended by cowboys, roamed the unfenced prairie (#144). They cropped the green tuft and buffalo grass throughout the year, for the heat of summer cured it to nutritious hay as it stood. Since winters were mild and the snow covering thin or absent, the cattle, horses, and sheep, with little attention, grazed on this dried fodder out of doors throughout the winter. But the fences of railways and farms have so destroyed the freedom of movement that the open ranges, except for a few in the foothills of the Rockies, are replaced by ranches enclosed by fences, or by farms.

Irrigation in Alberta. Large areas of this region, too dry for farming, have been supplied with water from rivers. Dams thrown across the rivers, such as the Bow at Bassano, store the water which flows down the mountains, and through the coulee-like valleys. Main channels from this reservoir lead the water to flat land at lower levels. Water is drawn off from these channels, to the farms, and by means of ditches and furrows, spread over the surface like the veins of a leaf, watering all parts of the fields. This irrigation works wonders in the hot sunshiny climate of south Alberta. Good soil, abundant heat, a strong sun from a clear sky, and water, as much and as often as required, transforms the desert into splashes of greenery. Thousands of acres around *Lethbridge*, *Bassano*, and *Raymond* (#139) grow three heavy crops of alfalfa annually, every stem of which depends on water spread over the fields from irrigation ditches. This clover is as nutritious to stock as milk is to babies, and some of the finest dairy and beef herds in Canada are found in this district. Sugar beets are grown on irrigated land at Raymond (#139), by industrious Mormons, who learned the art in the dry lands around Salt Lake City (p. 483); and a factory for extracting sugar is in full operation in this thriving town. Corn, tobacco, fruits, and vegetables, which do not thrive in other parts of the prairies, come to perfection under the mild climate, and controlled water in the irrigation area. In

#145 CHASING THE BUFFALO

these areas, cultivation is intensive, farms are smaller, houses are closer together, and life is less lonesome and has more social advantages than in other parts of the broad prairie.

People

Selkirk settlement. Until 1811 only Indians roamed over the broad lone land of the prairies; with bow and arrow they shot the buffalo (#145), whose body was as big as that of a cow, and whose numbers were countless. The Blackfeet Indians in Alberta chased their prey on horseback, but the eastern Indians chased the buffalo on foot. They would dry the buffalo meat in the sun, pound it into a fine powder and compress it into cakes, or pack it in skins in melted fat, as *pemmican*. As this was very nutritious and would keep for long periods, they were in no fear of hunger. In 1811 *Lord Selkirk*, a Scottish nobleman, who had read of the explorations of Alexander Mackenzie (p. 382), and who saw the poverty of many of the small Scottish farmers, started a settlement on the prairies (#86, #135). He obtained an immense grant of the finest land around Winnipeg, and brought out groups of sturdy Scots year after year. Everything at first seemed to go wrong. They suffered from cold, lack of seed, and farm tools; the hungry grasshoppers stripped their fields bare for three successive years; early frosts blighted the remnants of the crop that swarms of gophers had not destroyed, and the Red River suddenly rose in spring to flood their newly-built cabins. The North-West Company, and to a less extent the Hudson's Bay Company, believing that furs and farms do not mix, worried them without ceasing, so that progress was slow and even their Scottish stubbornness was almost broken. But finally they succeeded. They grew their first crop of wheat in 1815, and at last after 1826 came abundant harvests. Soon French half-breeds, the descendants of fur traders, began to settle with them around *Fort Garry* at the point where the Assiniboine empties into the Red River. The Scots kept to the west of the Red River, the half breeds to the east.

The North-West Territories in 1870. At 1870, when the North-West Territories was handed over to the Dominion government by the Hudson's Bay Company (#86), there were probably fewer than 25,000 settlers in the whole region between Hudson Bay and the Rocky Mountains, more than 10,000 of whom were gathered along the Assiniboine and Red Rivers between Winnipeg and Portage la Prairie. There were also small settlements at Prince Albert and Edmonton. About nine-tenths of these settlers were *Metis*, or *French half-breeds*. The only means of transport were boats and canoes on the rivers and lakes, and two-wheeled carts on the land. The wobbly, creaking *Red River carts*, as they were called, were made entirely of wood, and drawn by oxen or horses. With these one could go anywhere over the prairie.

Butchering the buffalo. The great events of the year were the spring and summer buffalo hunts, when thousands of settlers and

Indians would migrate to the haunts of the buffalo, and pitch their tents in a space surrounded by Red River carts. Mounting their horses, and armed with rifles, they would proceed to cruelly and criminally butcher in thousands those majestic monsters, the noblest of all North American animals. Many of these animals were slaughtered merely for their hides, which sold for one dollar apiece, often only for their tongues, and all too frequently

#146

RED RIVER CARTS

to satisfy the lust to kill. No animal could survive such sinful slaughter. After 1870 the thundering herds were alarmingly reduced, in a few years the great stream had dried to a mere trickle, and by 1900 the great problem was to prevent the total extinction of this noble animal. The building of fenced railways, and fenced ranches and farms, cutting off their regular migration paths, the terrific slaughter due to the Indians learning to use rifles, and the bringing of the land under cultivation—all played their parts in this ghastly destruction of the Indians' best friend.

Nomadic Indians settle on reserves. The carcass of this animal had supplied them with clothes, houses, boats, bridles, saddles, bags, weapons, fuel, meat, and their very life. By 1870 this monarch of the prairies and mainstay of the Indians was nearly gone. Their self-dependence went with it; with rifles they had knocked out the main

#147

NEW CANADIANS TRAVELLING WEST

prop of their life. These free nomads of the plains had to be placed in reserves and partly supported by the Dominion Government.

The beginning of a flood. The Red River rebellion of 1869-70, and the march of an expedition of soldiers from Ontario to Manitoba stimulated an interest in the country. The first little rivulet of immigrants, that so soon was to develop into a surging Amazon, began to come down the Red River from the United States. A splendid band of sturdy God-fearing *Mennonites*, a sort of German Quaker, came out from Russia and settled east and west of Emerson (#139) near the border. About the same time a group of Icelanders began fishing and farming at *Gimli* on the west coast of Lake Winnipeg, and soon became leaders in every department of intellectual and mechanical work (#139).

The rising tide of settlers. The first railway from Winnipeg to St. Paul's, Minn. was completed in 1879 and might have deflected trade and immigration north and south had not a far greater event been already planned. The Canadian Pacific Railway was even then crystallizing, a short line here and another there; in 1881 over 160 miles were built across the plains, in 1882 it gathered force and two and one-half miles were built every day; by 1886 the road was complete, a railway spanned Canada from east to west and through trains from Montreal to Vancouver ran a regular schedule. The greatest event in the history of Western Canada had taken place. Then began a steady stream into the prairies of settlers from Ontario, the Maritime Provinces, and the British Isles. From that day till the Great War broke out, the stream never ebbed. It flowed moderately at first, but after the best lands in the United States had been taken, the stream became a torrent. By 1896 they were surging across the ocean from almost every country of Europe. Between 1901 and 1906, the population doubled. The people from Ontario spread along the main line. English, Irish, and Scots often settled in communities; and when the Scandinavians, Jews, Hungarians, Germans, Ruthenians, and Doukhobors came they collected in national groups, each having its own schools, continuing its own customs, and speaking its own language (#147). But the children as they went to school were taught in English, and the new generations are becoming more and more Canadian.

The United States joins the procession. As a second generation grew up on the prairie farms of the Central United States, and available land was scarce and dear in their own country, these attractive immigrants began to flow over the border to the cheap, fertile farms of the prairies. During the present century they have come in ever increasing numbers. As they were well-to-do and brought stock and a good deal of money with them, they were a very valuable addition to the population, especially as they were largely of the same blood and language as native Canadians, or else our next of kin, the wonderful Scandinavians.

Blending the people. A map of the Prairie Provinces showing in different colours the nationality of each district looks like a great sea of British stock with an archipelago of islands of French, Scandinavians, Germans, Ruthenians, Russians and Jews scattered over the sea, the Ruthenian and Russian islands being especially large and numerous in the north. On every one of those islands you would see habits, houses, language, and religion somewhat similar to those in the part of Europe from which they came; nevertheless, the great sea of British customs, that beats ceaselessly on the shore of every island, rules the whole.

As settlement increased, railway building more than kept up with the invading population, and to-day every settled part of the country is enmeshed by a network of railways that would be hard to equal anywhere else in so young a country. But the Great War destroyed immigration and railway building, and neither has yet recovered from that terrific shock.

Manufactures

Hydro-electric power. Manufactures usually thrive where there is sufficient population to give a valuable home market, where raw material and cheap power are available, where the people have stable enough government to attract capital, and have intelligence to work the factory machinery. The people of the Prairie Provinces are second to none in intelligence and daring, and their governments are steady and progressive, but because the population is spread out in a very thin layer, manufacturing is but in its infancy. Besides the abundant coal deposits of Alberta that create power to run machinery, all the provinces have available water-power to make electricity. Many streams, tumbling down into Alberta from the Rocky Mountains, will some day be harnessed for that purpose. Already the Bow River west of Calgary (#139), distributes electricity to that city and as far south-east as Lethbridge and Magrath. Saskatchewan's available sources of water-power are unfortunately mostly in the north, on the Churchill and Reindeer Rivers. There is also available power on the Saskatchewan below Prince Albert. The only large development, however, is at *Island Falls* on the Churchill River which is used in the Flin Flon mining region of

Manitoba (p. 346). Manitoba is very fortunately situated for hydro-electric power as so much of it is occupied by the Canadian Shield (p. 135), which vibrates from end to end with rushing rapids and plunging cataracts. There is a very large development at several points on the Winnipeg River (#139), where electricity is distributed over much of southern Manitoba. The largest distribution is in the city of Winnipeg, where it is so cheap that it has made this great city the industrial centre for the west. Other electric power developments are supplying the gold mines to the east of Lake Winnipeg.

Growth of cities. The first towns of the Prairie Provinces grew at prominent points on the rivers, either at or near the point where a tributary joins the main stream, as at *Winnipeg, Fort Ellice* and *Prince Albert*; or at a ford on a river, as *Battleford* and *Portage la Prairie.* Others sprang up at the gateways to low passes through mountains as at *Edmonton, Calgary,* and *Macleod.* But with the coming of the railways, steamers vanished from the rivers and these advantages were almost entirely swept away. The size of the city now depends largely on the numbers and importance of railways crossing at that point. Every city is a distributing centre and has large wholesale and retail stores. The manufactures are of local raw materials and consist chiefly of flour, cereals, brick, cement, leather, butter, and the curing of meats (#139); lumber is made in the towns of the forested region; usually in the larger towns there will also be a brewery, a foundry, one or more assembly plants for farm implements, a soft-drink manufactory, and, of course, bakeries. The towns and cities, during the years of rapid growth, when money flowed freely, built the finest schools, churches, and other public buildings, and greatly improved the streets.

Winnipeg, the fourth largest city in Canada is the Chicago of the west. Between Lake Winnipeg and the Lake of the Woods there is a bottle neck through which all traffic between east and west must pass — Winnipeg, fills the mouth of the bottle. It is the greatest wheat market in the world, the centre of banking, insurance, wholesale distribution, and manufacturing for the whole of the Prairies. It has large meat-packing houses, over one thousand manufacturing establishments, and is one of the great fur-trading centres in Canada. Besides being the capital city, it is also the educational centre of the province. It has many fine buildings devoted to the cause and work of education, several modern collegiates and the University of Manitoba with its surrounding cluster of Church Colleges are located within its borders. A beautiful structure, of which the residents may well be proud, is the provincial Parliament Building, the seat of Manitoba's government.

Brandon and Portage la Prairie, two enterprising cities, are situated at the very centre of the flattest and most fertile prairie and have the advantage of being on the main line of the Canadian Pacific Railway. Brandon has an important Baptist college. Portage la Prairie is also on the main line of the Canadian National Railway. St. Boniface, across the river and a separate municipality, is really a part of Winnipeg.

Saskatchewan, though the most populous of the Prairie Provinces, is the most strictly rural and has few large towns and cities. *Regina,* the capital and largest city, is nearer to the centre of population of the Prairie Provinces than any other city, it has large departmental stores, assembly plants for motor cars and farm implements, and over 100 manufacturing establishments (#139). It is the only large city on the prairies not located on water, and grew largely because it was the capital of the North-West Territories, with headquarters for the mounted police. It was later selected as the capital of the Province. *Saskatoon* on the South Saskatchewan, 160 miles north of Moose Jaw, is a city of recent growth, whose welfare depends on its excellent railway service, and the agricultural productiveness of the surrounding regions as indicated by the relatively dense rural population. It has a large government elevator, two large flour and cereal mills, and is the centre of the tanning industry (#139). It contains the provincial university, a most elaborate Normal School, and the provincial agricultural college. *Moose Jaw* in the south, is an important railway centre. *Prince Albert,* on the North Saskatchewan, is one of the oldest towns and a centre of trade in fish, furs, and timber with the north of the province.

Edmonton, the capital of Alberta, is at the head of navigation on the North Saskatchewan River and at the cross-roads of two very important lines of traffic; one, west through the Yellowhead Pass to Prince Rupert and Vancouver, and another north through the Peace River to the Peace River farming district, thence through the Mackenzie River to the Arctic. It is the southern terminus of a line extending north to the head of navigation on the Mackenzie River. It is surrounded by a well-watered agricultural region, has coal mines at its door and has a pipe-line for natural gas from Viking. It contains both the Parliament Buildings and the provincial university, manufactures flour, cereals, leather, lumber, etc., is one of Canada's chief fur markets, and is the fountain head of traffic with the north.

Calgary, the largest city in Alberta, has the advantage of being near the entrance to Kicking Horse Pass on the main line of the oldest through railway. This makes it the spear head for traffic with the population of British Columbia, who are located along the Canadian Pacific Railway. Besides, it is the natural trade centre for the rich irrigated and ranch lands of the south. As has been already mentioned it has hydro-electric power from falls on the Bow River, which with cheap natural gas from Turner Valley and Bow River, contributes to its many manufactures.

Lethbridge has the same relation to the Crow's Nest Railway that Calgary has to the main line of the C.P.R. It is also in a coal mining district and the centre of irrigation for the south.

Medicine Hat, because it has exceedingly cheap and abundant natural gas, is an industrial city. It has a very large flour mill, makes tile, linseed oil, pottery and metal goods.

CHAPTER XVI

BRITISH COLUMBIA

A masterly location. British Columbia holds a masterly position. It has the same relation to the Pacific Ocean as the British Isles have to the Atlantic; their latitudes are the same, their climate and vegetation are similar, and people of good British stock are in control. This western province of Canada is as large as the combined areas of the British Isles, France, Belgium, and Holland, which support more than one hundred million people. While these European countries, in their youth faced an empty ocean (for America had not yet been discovered), this vigorous young province of British Columbia now faces the swarming multitudes of South Asia, who are just rousing themselves from a thousand years of economic sleep, eager to satisfy their keen hunger for all the commodities that the West can offer. The cities of Vancouver and Prince Rupert are closer to these hungry hordes than is San Francisco. The cutting of the Panama Canal has also brought Europe and the eastern parts of Canada and the United States so close to Western Canada that a steady procession of steamers loaded with Alberta and Saskatchewan wheat and British Columbia lumber are ever streaming through this little artery and spreading fan-like over the broad Atlantic (#143).

A gifted province. British Columbia is smaller than Ontario, and much smaller than Quebec, and its population is less than that of any of the Prairie Provinces. Yet in varieties of climate, industry, and surface features it stands alone. Another outstanding fact is that each person in British Columbia on the average produces more, spends more, and is worth more than in any other province of the Dominion. It is also interesting to notice that it is fifty thousand square miles larger than its neighbours, Washington, Oregon, and California put together.

Surface

A province of mountains and trenches. The surface of British Columbia is somewhat against it. It has been described as a sea of mountains, but it is not nearly so complex as that. Indeed there is an orderly disorder in the series of deep parallel valleys, partly gouged out by water and ice, and divided by a succession of stubborn, high,

BRITISH COLUMBIA
ITS SURFACE, WILD LIFE, EARLY EXPLORATIONS TRAILS AND SETTLEMENTS
#148
LIARD R.
STIKINE R.
SKEENA R.
BABINE L.
PEACE R.
PINE R.
PARSNIP R.
FRASER R.
N. THOMPSON R.
FRASER R.
OKANAGAN L.
KOOTENAY L.
COLUMBIA R.
PACIFIC OCEAN
OKTA SOUND
AINCOUVER I.
AN DE FUCA STRAIT
YELLOWHEAD PASS
HOWSE PASS.
KICKING HORSE PASS
CROW'S NEST PASS
FRASER IN "HELL'S GATE"
MACKENZIE TASTES THE SALT OF THE PACIFIC
BOUNDARIES
S = SELKIRK MOUNTAINS
M = MONASHEE " ROCKY = R
C = CARIBOO " MOUNTAINS
ST = STIKINE " PURCELL = P
B = BABINE " MTS.
MOUNT WADDINGTON
MOUNT ROBSON
ROCKY MOUNTAIN TRENCH
WILD LIFE IN BRITISH COLUMBIA
BEAVER
OTTER
JACK RABBIT
GRIZZLY BEAR
MOOSE
BLACK AND OTHER BEARS
CARIBOU
MOUNTAIN SHEEP
WAPITI, OR ELK
VARIETIES OF DEER
ROCKY MOUNTAIN GOAT
WOLF
FOX
EARLY FORTS ESTABLISHED
ROCKY MT. FORT RM 1806
FORT McLEOD M 1806
FORT FRASER F 1806
FORT ST. JAMES SJ 1806
FORT GEORGE G 1807
KOOTENAY HOUSE K 1807
FORT VICTORIA V 1843
EARLY GOLD SETTLEMENTS
HOPE H
OTHERS
LYTTON L
QUESNEL Q
BARKERVILLE B
OMINECA CREEK O
ARIBOO TRAIL
EWDNEY TRAIL
ACKENZIE'S JOURNEY 1793
RASER'S JOURNEY 1808
HOMPSON'S EXPLORATIONS 1807-1811
SHD

parallel ridges (#148). The main mountains, and valleys, so straight and deep that they are called *trenches*, run nearly parallel with the coast, and the Geographic Board of Canada after careful study has divided them into three belts called (a) *Eastern*, (b) *Interior*, and (c) *Pacific*.

The Rocky Mountains. The eastern belt is very distinct; it is formed by the most bulky and continuous of all the ranges, the *Rocky Mountains*. They are well named. Their naked edges of limestone, far above the tree line, bristle with sharp ice peaks and smooth snow domes of every shape and hue. The approach from the east is a series of platforms or foothills. This backbone is highest and most continuous in the southern third of British Columbia, Mount Robson (12,972 feet) forming the crown. In the middle third of the province it is so much lower that the *Peace* and *Liard Rivers* rise on its western slope and are able to break through and drain into the Mackenzie system. Here are found the lowest passes. The north part of this continuous belt in the Yukon is no longer called Rocky Mountains, but named *Mackenzie* and *Franklin*, after famous explorers, who fought famine and frost to explore these remote areas of Canada.

Passes through the Rockies. This formidable barrier, thrown across the main street of Canada, fortunately has a number of moderately low passes, through which first, explorers and fur-traders went on foot or on mule back; and next, three railway lines were built. The following are the main passes from south to north, with their height, the streams that are still wearing them down, and the railways that traverse them (#148).

NAME	HEIGHT	RIVER VALLEY	RAILWAY
Crowsnest	4438	Crow's Nest River (Branch of Old Man River)	C.P.R.
Simpson	6650	Simpson River (Branch of Kootenay River)	
Vermilion	5264	Vermilion River (Branch of Kootenay River)	
Kicking Horse	5296	Bow River (Branch of Saskatchewan River)	
		Kicking Horse River (Branch of the Columbia River)	C.P.R.
Howse	4800	Blackberry River (Branch of the Columbia River)	
		Miette River (Branch of the Athabaska River)	
Yellowhead	3738	Yellowhead Creek (Branch of the Fraser River)	C.N.R.
Pine River	2850	Pine River (Branch of the Peace River)	
Peace River	2000	Peace River	

The longest, deepest, and straightest trench. The western limit of the Rocky Mountains is indelibly marked by the *Rocky Mountain trench* (#148), one of the deepest, straightest, most nearly level furrows that ploughs the face of the earth. It varies from one to fifteen miles wide, can be traced 1,500 miles from Montana to the

Yukon-Alaska border, and is so nearly level that the streams which traverse its bottom are for the most part navigable. These rivers are its most interesting feature. The *Kootenay* (#148), bordered by farms of luscious fruits and rich grasslands, occupies its southern part. Then four rivers in succession, whose main courses are more or less at right angles to the direction of the trench, throw a right and a left arm to flow in opposite directions along its course. The Columbia rises in the same marsh as the Kootenay but flows north-west along the trench, and, where it breaks through its eastern slope at the *Great Bend*, gives off the Canoe River to the north. Near the source of the Canoe the Fraser River takes up the path to the north, and where it breaks through, again gives off a small branch to the north. Then farther north, where the Rocky Mountains are lower, the Peace and Liard in succession have pierced the Rockies, and each throws off a right and left arm to occupy the trench as far as the Yukon, where the course becomes more obscure.

Interior belt. The interior belt (#148), is much older, more deeply eroded, and much more complex than the eastern belt. The southern part is about four thousand feet high and is divided into long ranges with deeply eroded trenches between them (#148). The central part is lower and completely dissected by transverse trenches, many of which form exquisitely coloured long deep, narrow lakes. In the north the belt is higher, and more continuous, the whole central part being occupied by the *Stikine Mountains* (#148).

Parallel trenches and mountains. In the south, three distinct deep trenches, with several intermediate ones, run at an acute angle to the Rockies and between each two trenches is a range of mountains. Not so many thousand years ago these trenches were filled with immense glaciers, which ploughed them deep and U-shaped as the ice moved to the south; as the glaciers finally melted, they often deposited dams across the southern end of the valleys, so that in part they are now occupied by deep mountain lakes, with high sloping sides; through the bottom of the remainder of these trenches run streams, which are continuing the work of erosion. These narrow valleys are of exceeding fertility. The shelter of the overhanging slopes and the foaming mountain streams, when brought down from above, cause the growth of choice fruits and smiling meadows. The most easterly is the *Purcell trench*, containing the picturesque *Kootenay Lake* and *Kootenay River* in its south end and the *Duncan* and *Beaver Rivers* in the north half. Between the

Rocky Mountain trench and the Purcell trench are the pear-shaped massive *Purcell Mountains,* bursting with their bounty of gold, silver, lead and zinc ores. The *Selkirk trench,* occupied by the *Columbia River* and its beautiful silver-like streaks, the *Arrow Lakes,* lie next to the west, and the rugged snow capped *Selkirk Mountains* sternly separate the Selkirk and Purcell trenches; the great bend of the Columbia River bounds these mountains on the north. The Canadian Pacific Railway found the greatest difficulty in traversing the frowning front of the Selkirk Range. *Roger's Pass* was selected, but so high and rugged was this found to be that the *Connaught Tunnel* had finally to be bored for seven miles through the mountain. The northern part of the Selkirk Mountains is a panorama of vast snow fields, steep mountain peaks, and blue, glimmering glaciers, worming their way through the valleys, whose lower courses vibrate with the rush of foaming, tumbling streams. Indeed, here are found some of the finest tourist playgrounds in the world.

The *Okanagan trench,* still further to the west is occupied successively from north to south by a part of the *North Thompson River, Adams* and *Shuswap Lakes, Okanagan Lake* and *Okanagan River.* Between it and the Selkirk trench lies a series of ranges, sometimes vaguely called *Gold Range* but more properly the *Monashee Mountains.*

Still further west is the valley of the *Fraser River,* which, after flowing north-west along the Rocky Mountain trench for 240 miles, loops around the *Cariboo Mountains,* cuts diagonally across the whole central belt, and finally in a succession of yawning canyons bursts in fearful surging rapids right through the Coast Range, which we are about to describe.

Western belt. The *Western belt* consists of two ranges: the first is an impressive ridge of ancient mountains, one hundred miles wide, running throughout the province along the coast, and appropriately called the *Coast Range*; the second ridge, which may be called the *Insular Range,* is partially hidden beneath the sea but appears as a long row of islands, including *Vancouver, Queen Charlotte,* and that complex tangle, which forms the frayed outer border of what is known as the *pan handle of Alaska.* Between these two ridges is the most western of the longitudinal furrows, that plough the rough surface of the province. It is called the *Coastal trench,* is filled with sea-water, and forms a sheltered channel 800 miles long for ships, and a bountiful feeding-ground for salmon, halibut, and countless swarms of lesser fish.

Coastal Range. The Coastal Range is high, wide, flat-topped, and ascends steeply from the sea in slopes which are enrobed in the most magnificent forests of giant firs, spruces, and cedars, and still higher by snow and ice. *Mount Waddington* (13,260 feet), the highest peak, towers even above Robson, the highest peak in the Rockies of British Columbia. While the Rockies are made largely of water-formed rocks in layers, immense masses of volcanic rock are laid bare everywhere in the Coastal Range, and along the line where the volcanic and layered rocks meet, veins of metals are frequently found; so that some of the most valuable mines are found in these mountains.

An embroidered coast. During what is called the glacial period, the whole of British Columbia, and in fact almost all of Canada, was crushed under a thick sheet of ice and snow called an ice-cap. As this terrific ice-plough pressed from above and behind by millions of tons of ice, and studded beneath with lumps of hard rocks, rasped its slow journey down the valleys to the Pacific Ocean, they were furrowed deeper and deeper and polished smoother and smoother even beneath the sea. When this sheet of ice, thousands of feet thick, and covering most of Europe and more than half of North America, finally melted, the immense volume of water set free, raised the level of the oceans hundreds of feet, so that it ran up into these long narrow valleys, which to-day make the west coast of British Columbia the most tattered in the world. It is everywhere cleft by these gigantic furrows, very narrow and winding, into which the sea-water flows (#148). In British Columbia they are called *inlets*, *channels*, or *canals*. In Norway, similar winding coastal channels are called *fiords* and in Scotland *firths*. They often push their worm-like course deep into the volcanic rocks of the Coast Mountains, and usually send out branches like the toes from the foot of a giant bird. The steep sides and heads of these inlets are braided with waterfalls, varying from trickling shining ribbons and veils of white foam to full-sized rivers. Not only did the rising water of the ocean drown these valleys, but it also cut off projecting elevations from the mainland to make an embroidered coast of thousands of rocky islands of all shapes and sizes. Consequently British Columbia can boast an array of deep and perfectly protected harbours along its eight hundred miles of coast, which contrasts markedly with the rugged, repelling coast of the United States to the south, with hardly a dint in its rigid outline from Juan de Fuca Strait to San Francisco

Bay. It is hard to conceive of a more perfect feeding and spawning ground for fishes and every other creature of the sea than these inlets, countless in the variety of their shape, size, and depth, which nestle among the islands, and wander up into the clefts in the mountains.

Climate

Ocean winds on Coastal Range. The climate of the west coast of British Columbia and especially of Vancouver Island is very similar to that of the British Isles. The latitude is about the same, the mild, moist westerly or south-westerly winds blow in from the ocean, and a warm ocean current laps the shore of each. But the high

#149

barrier of the Coast and other ranges spreads right across the path of these winds, intensifies their effect on the coast, but largely cuts off their kindly influence from the interior (#149).

Ascending and descending air currents. Just as the bottom book of a pile is pressed by all the volumes above, so the different layers of air are compressed by all the layers above; consequently as air rises, it expands because of the decreasing pressure. As heat is used to push the particles of the expanding air apart, the atmosphere gets colder and colder the farther it ascends. If it contains water vapour, and it always does, the chilling in time will cause part of the vapour to be condensed into little globules of water or crystals of ice, depending on the temperature. Such particles floating in the air are called *fog* and *mist* when near the surface but *clouds* when high in the air. If the chilling and accompanying condensation go far enough, rain or snow will be produced. It is these ascending air currents that usually cause the clouds to be formed. Exactly the opposite results are produced by descending air currents. They become warmer and drier; clouds will be dissolved, leaving a clear sky.

The dripping coast. As these west winds, which have been licking up moisture from the Pacific Ocean, strike the west slopes of the mountains of Vancouver Island and the Coast Range, they are deflected upward and chilled; the moisture is condensed, and dripping rains deluge the land (#149). It is no wonder the trees grow so large, and the forests so dense. Almost every slope directly facing the Pacific receives more than 100 inches of rain annually, and at some parts, when nothing interrupts, and perhaps the winds are concentrated by a funnel-shaped valley, there may be 200 inches. *Henderson Lake,* near Alberni, with an average of 262 inches of rain per annum, seems to be not only the wettest spot in Canada but the wettest in the world, outside the region of the trade-winds and monsoons. As the westerly winds blow more strongly and more steadily during the winter, and the land at this season is much cooler than the water, the heaviest rains fall at this period of the year.

A great contrast. When the winds pass over the crest of the Coast Mountains and descend on the eastern slope, their whole action is reversed; they become warmer and drier, and less rain falls. The contrast between the vast, compact, green forests on the west slope and the scattered, stunted trees and semi-arid scrub of the east slope of the Coast Range is like that met in passing from an oasis to a desert. The same contrast is found between the east and west slopes of the Selkirk and Rocky Mountains (#149).

The dry trenches. Down in the trenches of the Fraser, Thompson, Okanagan, Columbia, and Kootenay Rivers, there is never more than thirty inches of rain and often much less. The driest spot in British Columbia is at *Clinton,* in one of these river trenches, where an average of less than six inches of rain falls annually. Fully nine-tenths of British Columbia is a high plateau, whose rainfall is much more copious than the niggard moisture of the valleys or trenches.

Moderate climate on the coast. As the prevailing west winds sweep the air of the ocean over British Columbia, the lines of equal average temperature, called isotherms, run more nearly north and south than east and west. As the water of the ocean never gets very warm in summer or cold in winter, the air over it, being continually carried in, gives the coastal regions of British Columbia cool summers, and winters so mild that in spite of the heavy precipitation little of it falls as snow. Consequently sleighs are never used for transporting logs to the inlets as is done in eastern Canada. Although Prince Rupert is 250 miles farther north than Winnipeg its winter temperatures are 25° warmer and its summer temperature 10° cooler.

Hot summers and cold winters of interior. Of course further inland these moderating winds have less influence; summer becomes warmer and winter much colder, and more of the precipitation falls as snow. Indeed, on the eastern slopes of the Selkirks, as at *Glacier*, there may fall over 100 feet of snow, which makes a serious problem for the railways. Down in the trenches, or river valleys, the cooling influences of the westerly winds is felt least of all, and in these the summer days are often uncomfortably hot, though the radiating of the heat to the cloudless sky gives some relief at night. One characteristic of almost the whole province is a lack of gales, blizzards, and thunder-storms. The hottest temperature ever experienced in British Columbia was 110° F at Griffin Lake in a well-protected valley on the C.P.R.; the thermometer at *Kamloops* occasionally goes over 100°.

The Chinook wind. When, in winter, the westerly winds slide down the eastern slopes into the trenches, they play the part of a good fairy (#149). Like magic, zero weather may be turned into balmy spring. The thermometer has been known to rise 60° in five minutes. A foot of snow may be licked up from the frozen ground by the warm, dry air so quickly and quietly that in three or four hours the hungry ranch cattle, which before had to be fed in stalls, can now once more find pasture on dead bunch grass. Though there is no escape for the melted snow into the ground, for it is frozen, so completely does the thirsty air drink up the snow, that there is none to run off. These delightful winds, bringing a relieving break in the monotony of cold, are called *Chinooks*.

Exploration of Coast

Captain Cook at Nootka Sound. British Columbia was the last part of Canada to be traversed. Though explorer after explorer had failed to find such a stream, Western Europe for long was haunted by the belief that there was a great river connecting the Pacific with the Atlantic. The search for this mythical river or strait from the east by Cartier, Champlain, La Salle, and a host of others had failed (p. 252). Wild rumours gradually leaked out that Spanish explorers had discovered such an opening from the west, and it had even been called the *Strait of Anian.* Now that Great Britain had conquered Canada (1763), a new motive was added to seek this illusive strait. The British government sent that prince of sailors, *Captain James Cook,* to find whether this mystery river was a fact or a fiction. He sailed by Cape of Good Hope, New Zealand, across the Pacific Ocean until he reached North America in 1778, and then sailed north along the coast (#148.) Skirting Vancouver Island he dropped the anchor of his good ship, the *Resolution,* in *Nootka Sound* (#148), where, owing to his skilful management, he was favourably received by the Indians, and a brisk trade in beaver skins and sea-otter furs took place. Needless to say, Cook discovered no sign of the mythical Strait of Anian, though he searched the whole coast to Behring Sea (#148). Unfortunately this brave, capable sailor was unable to continue the search for a north-west passage through Behring Strait to the north of Canada, as a few months later he was murdered by the spears and stones of natives of the Sandwich Islands (Hawaii).

Captain Vancouver explores the coast. Following Cook's visit, the Spaniards under Quadra had occupied Nootka Sound. At that time there existed a brisk demand for the skins of the sea-otter. To secure these skins there was considerable rivalry shown between the American and Spanish traders who came to Nootka Sound and other inlets in their search for the mammals. Two sea-otter skins, with their luxurious, soft brown fur, frosted with the hoary tips of longer hairs, would be sufficient to make a long fur coat. These coats were much sought after by the rich mandarins of China.

The occupation of Nootka Sound by the Spaniards led to many disputes and some fighting between the British and the Spanish. When finally this question was settled with the Spanish Government who agreed to surrender all rights to the British, *Captain Vancouver* was sent out to explore the whole west coast which was later to become British Columbia. With two ships he sailed through Juan de Fuca Strait (#148), explored and named every important island and inlet along the maze of channels separating Vancouver Island from the mainland, proved for the first time that Nootka Sound was on a great island, buried for all time the myth of the Strait of Anian, but failed to find the mouth of the Fraser River, though his ships noticed its current, the discoloration of the sea-water, and the low swampy delta formation.

Interior Exploration

Cut-throat competition for fur trade. While the lure of the glorious silky fur of the sea-otter, was leading to minute inspection of the winding inlets of the tattered braid of the western coast, the powerful lodestone of profitable fur trade was leading brave adventurers toward the interior of British Columbia from the east. When Canada was conquered by Great Britain, English merchants of Montreal lost no time in seizing their share of the fur trade which had been so profitable to the French, and pushed their buyers deeper and deeper into the western wilderness to bargain with the Indians before they reached their rivals, the Hudson's Bay Company. The competition was so keen that danger of destroying all their trade led these merchants to unite in the *North-West Company*, with western headquarters at *Port Arthur*. It had become a habit of the Hudson's Bay Company to have the Indians bring the furs to them rather than for their company to penetrate further and further into the interior to open new trading posts. The aggressive North-West Company knowing this, endeavoured to deflect much of the fur trade which went to the Hudson's Bay posts at the mouths of the rivers emptying into Hudson Bay.

Explorers enter British Columbia from the north-east. It was the North-West Company, consequently, that furnished the explorers, many of whom they attracted to their service from the Hudson's Bay Company. The exploration of British Columbia was made from the east rather than from the coast, where the continuous rampart of the Coast Range formed a forbidding barrier. The western entrance to the province was in the north, where mountains are low and where northern rivers of the plains have cut their channels right through the Rocky Mountains, rather than from the south where the rugged Rockies rise in all their majesty to awe the man who dares to traverse them.

In 1792 *Alexander Mackenzie*, still drawn by that mythical Strait of Anian, turned his face doggedly westward to reach the Pacific Ocean. To be ready for an early start in 1793 he spent the previous winter at a fort which he built near where the Smoky River joins the Peace (#148). Almost before the ice had left the river, he with six French-Canadians and two Indians, in a 25-foot bark canoe, faced the swift current of the Peace River to search for the western sea. They continually met great flocks of wapiti (elk) and buffalo. So numerous were the latter that many open parts of the country looked like farm-yards. In the beginning of the gorge through the Rocky Mountains, the rapids were so swift that they were constantly in great danger as the frail canoe was buffeted from rock to rock; many times they were wading to their armpits in the icy, turbulent water, tugging the canoe by a rope; sometimes they had to carry their whole equipment up rasping slopes, so steep that steps had to be cut in the solid rock. More than once they had to hack a path through the dense forest. But nothing daunted the Scottish stubbornness of Mackenzie, who had to coax, bribe, and threaten his less eager companions. At last they reached the forks of the Peace and decided to turn south up the Parsnip River (#148), swollen with the spring melting of the snow on the mountain slopes. Never had they seen beaver so numerous; everywhere countless acres of logs cut down by these patient animals were found. But Mackenzie was an explorer rather than a fur trader, and pushed on. At last by a portage they reached the Fraser, and travelling was easier. On the advice of Indians, who knew the district, he left the Fraser River near *Quesnel*, turned west, and travelled up the *Blackwater River*. The last part of the journey was made on foot in a race against starvation, as they had only slim rations for four or five days. On July 19th, when his food was nearly gone, his men exhausted, and with only his own untiring energy as a driving force, he was overjoyed to reach an inlet at *Bella Coola*. As he tasted the salt water he knew that at last the search for the western sea, which began with Jacques Cartier in 1534, had been accomplished and that he had reached the Pacific Ocean (#148). It seems strange that two Britishers, James Cook and Alexander Mackenzie, both started out from Britain only a few years apart, one went westward around two-thirds of the world, the other eastward and they almost met on the edge of the Pacific.

Simon Fraser builds the first trading posts. The mouth of the Columbia River had already been discovered from the ocean, when *Simon Fraser*, a seasoned fur trader of the North-West Company, followed Mackenzie along the Peace and Parsnip River in 1805 (#148). He built *Rocky Mountain Fort* (#148) at the mouth of the Parsnip and *Fort McLeod* on a delightful lake, set like an emerald in the midst of the opal mountains (#148), on a branch of the Parsnip. These were the first two fur-trading posts of British Columbia, but they were soon followed by others; by 1806, this courageous Scot was on the Fraser River, followed the Nechako westward, and before the summer was over, Fort St. James on Stuart Lake and Fort Fraser on Fraser Lake were ready for trading. The next spring, 1807, he erected another post where the Nechako joins the Fraser, naming it Fort George after the British king. Now the race was on between the United States and the North-West Company as to which would be first to explore the Columbia River. An urgent message came to Fraser at Fort George to prepare to descend this river, which he firmly believed was the Columbia, and to obtain its fur-trade for his energetic company.

The dangers of the Fraser River. In May 1808, Simon Fraser with four canoes and about twenty men swung out from Fort George and was soon whirling through tumultuous currents of the upper canons of the Fraser River (#148). As they advanced, they were cheered by occasional stretches of merely swift, muddy currents, for the waters were swollen by the melting of the winter snows covering the mountains, which rose steeply from the river on both sides. The whole trip was a succession of hairbreadth escapes. Now their canoes were tossed from side to side like eggshells, or almost ripped in two on rasping rocks, as they plunged through a gloomy throat in the river, contracted to forty yards, and arched above with rocky walls, a thousand feet high. Again the currents, cataracts, whirlpools, and torrents became so fearful that even the dauntless courage of Fraser dare not face them. Canoes and baggage had to be carried or dragged over wooded portages, where deep ravines, steep hills, and yawning chasms offered such overwhelming resistance that only the dogged Scot, spurred by the race against the United States explorers, faced them without flagging. They were constantly meeting Indians, many on horseback, who were not always friendly; sometimes they deceived them, and occasionally they were quarrelsome. At last the currents became so furious and the rocks so threatening that they had to hide their canoes and proceed along canyons on foot, each man with nearly 100 pounds on his back.

Through Hell Gate. But the chief dangers, the most desperate currents, the steepest canyon ramparts were still ahead, where the Fraser River has battered its fierce course through the coast range. A canoe was cut right in two as it swung against a rock; at one point the empty canoes were allowed to take a chance against the deadly waters, as the men packed along the ragged shore. Ladders, made of poles bound end to end, with twigs tied across, and which swung timidly hundreds of feet down the side of the canyon had to be used to traverse the side of the canyon. Though Indians, through long experience, could use these dangling ladders safely, it was a deadly peril for these fatigued voyagers, each with a heavy pack on his back, to climb hundreds of feet up these strange shaking stairs. But by skill and luck they all came through this awful canyon, one of the most dangerous parts of which is well called *Hell Gate.*

Success and defeat. At last they were drifting placidly on the tranquil waters of the lower Fraser, with the level stretches of its fertile delta spread out before them. But even in the sweetness of success there was the bitterness of defeat. When Fraser measured his latitude and longitude he found that the mouth of the river he had explored was nearly two hundred miles north of the mouth of the Columbia, which he, in a race with a fur-trading company of the United States, had set out to possess. It is rather remarkable that while Fraser was feverishly plunging down the Fraser, thinking it was the Columbia, away back in the Rocky Mountain trench another great Canadian explorer was paddling upward on a river which he never suspected to be the Columbia. How could he consider it to be this river, flowing in the opposite direction to which the Columbia was supposed to move?

Thompson unravels the coils of the Columbia. That man was *David Thompson,* the most thorough explorer of the three giants who traced the confusing water courses of British Columbia. While Mackenzie and Fraser crossed the Rockies through the Peace River gap in one of its lowest passes, Thompson assailed this barrier in the south, where it throws across the path not one but a whole series of confusing crests, with towering peaks and appallingly high passes. He attacked these one after the other; in 1800 he gazed at the glories of Kicking Horse Pass, now treasured in Banff National Park; in the same year he traced the North Saskatchewan to the *Howse Pass,* studded on both borders by a dozen of the highest peaks in the Rockies, every one of them clad far down with perpetual snow. It was through this pass that he descended a swift tributary in 1807 to catch his first view of the Columbia, whose mazes he mastered in the next five years. He poled southward up this stream in a canoe, and near its source built *Fort Kootenay* (#148), which became the first trading-post in south British Columbia. After going around the loop of the Kootenay River in 1808, he was called back to the east; but again in the late autumn of 1810 he was fighting his way against snow, starvation, and below-zero weather over the ice of the Athabaska River, since unfriendly Indians closed the Howse Pass. By Christmas, he and his men were floundering through the deep snows of Athabaska Pass; he pushed forward and soon was again on the Columbia, which he followed through the Rocky Mountain trench, crossed over to the Kootenay, descended it to Montana, continued westward on horseback until he reached the Columbia, then paddled with the current to its mouth. On the

#150

A HUDSON'S BAY POST

return trip he discovered for the first time part of the Columbia going through the Selkirk trench and arrived back at the Great Bend (#148). He thus had explored every part of the course of this confusing river as well as its confusing tributary, the Kootenay.

Other explorers. Space does not permit a description of the equally thrilling adventures, dangers, and hardships of *D. Stuart*, as in 1811-13 he pushed up from the mouth of Columbia, through the beautiful Okanagan Lake to Kamloops; and of *J. McLeod* and *R. Campbell* who in 1834-38 paddled, packed, and drifted along the cold

Liard River to *Dease Lake* and across to the *Stikine*; or of *Campbell* who in 1840-48 went up the north branch of the Liard and down the *Pelly* to the *Great Yukon. William Dionne* in 1859 conquered the canyons of the *Skeena* as far as Lake Babine.

Founding of Fort Victoria. These explorers were all interested primarily in the fur trade and not a settlement was formed. After a long dispute, an arbitration court gave what are now the States of Washington and Oregon to the United States. The Hudson's Bay Company, which had absorbed the North-West Company, expecting such a decision, had in 1842, changed their headquarters from *Fort Vancouver*, near the mouth of the Columbia River, to *Fort Victoria*, on Vancouver Island (#148); here the first settlement in British Columbia grew up around the trading-post.

Present state of fur trade. The fur trade of British Columbia has continued until to-day but is not of such relative importance as formerly. Indeed, the exploiters of furs and fish of this province have been blind to their interests. The superb sea otter (p. 381) was slaughtered almost to extinction more than fifty years ago; the fur seal of the north Pacific was shot on the open sea so continuously from fifty Victoria schooners that an international commission had to prohibit the trade entirely. But throughout the province, and especially in the north, trappers still catch the beaver, mink, marten, muskrat, ermine and lynx.

Mining

Gold on Queen Charlotte Islands. While the fur trade repelled the settler, since furs and farms do not readily mix, the rush for gold, which spread like a prairie fire over British Columbia for fifty years, brought permanent settlers to every part of the province. The golden fire was lit in 1851 on Queen Charlotte Islands, when two Indians stumbled on nuggets weighing two pounds and three pounds; but for seven years the fire only smouldered, as the search in these islands did not succeed.

Gold rush to the Fraser. In 1858 what were previously vague, whispering rumours of gold in the sand-bars on the Fraser River became a deafening roar, with streaming headlines in all newspapers; one of the greatest stampedes in history began through the wilderness toward the treacherous canyons of the Fraser River. The rich, the poor; the proud, the humble; the young, the old; the sturdy, the feeble; all joined the furious race to get first possession of the golden sands. Small boats, canoes, every species of craft that would float, poured their human cargoes across Juan de Fuca Strait into the mouth of the river. Hudson's Bay Company employees left their posts, United States soldiers deserted their regiments, and sailors their ships. Indeed, shipping was paralyzed in Puget Sound because there were no sailors left. Every kind of craft, sea-worthy or rotten and ready to sink, that sailed from San Francisco, was

#151

WASHING SAND FOR GOLD ON THE FRASER RIVER

packed with a motley mass of excited humanity, buoyant with golden hopes. Seventeen hundred left the city in one day. Soon 17,000 miners, carried on the eager tide, were fighting their way through the Fraser canyons that had almost conquered the tried and tested Simon Fraser. A good number were drowned in the whirlpools and rapids, some almost starved to death, a few were killed in quarrels with Indians, many turned back, but thousands were soon washing the sands for gold in sieves and rocking cradles (#151). Each man was given a patch of sand-bar twenty-five feet square as his domain.

Only a few struck sands rich enough to make a hundred dollars a day; but most, if willing to work, could make a fair wage of ten or twelve dollars a day. The majority of these restless prospectors, as they impatiently filtered out the yellow dust that had been washed down by the current, had visions of the massive *mother lode* farther back, from which these particles had been worn off. This was the lure that dragged them upstream through the yawning canyons, over the whirling waters, across the heart-breaking mountains, to seek richer and richer gold beyond. Only the Chinese, large numbers of whom were attracted from California and China, were willing to work over the sands on the bars thoroughly until they had delivered their full quota of gold. Many miners reached as far as Lytton (#148) the first year, though there was no safe transport through those grand, but forbidding and awe-inspiring canyons. In November 1858, 10,000 miners were washing the sands of the Fraser River between Hope and Lytton.

On to the Cariboo. By 1859 the search for the mother lode had drawn one thousand men to *Quesnel.* Then the real thrills began, as the most daring went back to explore the creeks in the Cariboo Mountains. *Williams* and *Antler Creeks,* near *Barkerville* (#148), were as well known names as Klondike Creek in the Yukon became later. For a few, to whom luck gave the choicest positions, these creeks poured out the yellow dust as though it were water. One man in Williams Creek took out 900 ounces, 500 ounces, and 300 ounces of pure gold in three successive days; others made $1,200 a week; one pan gave $800 to a lucky miner. John A. Cameron, better known as "Cariboo Cameron," took out $150,000 worth of gold in three months. In 1863 there were 4,000 men at Williams Creek and the Cariboo was beating California in its palmiest days. Quesnel, now a sleepy village on the Fraser River, pulsated with life as the streaming thousands of dare-devil men made its streets hilarious as they pushed nervously onward to the mines. Its shops did a thriving business, goods were paid for with gold dust; men from every nation, of every type, of the titled aristocrat of Great Britain, the middle class farmer of Ontario and the United States, to the offscourings and gamblers of the cities who flocked like vultures after their easy prey, all jostled one another on the streets of Quesnel and Barkerville.

The gold rush spreads. In 1864 the best days of the Cariboo were over. During this year a new gold rush to the *Kootenay* (#148) at

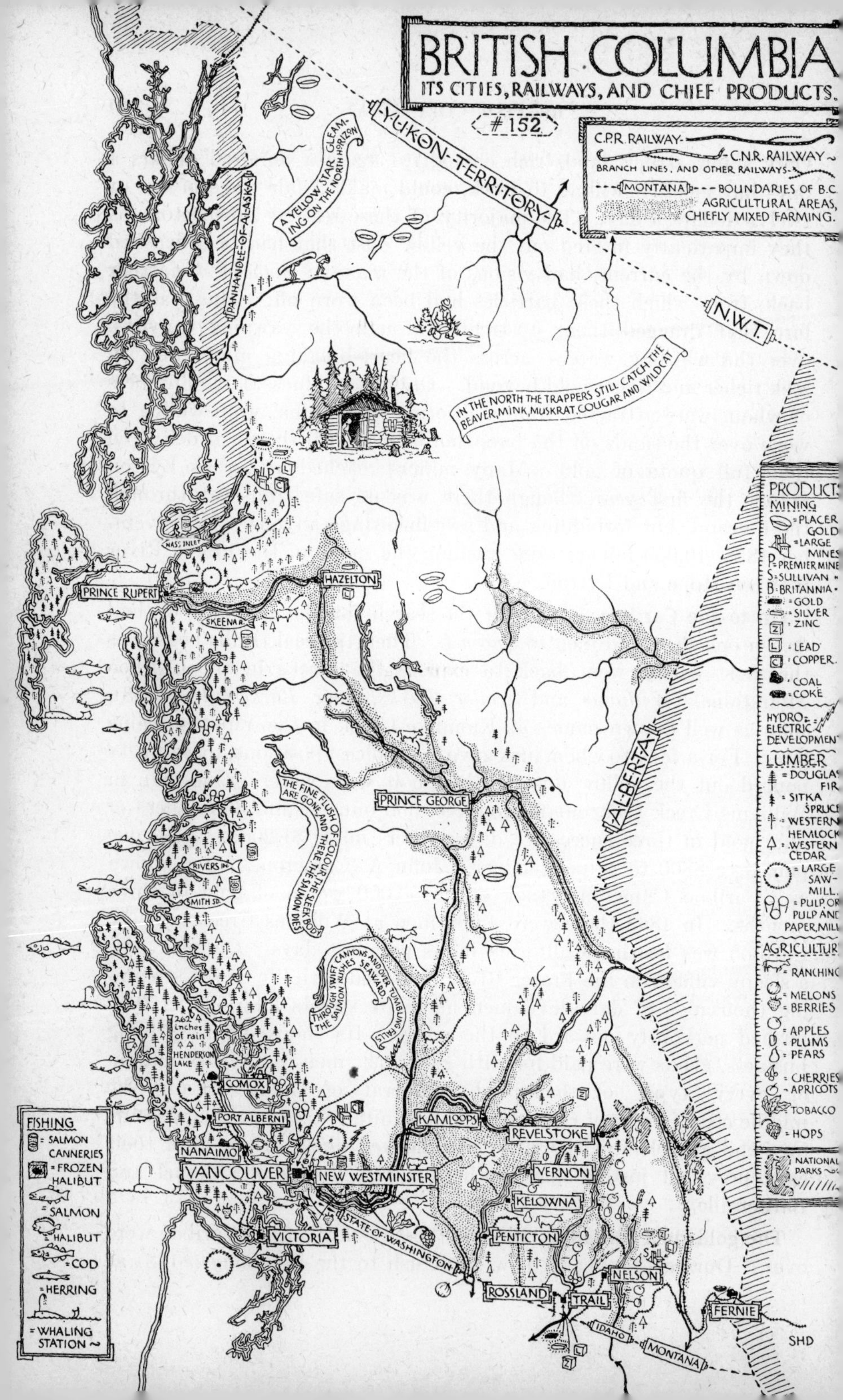

BRITISH COLUMBIA
ITS CITIES, RAILWAYS, AND CHIEF PRODUCTS.
#152
C.P.R. RAILWAY
C.N.R. RAILWAY
BRANCH LINES, AND OTHER RAILWAYS
MONTANA --- BOUNDARIES OF B.C.
AGRICULTURAL AREAS, CHIEFLY MIXED FARMING.
YUKON TERRITORY
N.W.T.
ALBERTA
PANHANDLE OF ALASKA
A YELLOW STAR GLEAMING ON THE NORTH HORIZON
IN THE NORTH THE TRAPPERS STILL CATCH THE BEAVER, MINK, MUSKRAT, COUGAR AND WILDCAT
THE FINE FLUSH OF COLOUR THE SLEEK BODY ARE GONE AND THERE THE SALMON DIES
THROUGH SWIFT CANYONS AND OVER TUMBLING FALLS THE SALMON RUSHES SEAWARD
NASS INLET
PRINCE RUPERT
HAZELTON
SKEENA R.
PRINCE GEORGE
RIVERS IN.
SMITH SD.
262 inches of rain
HENDERSON LAKE
COMOX
PORT ALBERNI
NANAIMO
VANCOUVER
NEW WESTMINSTER
VICTORIA
STATE OF WASHINGTON
KAMLOOPS
REVELSTOKE
VERNON
KELOWNA
PENTICTON
ROSSLAND
TRAIL
NELSON
FERNIE
IDAHO
MONTANA
SHD
PRODUCTS
MINING
= PLACER GOLD
= LARGE MINES
P = PREMIER MINE
S = SULLIVAN
B = BRITANNIA
= GOLD
= SILVER
= ZINC
= LEAD
= COPPER
= COAL
= COKE
HYDRO-ELECTRIC DEVELOPMENT
LUMBER
= DOUGLAS FIR
= SITKA SPRUCE
= WESTERN HEMLOCK
= WESTERN CEDAR
= LARGE SAW-MILL
= PULP OR PULP AND PAPER MILL
AGRICULTURE
= RANCHING
= MELONS
= BERRIES
= APPLES
= PLUMS
= PEARS
= CHERRIES
= APRICOTS
= TOBACCO
= HOPS
NATIONAL PARKS
FISHING
= SALMON CANNERIES
= FROZEN HALIBUT
= SALMON
= HALIBUT
= COD
= HERRING
= WHALING STATION

the other end of the province, drew away thousands of the more restless spirits, and the next year they came surging to the *Great Bend* of the Columbia River at the north end of the *Selkirk Mountains* (#148), where only a little gold was found. By 1869 no part of the province was too distant or too difficult for these hungry prospectors, who had tasted gold. They pushed into the endless sea of mountains of the north, from the bases of which plunged the creeks, which rolled their noisy waters over the yellow sand. By 1871, more than 1,200 miners were at work on the branches of the lonely *Omineca*, which empties into the Findlay River. By 1874 these streams were nearly deserted, for a new yellow star was gleaming on the north horizon; and by 1876 more than two thousand men were digging the sands of the streams from the *Cassiar Mountains.* Out of one of these creeks the largest gold nugget ever discovered in British Columbia and worth $13,000 was picked up.

The next mineral discoveries were in the south and were no longer in the sands of the streams, but in the rocks, from which they had to be blasted. Almost every town along the south was founded around a mine. In 1888 the charmingly-situated town of *Nelson* (#152) sprung to life in the midst of a number of gold and silver mines; in 1889, the first gold was found in *Rossland* (#152). The same year was a notable one when the first smelter was erected in *Trail.* This has now grown to vast proportions, and to-day it has one of the largest and most efficient chemical works in the world, in which vast quantities of zinc, lead, gold, silver, cadmium, acids, and fertilizers are produced. 1892 was a red-letter year for mining in British Columbia, since the *Sullivan Mine* (#152) in East Kootenay near *Kimberley* was discovered. To-day it is the greatest zinc and lead mine in the world and is largely responsible for making Canada the second largest producer of zinc and the fourth largest producer of lead. This mine is also Canada's chief source of silver. In 1898 the *Britannia Mine* (#152), on Howe Sound, near Vancouver, was discovered, and ever since it has been one of the chief producers of copper in this province. The third great mine of British Columbia is the *Premier* (#152), discovered in 1918, and situated on Portland Channel. From it gold, silver, lead, and copper are obtained. The famous Granby Mine at *Anyox* on Portland Channel, after being Canada's chief producer of copper for nearly forty years, closed down in 1935.

The coal mines. The production of coal in British Columbia has been decreasing on account of the increasing use of substitutes such as fuel oil, gasoline, and other petroleum products. In 1850, miners were brought from Scotland to open the thick seams of coal at *Nanaimo* (#152), on Vancouver Island, and twenty-five years later the colliery farther north at *Comox* was started, and both have been steady producers ever since. When the Crow's Nest Railway was built through south British Columbia, a new coal field of astonishing thickness and richness was found and has become easily the chief producer in the province. These three are still the chief coal mines. Large ovens, established at *Michel* in the Crow's Nest region, produce coke, which is largely used in the province as fuel, instead of coal and wood.

The future of mining. Undoubtedly mining has just begun in British Columbia. The rocks that seam this province from end to end have proved richly productive in the corresponding parts of the United States and Mexico; while each mile of the Cordillera in the United States has already yielded $3,200,000 in minerals and in Mexico $3,400,000; in British Columbia it has yielded only $103,000 per mile. Even the sands of the creeks and bars, which one might think had been combed clean of gold by the tireless Chinese, still contain by conservative estimate, sufficient gold to yield $300,000,000 if efficient, large-scale machinery were brought into operation.

Agriculture

The beginning of farming. Mining has done much more for British Columbia than only to pour out its mineral wealth. The fur-traders merely revealed a skeleton of rivers, mountains, valleys, and passes; the mining industry covered this skeleton with flesh and blood and made the province a living colony. It swept like a storm along the rivers and valleys and strewed them with men, who, when the gold excitement was over, settled down to make a living off the land. To feed the onrush of miners was an almost impossible task; prices of flour, meat, and milk were so extravagantly high, that some of the wiser among the invaders realized that there was more profit to be made from raising wheat, vegetables, meat, and milk than from the more exacting and less certain process of washing sand for gold. Wherever was found a little patch of suitable soil or a range of grassland, some of these men settled down to become farmers and ranchers. Many, who found no gold, or gambled their wealth, were left stranded and had no choice but to stay where they were; they became a type of shiftless farmers. *Barkerville*, *Lillooet*, and *Kamloops* soon became farm centres and general markets, and the Fraser delta a chequer-board of grain fields, orchards, fruit and vegetable gardens. Even in 1862, *Langley*, *Hope*, *Yale*, and *Lytton* were towns.

The first farms and ranches. In 1811 *D. W. Harmon* dared to plant potatoes, barley, and turnips as far north as Fraser Lake in the centre of the province. This was the first garden west of the Rocky Mountains, and probably succeeded, for in 1816 he states that 41 bushels of potatoes were obtained from planting one bushel, and 72 bushels of barley were grown to the acre. By 1830 the Hudson's Bay employees had cattle at Fort George at the bend of the Fraser River and nearly six hundred horses at Kamloops and Fort Alexandra. As early as 1851 Governor Douglas had fruit trees growing at Victoria, and in 1874 the *Okanagan district* saw the beginning of the fruit industry, when *Thomas Ellis* planted trees at *Penticton*. In the eighteen sixties herds of cattle were driven from Oregon to Kamloops and Lillooet.

Climate suitable for farming. The climate of British Columbia is very favourable for farming. The summer warmth on the coast is not so intense as in the interior but is so prolonged that vigorous growth begins in February and continues without a break until November. Though rainfall in this region is greatest in the winter, sufficient falls in the summer for healthy growth. The great summer heat from a cloudless sky in the interior is ideal for many crops, but summers are so dry that farming is only possible where irrigation water can be brought down from the mountains.

Farming the river benches. Though climate is so friendly, the surface of the province is so largely a sea of mountains that farming must always be confined to the 20,000 square miles of river and lake valleys, along which nutritious sediments have been gathering for millions of years. Many of these valleys have a succession of long benches, or level plains, like a series of platforms, rising one by one from each side of the river. These show former levels of the stream and are composed of fertile sediment settled out of the river itself. These benches are the chief farming regions. The upper ones receive enough moisture to grow bunch grass and other pasture crops for cattle, which the hot sun cures in the autumn to excellent fodder. The cattle are able to browse outdoors throughout the winter on this dry herbage. When water is brought down from the slopes above, which are high enough to catch rain and snow, to the lower benches, the thirsty land blossoms with a vivid freshness against the drab monotone of the dusty sage-brush.

Fruit perfect in form and colour. The two most notable of these valleys are the Okanagan and Columbia-Kootenay, where are grown apples (#153), plums, pears, apricots, and cherries (#152), which are pleasing to the eye and luscious to the taste. Nowhere in Canada

#153
PICKING APPLES IN THE OKANAGAN VALLEY

is the fruit more scientifically graded and more carefully packed for shipment to Great Britain, Prairie Provinces, and east Canada. One may be almost sure that when a diner in Toronto or Montreal sits down to a notable banquet, the beautifully coloured, perfectly formed, and delicious apples on the table are from the Okanagan valley. The small fruits, such as raspberries, loganberries as big as plums, and strawberries as big as small eggs, are more abundant on the delta of the Fraser and on Vancouver Island (#152).

The progressive British Columbia people, so far from a market, in 1935, actually shipped successfully to Great Britain nearly four thousand barrels of strawberries in airtight containers filled with sulphur fumes.

Hops, tobacco, melons. Special crops are being now featured in British Columbia. Many acres of hops on high trellises can be seen at *Agassiz* near the mouth of the Fraser River; tobacco is being tried in the same region, and at *Oliver*, south of the Okanagan, honey-sweet *musk melons* for the whole province are being grown (#152).

Ranching. Cattle ranching and mixed farming, two branches with a great future, are steadily increasing on the bench lands of the river valleys. Around *Kamloops*, large ranches have flourished for seventy years, and in the valleys of the *Kootenay* in the south end, in *Nechako* and *Bulkley* along the Canadian National Railway, and in the Cariboo district the output of beef cattle is steadily increasing as markets are found (#152).

Take courage from the Swiss. Since British Columbia can grow far more wheat, oats, barley, potatoes, or hay to the acre than any other Canadian province, its future as an agricultural district is assured. To show its possibilities, let one small region, the Columbia-Kootenay valley in the Rocky Mountain trench be compared with Switzerland. This valley contains two-thirds more cultivable land and much more timber and pasture land than the whole of Switzerland, and in addition possesses minerals. Yet Switzerland supports 3,500,000 people, most of whom make their living from the land (p. 154).

The harvest of the sea. The waters washing the west coast of British Columbia should be a veritable Garden of Eden for fish. Countless islands, the peaks and plateaus of a drowned mountain range, form a breakwater to beat back the waves of the Pacific Ocean, and give thousands of miles of safe and sheltered waterways fringed with a frayed embroidery of countless branched inlets of every shape and size. These snug inlets, growing every conceivable type of fish food, are ideal places for the fish to spawn and for the hatching of their young. Indeed, they contain such a variety of conditions, that every species of fish, no matter how odd its wants, should find food, depth, temperature, and movement of water just suited to its needs. One is not surprised that these complex waters are populated by myriads of flat giant *halibut*, sleek little *cod*, and shoals of *herring* as numerous as a plague of locusts, and which come and go as mysteriously. Vast schools of *pilchards* as fat as butter are found in the waters of this bountiful fishery which are also visited regularly by six species of *salmon,* where on the rich pastures of the sea they rapidly grow to beauty and maturity (#152).

The life of the salmon. The salmon is the prince of Pacific fishes (#154). There are five important species: the *sockeye*, *spring*, *coho*, *pink*, and *chum*, all of which resemble one another in habits and are valuable in the fishing industry. The beautiful sockeye with the firm body of rich red flesh, and savoury taste, is prized above all the rest. It is born far up the rivers, often in the mountain lakes, where

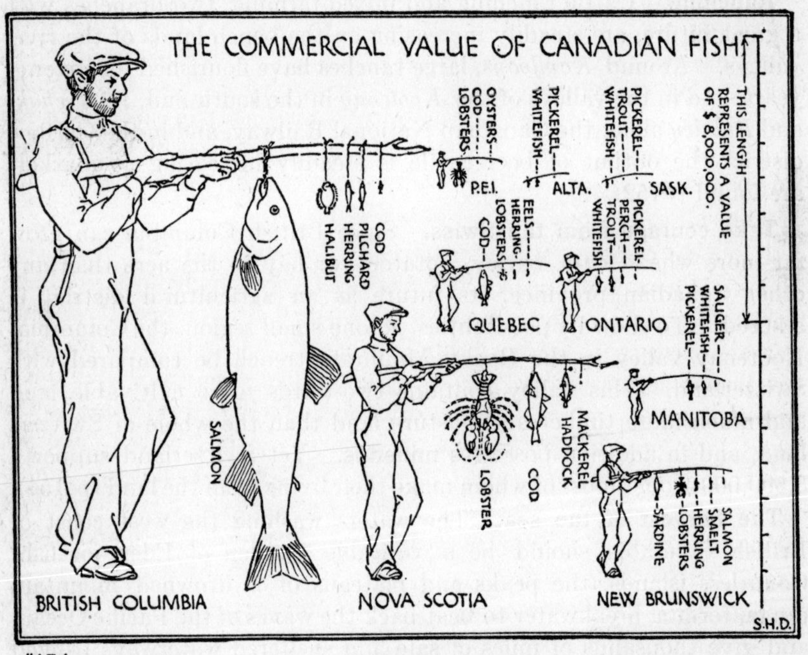

#154

the male and female before they die cover the eggs with gravel. Almost a year later, urged by a longing it does not understand, the young salmon starts the rough journey for those luxuriant mazes of the sea, previously described. Through the swift canyons and over the tumbling falls it rushes seaward, and then it is lost for three years; but the rapid growth indicates that it is finding abundant food among the smaller fish and shellfish that abound in the coastal trench of the Pacific. It is at last mature and is fat and strong. It starts back on its last journey to the little lake where it was born. Among the thousand gaping mouths in that tattered shore line, how it finds the right one is the sockeye's secret, but it seldom or never misses; and how, as it passes up the river, hundreds of miles, it knows just which of the hundred branches to enter is a great mystery. But it knows its street and number, though it has travelled the route only once before and that three years ago when it was still an infant. At last it is back on its home lot in its own lake; the fine flush of colour, the sleek body, the firm flesh are all gone; it has eaten nothing since it entered the river; it is worn out; when it lays its eggs its life is complete, and there it dies where it was born, the whole district often being foul with the smell of the decaying bodies. This remarkable story has been worked out by a study of the lines of growth on the scales. By studying a single scale the skilled eye of the scientist can tell not only how old a salmon is, but whether it came from the Fraser or Skeena River, and even the particular branch of the river. Other species of salmon may spend only two or three years in the sea but all come back to their home to die.

Catching salmon. It is chiefly while coming up the inlets and rivers on their last journey that these fish are caught. A net hung vertically in the muddy water from a small boat across their path catches them in its meshes behind the gills. In a few localities traps are placed across their path. They enter the large funnel-like opening of this net and soon work backwards through the throat to the heart from which they can never hope to escape. Out in the sea the spring and coho are caught in great numbers on troll lines. Small boats can be seen by hundreds at the mouth of the Fraser River and along the Skeena. These fish are also caught in Nass and River Inlets (#152).

The "Iron Chink." Along these fishing grounds are nearly fifty canneries to which the fresh fish are rushed. They are fed one by one into the mouth of a great machine, called the "*Iron Chink.*" It fits itself to the size of the body, cuts off the head and tail, in a flash removes the six fins, then slits the fish open, removes the insides, and sends the clean carcasses rushing forward on a carrier at the rate of one a second. Along the carrier they are thoroughly washed by sprays of water and cut by great knives into pieces of the proper size. Cans come along one carrier, cuts of salmon along another, and as they meet a machine rams the salmon into the can, which rolls down to another chamber, in which air is removed, salt added, and the can sealed. Then on trucks they are wheeled by thousands into big chambers where they are cooked by superheated steam until all the bones are soft. Machines shellac and label the cans. Man's hand does not touch the fish after it enters the cannery. This nutritious food is shipped to many countries but especially to the more advanced parts of the world such as eastern Canada, Great Britain, and France. Sockeye, cohoes, and pinks are canned; spring and cohoes are frozen and shipped fresh to eastern Canada. Though dog salmon are sometimes canned, they are largely salted and shipped to China and Japan.

History of salmon industry. Salmon, dried on raised platforms, were the chief food of the Indians along the coast and the rivers from the earliest times. The Hudson's Bay Company as early as 1835, shipped 3,000 to 4,000 barrels of salted salmon to Hawaii. The first

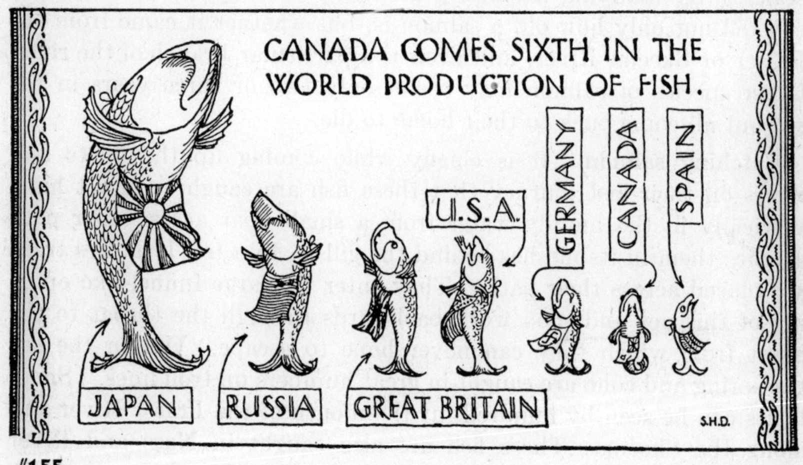

#155

commercial cannery was started in 1870 near New Westminster, and the industry after a few hard years, romped ahead by leaps and bounds. It is not likely to expand much beyond the present level, since no greater number can be captured than at present without depleting the supply.

Halibut, herring, pilchards. Halibut are caught by Japanese and Indians, largely with hook and line, and rushed to Prince Rupert, where they are frozen and shipped to eastern markets (#152). This industry only developed after the Canadian National Railway to Prince Rupert was finished. Herrings and pilchards swarm in immense numbers off the coast of Vancouver Island (#152) and in some of the inlets and could be captured in far larger quantities than they are at present if a suitable market could be found. The pilchard is caught largely for its oil and to make fertilizer. The herring is also used for these purposes, but may be cured by drying, salting, and smoking, and finds a market in China and Japan.

Forest Wealth

The "big four". The heavy rainfall and mild weather of the western slopes of the Coast Range have clothed them with a forest

#156 LOGGING IN BRITISH COLUMBIA

of trees, whose noble trunks, four or five feet in diameter, may tower up one hundred feet before the first branch is reached. These trees are crowded so close together that often they yield 100,000 feet of lumber per acre. The *Douglas fir, western cedar, western hemlock,* and *Sitka spruce* are among a dozen trees that grow on these slopes (#156). The Douglas fir dwarfs all others in importance, and its beautifully-grained, light, tough lumber is found in structural work, ornamental panels and in masts all over the world. It would be easy to cut boards from it one hundred feet long and four feet wide. The western cedar is so resistant to decay, that lumber can still be cut from logs that have fallen one hundred years ago. Though too soft for many purposes, it is safe to say that most of the shingles used in Canada and the United States were cut from this tree. The giant canoes manned by fifty Indians, which astonished the early explorers on the Pacific coast, were shaped from a single trunk of this huge tree. The western hemlock has been neglected for lumber not because it is not strong, light, and beautifully figured, but because Douglas fir is better and more readily obtained. It is the chief wood used in the six pulp and paper mills along the coast (#152). Sitka spruce is most common on the Queen Charlotte Islands. Because it is strong, light, and elastic, and can be obtained in large, clear, straight-grained planks, it is the best wood in the world for aircraft manufacture. It is also used for pulp.

Logging. The timber felled and trimmed on the slopes of the inlets is cut into suitable lengths and then dragged down the slopes, through the woods, or on overhead cables by donkey engines, to rough logging railways (#156). These transfer it to the water edge, where the logs are joined into rafts which are towed by tugs to *Prince Rupert, Port Alberni, Victoria, Vancouver,* and *New Westminster* (#152), where, in some of the largest sawmills in the world, they are cut into lumber.

A monstrous machine. A continuous stream of giant logs is fed from the mill pond along an endless moving belt to a carriage made of a maze of wheels, cogs, belts, platforms, bars, and shafts (#157). The alert sawyer stands at his post, with numerous levers in front of him. By means of these he can guide every movement to his purpose as though the carriage were a living thing. A bar presses the log against one side and at the same instant the carriage and log rush forward to meet a fierce, whirling, circular saw; in an instant it swings back, a slab falls away from the side; it is rolled over as though it were a fence post and again rushed forward to have a second slab removed. After four cuts it is square, and then the marvel of machinery takes place. On the fifth mad rush forward a series of parallel band saws cuts it so that it falls apart into a pile of great boards, which, before one has time to think, are being carried away on belts, sorted, and put in loose piles to dry. A steady line of logs enters the hungry maw of this monstrous machine at one end, and a continuous stream of lumber flows away from the other.

#157

SAWMILL

Lumbering in the interior. In the interior of British Columbia, where there is less rain, and the growing season is shorter, the trees are smaller and more scattered, and are in some cases of different species. But lumbering is carried on along all the rivers, and smaller sawmills which mill largely for local use, are found throughout the province (#152).

History of lumbering. The first sawmill in British Columbia operated in 1846 at Parson's Bridge, near Victoria. With the discovery of gold and the demand for buildings, many sawmills were built. In 1861 the first large mill was constructed at *Alberni*, also on Vancouver Island. In 1873 there were fifteen mills. At this time long rows of oxen were used to transport the logs, but by 1890

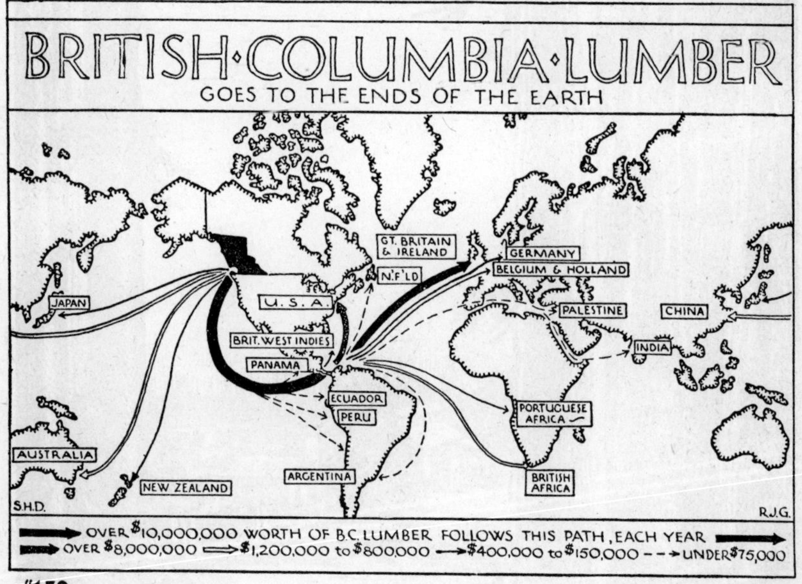

#158

the donkey engine had entirely replaced these draught animals (#156). By 1915 the use of high level cables attached to the tops of trimmed Douglas fir trunks were in common use. To-day there are 75 logging railways to bring the logs to tide water and 60 tugs to tow the rafts to the mills.

The Cariboo Road. With its surface billowed like the storm-tossed sea with high mountains, many covered with ice and buried in snow, its deeply furrowed trenches, its gaping canyons, its swift rivers tormented with tumbling rapids and tossing waterfalls, British Columbia has always been a difficult country for the transportation of man and goods. When thousands of miners suddenly swarmed into the country, trails had suddenly to give place to roads. *Governor Douglas* at once undertook to build a first class road from Hope

#159

ON THE CARIBOO TRAIL DURING THE CARIBOO RUSH

How many kinds of transportation are using the road?

right into the Cariboo country, over four hundred miles away (#148). It had to conquer all the obstacles that had brought Simon Fraser and his brave band face to face with death many times. The road was to be smooth and wide enough to take heavily loaded wagons at all seasons. In a new colony, with scarcely any settlers, with small and uncertain revenues, the courageous governor undertook this great work. A group of royal engineers built the first few miles, contractors completed the task, but, much of the hard, back-breaking work was done by Chinese, brought in for the purpose. In 1863, two years after it had been begun, the great *Cariboo trunk road* was completed (#148). Fifty years after the excitement of the rush to the Cariboo had been almost forgotten, this wonderful trunk road still stood, a monument to the engineering skill and political courage of the pioneers of British Columbia. Within recent years it has been repaired and is now a first-class motor road.

The Cariboo Trail. In the early days, even before the completion of this road, long lines of mules, each with a two hundred and fifty-pound pack on its back, swung along the tortuous trail (#159), close to ledges looking down a thousand feet, over mountains, and down through ravines, where these safe-footed animals never stumbled. One speculative trader introduced 21 camels on the route, but their spreading toes, adapted to sand, were quite unsuited to the rough rocks of the Cariboo, and he finally left them to their fate in the mountains of British Columbia. Many a man for years after, who did not know the story, was startled to see a group of desert camels in the midst of the mountains. The last one died in 1905.

The Dewdney Trail. Sir Edgar Dewdney had already built a trail eastward from Hope to the Similkameen River (#148), and later, 1864, when gold was discovered in the Kootenay country, this was continued right across the south end of the province through *Princeton*, *Hedley*, *Greenwood*, *Rossland*, *Trail*, and *Cranbrook*. It was known as the *Dewdney Trail* and played a big part in the opening of the row of mining towns, strung like a necklace across the south of the province. The *Kettle Valley Railway* and a good motor road now follow the course of this trail.

Canadian Pacific Railway. But the repelling ramparts of the Rocky Mountains hung as a dark curtain between this province and the rest of Canada. Until it was pierced by a railway, full, prosperous life could not come to this colony; consequently the building of the Canadian Pacific Railway was the rebirth of this western province. What the St. Lawrence was to Quebec and the

Great Lakes to Ontario, the C.P.R. became to British Columbia; it formed a silken cord binding the provinces into one Dominion; it gave British Columbia a market in the prairies for the fruit and lumber, and gave the Prairie Provinces an outlet for their wheat; it opened to all Canada the treasures of the Pacific Ocean and brought Australia, New Zealand and India near to the Dominion.

Vancouver. From the day of the completion of the C.P.R. in 1886, the growth of British Columbia crystallized along that line. *Revelstoke*, *Kamloops*, *Lytton* and *Hope* all increased in importance (#152), and at its terminus on the beautiful, commodious harbour on *Burrard Inlet* stands the city of *Vancouver*, which in fifty years has left behind it every other city in Canada except Toronto and Montreal. Its grain elevators rival those of Montreal; its commerce with Japan, China, Australia, New Zealand, and Singapore competes successfully with those of Seattle and San Francisco. As a manufacturer of lumber, shingles, etc., it has no Canadian rival, and it is the chief producer of manufactured goods for the whole province. *New Westminster* on the north arm of the Fraser River was the chief town of the mainland until the building of the Canadian Pacific Railway pushed Vancouver ahead. It had also at one time been the capital of the mainland before Vancouver Island was included in the union. It still is a thriving city, served by both the great Canadian railways, and has the only fresh-water port on the west coast of Canada. As boats drawing 30 to 40 feet can come up the Fraser River to this point, its commerce is of great importance.

British Columbia's second storey. When the Canadian National Railway found a low passage through the Rocky Mountains by the *Yellowhead Pass* (#148), and laid a line westward along the Skeena River to Prince Rupert; British Columbia began to build its second storey; already a group of thriving towns, such as *Prince George* and *Hazelton* are springing to life; and broad, well-watered valleys, such as the *Nechako* and *Bulkley*, have been opened to mixed farming and ranching (#152). But no railway can afford to pass Vancouver by; the Canadian National therefore extended their line along the North Thompson River to Kamloops and paralleled the Canadian Pacific for the rest of the way to Vancouver. Since one line can easily carry all the traffic, this duplication was wasteful.

The Kettle Valley and Pacific Great Eastern. A third railway follows very closely the Dewdney Trail across the south of the province, giving an outlet to the numerous towns, and passing the Rockies through the Crow's Nest Pass. A fourth railway, the *Pacific Great Eastern* (P.G.E.), built by the province through a desolate country, at great expense, starts at the none-too-good harbour of *Squamish* on Howe Sound and runs north-east to *Quesnel*. Since both terminals are of little importance, and since the line has very heavy grades, and the part of the plateau across which it passes is semi-arid and thinly populated, it does not pay its way and has become a great drain on the province.

In spite of the mountain roughness of the province, many of the rivers, especially where they flow through the trenches, or expand into lakes, are navigable for shallow boats. Many have been used in the past but today less frequently, due to the numerous railways available for transportation.

Settlement

The coming of the Chinese. The settlement of British Columbia spread along its valleys and coasts, which run north and south, contrasting with every other province in this respect. Since it was always far away and too costly for the working man easily to reach, labour has always been scarce. Even when the Canadian Pacific was completed, the temptation of the poor settler to drop off in the Prairie Provinces where smiling farms, ready for the plough, could be had for the asking, was so great that few ever got beyond the Rocky Mountains. Consequently, from the earliest settlement, this province followed the example of California, and allowed *Chinese* to enter the province to do the manual labour. We have already seen that they were among the first to sift the golden sand on the Fraser, and a large part of the strenuous, monotonous work of building the Cariboo Road was performed by these patient men, while the white man did the not-less-strenuous, but much-more-exciting work of gathering gold from the sand-bars. In 1864, there were 4,000 Chinese, and by 1882 their numbers had increased to 12,000. The number entering during the building of the Canadian Pacific Railway (16,000), was alarming, though without them the railway could not have been completed. After doing the hardest labouring work in the coal mines, in the canneries, and on construction work for some years, these energetic little men began competing with the white settlers at store-keeping, market-gardening, on trains, in hotels, etc., and a strong agitation arose to exclude them from the country.

After the Russo-Japanese war in 1905 the desire to exclude these eastern people greatly increased; for then the *Japanese* began to come into the province in great numbers, no less than 8,000 entering in 1908 alone. Most of these came from *Hawaii.* To add to the difficulties of this province, which was in the wildest state of excitement and vexation at the entrance of the yellow

#160

SPORT AND SCENERY FOR THE TOURIST

men in 1907, a wave of East Indians, citizens of the British Empire, swept into Vancouver. After much agitation a head tax of $500 has slackened the stream of Chinese to a mere trickle, and by arrangements made with the governments of Japan and India their citizens have almost ceased to enter the country.

The immigrants of British Columbia are chiefly from the British Isles, many of whom were attracted by the climate, so similar to that of old England. Most of these men were highly educated, and they have had a powerful influence in improving the cultural and educational influences of the province.

Tourist Attractions

Probably no part of Canada can offer more varied interests to attract the tourist (#160). The delights of the sheltered boat trip along the indented coast from Vancouver to Prince Rupert or Alaska cannot well be described in words. One passes through a perpetual maze of thousands of tree-clad islands, long, narrow, winding inlets that worm their way into the very heart of the Coast Mountains, and open stretches of water swarming with marine life. In the Rocky and Selkirk Mountains is a continuous panorama of filmy waterfalls, tumbling torrents, majestic rivers, lakes as glorious as emeralds in an opal setting, mighty mountains, ice-fields, thousands of glaciers never seen by man, and countless snow-capped mountain peaks waiting to be climbed for the first time (#160). Three national parks, Yoho, Glacier, and Kootenay (#152) have been set aside in these mountains for man's enjoyment. They have been made accessible by good roads, and are easy to traverse by well-kept trails. For its game-life, whether fish, waterbird, or wild beast (#152), in variety, size, and gameness British Columbia cannot be surpassed by any part of the world. Noble cariboo, moose as big as cows, graceful deer of several species, fierce grizzly bears, tiger-like cougars, mountain goats, and big-horn sheep are a few of the species that are widely scattered through the mountains and trenches.

CHAPTER XVII

NEWFOUNDLAND

The link between Europe and America. Newfoundland looks like a piece of North America which has become broken off and thrust itself out toward Europe. Perhaps that is the reason why it was the first part visited by Europeans. The daring *Norsemen* in their poor, crazy, open square-sailed rowboats, visited it in 1000 B.C., and *John Cabot* sighted its rocky shore in 1497 (p.163) and reported its waters to be choked with fish. Soon its coasts swarmed every summer with hundreds of fishing vessels from Europe. This harvest of the sea has never ceased from that day to this to attract these daring fishermen, and has been the chief cause of Newfoundland's growth. Its stand, like a beckoning watchman far out in the Atlantic ocean, has played a big part in the conquering of the barrier of the stormy Atlantic. The first cable from Ireland across the ocean was laid to Newfoundland, and it is still the great outpost for trans-Atlantic cables (#169). *Heart's Content, Harbour Grace, Bay Roberts* and *St. John's* all have European cables rising from the sea. These pass on messages to *Sydney, Canso,* or *Halifax,* in Nova Scotia. It was at St. John's in 1901 that *Marconi* placed his station and stood on its bleak shore to listen for signals that were being sent out from Ireland, and when the thrilling sound of the dots and dashes pulsating through two thousand miles of space struck his expectant ear, the new area of wireless telegraphy had been ushered in. Again in 1919 *Captain John Alcock* and *Lieutenant A. W. Brown* rose in their little aeroplane from St. John's, struck out over the foggy Atlantic, and were the first to traverse the ocean in a non-stop flight. From that day to the present this bleak coast has seen most of those adventurous fliers start on their eastern journeys.

A torn and tattered coast. The island is triangular, with *Cape Bauld, Cape Ray,* and *Cape Race* at the angles (#162). The last is the best known cape along the whole continent, as from the days of John Cabot its forbidding, rugged height has been the first welcome sight of the New World to sailors after a long and tedious trip across the stormy Atlantic. The coast is so torn with long, narrow, deep inlets that in all the early maps it was shown as a sea of islands,

It was the careful survey of the famous explorer, *Captain Cook*, which showed these tattered fringes to be all united at the centre into one island (#162). The bays are deep to the very shores, which

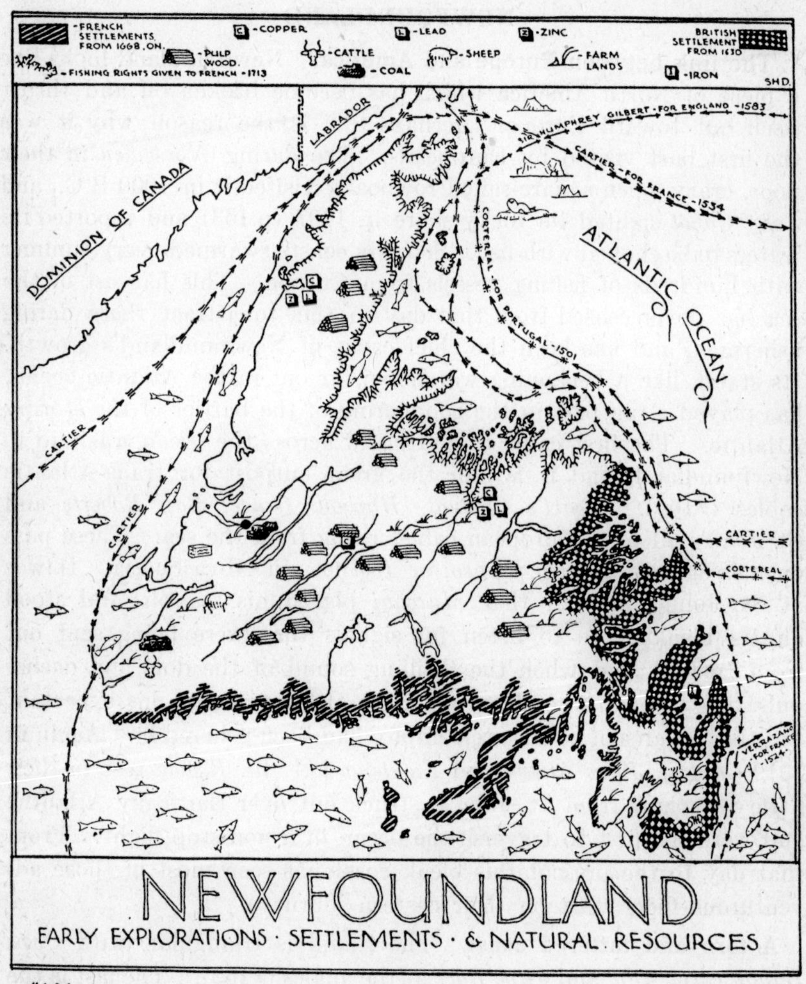

#161

are often high and rugged, though notched with many little harbours, each bordered by a fishing village. Many fragments of Newfoundland, in the form of rocky islands, mottle most of the bays and give them a wild, rugged beauty not surpassed by the fiords of Norway or British Columbia.

NEWFOUNDLAND
AND A SMALL PART OF LABRADOR
SHOWING PHYSICAL FEATURES, FISH, GAME, AND FORESTS
#162
CANADA
(PROVINCE of QUEBEC)
LABRADOR
ATLANTIC
OCEAN
LABRADOR CURRENT
BELLE ISLE
CAPE BAULD
STRAIT OF BELLE ISLE
GULF OF ST. LAWRENCE
LIMIT OF SHALLOW WATER
LONG RANGE
WHITE BAY
NOTRE DAME BAY
HUMBER
GRAND LAKE
GEESE
SPRUCE
BALSAM
EXPLOIT RIVER
GANDER L.
GANDER R.
BONAVISTA BAY
CAPE ST. GEORGE
ST. GEORGE BAY
TROUT
GROUSE
CARIBOU
TRINITY BAY
CONCEPTION BAY
CAPE RAY
FORTUNE BAY
PLACENTIA BAY
AVALON PENINSULA
SALMON
CAPE RACE
ST. PIERRE BANK
4800 SQUARE MILES
GREEN BANK
1450 SQ. M.
GRAND BANK
36,000 SQ. MILES

Rivers tumble through a rough country. The surface, from the very shore, is rough, rocky, and repelling. Most of the rounded ridges covering the surface run from south-west to north-east (#162) and are carried out into peninsulas before they sink beneath the sea. Long, deep, narrow lakes fill in the gulches between the mountains in

Copy of 18th Century Drawing of Fishing Stages in Newfoundland.

#163

FISHING FOR COD IN EIGHTEENTH CENTURY

A. Clothing of fisherman. B. Line, sinker, hook and bait. C. Fishing from barrels on side of schooner. D. Cleaning the fish. E. Trough for dressed fish. F. Salting boxes. G. Carrying the fish. H. Washing the cleaned fish. I. Press for squeezing oil out of cod livers. K. Barrel for refuse from livers. L. Barrel of cod-liver oil. M. Spreading fish out to dry.

the interior (#162) and are drained by rivers that tumble along the same valleys to the coast. All the largest, except the *Humber*, empty into bays along the north-east coast. The Humber has cut its way through *Long Range*, the highest ridge in the island (#162), which runs parallel with the west coast, and empties into the Gulf of St. Lawrence.

Pastures of the sea. Just why the food-fish population in Newfoundland waters is the densest and richest in the world has never been fully explained. The ragged coast provides ships with ideal harbours, fishermen with abundant bait, and fish themselves with the most varied breeding grounds, swarming with choice food for every appetite. The Banks of Newfoundland, the largest and most notable of which is the Grand Banks (#162), stretch all along the south and east of the island as a shallow, sandy platform. The water above this is a luxuriant pasture-field of billions upon billions of minute, jelly-like, animals and plants, on which swarms of small

#164

A FISHING VILLAGE

animals fatten. These become the prey of the cod and other food fishes, just as they in turn are devoured by seals and man. One cannot help believing but that the *Labrador Current* (#162), bathing the whole eastern and southern shore of the island, brings down from the Arctic vast multitudes of these minute animals and plants that love cold water. Just outside this cold current the nearly exhausted warm *Gulf Stream* flows in the opposite direction, and along the line where these opposing streams flow side by side there must be a stirring up of the waters that may be as stimulating to the growth of animals and plants as the churning of cream is to the forming of butter. Somewhere across these currents almost every organism of the sea should find suitable food, agreeable temperature, and comfortable movement of water.

Settlement of fishermen. The fisheries in these waters have been the bread and butter of the people of Newfoundland for four hundred years. Before it was settled, daring mariners from western Europe came over every spring in great numbers, caught cod on the Banks or near the shore, dried them on the beach, and carried them back to be sold in Europe. In time, a number of them, principally Englishmen, began to stay on the shore throughout the winter in

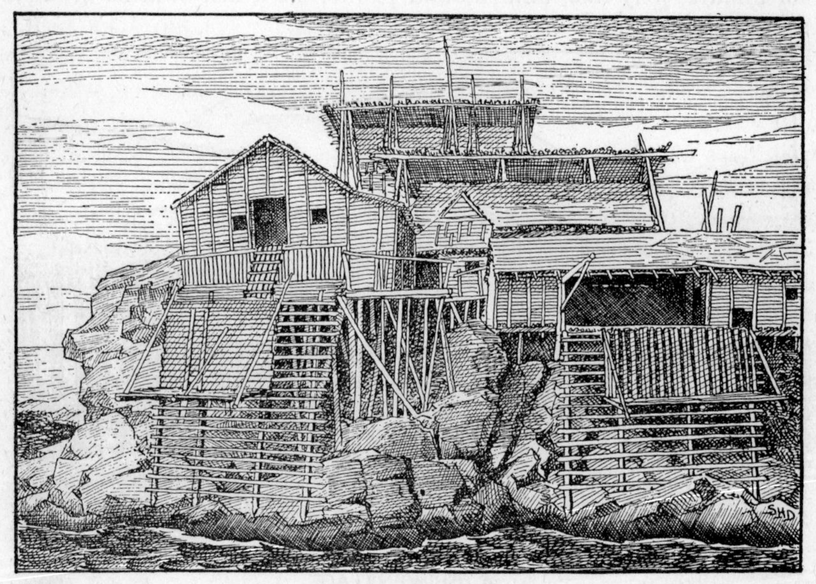

#165

FISHERMEN'S HOUSES AND DRYING RACKS

order to have an early start and get the best places in the spring for fishing and drying. This was greatly discouraged not only by other fishermen but by the British government, which claimed possession of the island. In fact settlement was, for a long time, forbidden and sometimes threatened with removal. But in spite of difficulties, men from the south-west of England and south Ireland gathered around the little inlets along the coast of the barren Avalon Peninsula on the south-east (#161) and crept up the east coast as far as *Bonavista Bay*. Later French settled along *Placentia Bay* (#161), and especially at *Placentia*. The same struggle between the two nations occurred here as in Canada, and the French were finally driven from Placentia but were left the two small islands of *St. Pierre* and *Miquelon* on the south coast. There they still remain, the last fragments of the continent, a large portion of which they once owned.

Inshore fisheries. The codfish is king in Newfoundland. Just as a beaver skin at one time took the place of the dollar in Canada, the codfish was the money unit in this island for centuries. Though there are nearly two hundred schooners fishing on the banks, the greatest quantity of fish is caught near the shore. Sixteen hundred fishing villages (#164), set in every inlet, form an embroidered necklace 6000 miles long right around the island. Though they fish with hook and line, they more commonly place traps, or weirs, near the shore. The fish are cleaned, salted, and dried on platforms, which cover every front yard (#165). The village covers a steep, rocky slope, lines of homes succeeding one another like rows of seats in a steep gallery (#165). The shore fisheries begin in June, because then swarms of little fish, called *capelin*, as numerous as flakes in a snowstorm, crowd up on the shore, chased by the greedy codfish. Fishermen scoop them out with nets in wagon loads to manure the land. After they depart, the cod are next attracted to the shore by countless schools of strange creatures, called *squid* (#11), which come and go most mysteriously, nobody knows where. These brown creatures, with tails like spearheads, have eight long pointed tentacles covered with suckers around their mouths and carry a pouch of ink, called *sepia*, which they do not hesitate to squirt in your face. They are almost as numerous as capelin but sometimes hard to catch. As the waves of squid are succeeded by waves of herring, the hungry cod are attracted near the shore well on into October, when the inshore fishing ceases. The catching of these small animals (capelin, squid, and herring), their preservation, and sale for bait to the Banks fishermen, is also an important industry.

Fishing in Labrador. The adventurous Newfoundlanders also fish on the coast of Labrador, seven hundred miles away. In Spring, more than eight hundred ships (#166) leave this region. They are loaded with household goods and fishing gear. Often the whole family goes along, taking the hens and even a cow. They catch and dry their fish throughout the summer and return in the autumn. They usually live in shacks on the coast, and often there are schools for the children. Some, however, live entirely in their ships.

Crisp salmon and toothsome lobster. The lordly cod is not the only fish of importance in Newfoundland. Crisp, fat *salmon*, made firm as rock and tasty as speckled trout by the icy waters of the Labrador current, is food fit for kings. In the inlets, all around the island,

#166

FISHING SCENES ON LABRADOR COAST

the salmon are caught in gill-nets as they are about to enter the rivers (#162); they are quickly frozen and, in a coat of icy armour, are shipped to Great Britain, where they are considered by leading London chefs to be equal in flavour to the best fresh salmon. *Lobster* factories, especially in the west, are strewed along the coast (#162), but too eager fishing has so decreased the number that the yearly catch has now shrunk to one-half of what it was formerly.

Seal hunting. The thrilling sport and hard work of catching *seals* attracts the most daring spirits from all over Newfoundland (#167). In the spring this dangerous work acts as a magnet, and men from

#167 SEALERS READY FOR THE SLAUGHTER

every direction come by train, sleigh, and on snowshoes hundreds of miles to St. John's. There are two kinds of seals: the *hoods*, which spend the summer under the shadow of the glaciers along the west coast of *Greenland* and the *harps*, which at the same time are on the opposite *Labrador* coast. But in the autumn, when the ice becomes thick in these regions, they swim south and spend the winter on the Banks of Newfoundland, fattening on the cod. As each seal (and there are probably a million of them) eats about three good sized codfish each day, they probably are the most successful fishers on the Banks. In January they move north, and by the beginning of February the ice fields covering the sea, both east and west of the north point of Newfoundland, are dotted with a million seals. In less than two days after they come up on the ice, 500,000 little white bundles of downy fluff are being nursed by as many proud mother seals. These seal pups, weighing seven pounds at birth, grow at a

#168

A CATCH ON THE ICE

quick rate, putting on several pounds a day, so that before they are six weeks old they are able to look after themselves and have a layer of pure, clear fat two and a half inches thick, just under their valuable skin. These buttery balls (#168) are the victims the seal hunters seek.

Battling through the ice. Eight steel ships (#168), each packed with two hundred of the bravest, toughest sailors the world possesses, begin in early March to batter their way out of the narrows of St. John's harbour; heavy field ice covers the sea as far as the eye can pierce; every foot gained has to be won by force. Often the men have to break a passage or push aside the ice in the biting gales. Occasionally dynamite has to be used to break this thick floor. These men never waver, however, but laugh at such difficulties. Though they are on the roughest corner of the most stormy ocean, they fight every inch of the way outwards, until perhaps the wind changes, blows the ice aside, and makes a wide opening.

By the middle of March they are near the hunting grounds (#162). Aeroplanes above search the boundless ice for those dark specks as thick as blackbirds in a field in autumn. At last the steamer arrives in their midst. Every man is out on the ice, an iron-shod club in his hand, a rope around his shoulder, and a package of food on his back (#168). The shrewd old seals drop down through breathing holes in the ice, but the innocent babies suffer. One whack on the head and a pup is laid low; with a sharp knife the belly is ripped open (#168), and in less than a minute, the skin, with lining of thick fat, is pulled from the quivering flesh; and away goes the hunter, dragging the skins behind him in a bloody trail on the ice (#168) and searching for another little victim. The skins are gathered in piles on the ice with a flag up to mark the place (#168). There is no pause, because every man's wage is measured by the number he kills; the slaughter never ceases until it is dark, or until the seals are all gone. Sometimes they kill six thousand seals in one day. The steamer comes along and lifts the piles on deck. This feverish slaughter goes on as long as there is anything to kill. If they have success, the ship may be so laden with skins that they overflow into the sleeping quarters of the men, who are packed anywhere for the rest of the voyage.

The dangers. Seal hunting is alive with dangers. The ice is always moving and carries the ship with it. It moves up and down with the waves as though it were made of rubber. Sometimes, after the men are on the ice, a fierce blizzard from the north-west blinds the air and drives the ship far away. The men cannot see more than a few feet in front of them and are lost; a night on the ice is deadly even for these tough men. Occasionally the ship is pinched between two great masses of ice and crushed like an eggshell, or it springs a leak and sinks to the bottom beneath the ice, leaving the whole crew on drifting ice, hundreds of miles from shore. Yet the trip has a strange fascination for these daring sailors. The man, who, at the end of the trip, says he will never go again, is one of the first to be drawn to St. John's by its magic the next spring.

Safely home. By the end of April the eight ships are back in St. John's with 200,000 smelly sealskins. The fat is removed and rendered into seal oil, which is shipped to the United States to be converted into soap. The skin, covered with coarse hair, is of little value as fur, but, when tanned, it makes the most beautiful *seal-skin leather.* If you have a rich, soft, flexible, leather bag, purse, or pouch, as you look at it, think that it once covered a fluffy little seal pup, and that brave men risked their lives to obtain it.

Pulp and paper. For more than three hundred years fish was to the Newfoundlander what wheat now is to the Prairie Provinces — its chief but uncertain wealth.

The forest is now rivalling the sea (#161). Since this island receives heavy rains at all seasons and has warm summers and only moderately cold winters, all the valleys of the interior are well covered with spruce, balsam, pine, and birch, though the backs of the hills are barren, having no better cover than moss and lichen. As soon as a railway across the island gave access to these forests, lumber mills were built and paper mills erected. Two very large mills, one at *Grand Falls* in the interior, and the other at *Corner Brook* near the West Coast, are well supplied with hydro-electric power (#169). The first sends vast quantities of newsprint to some of the largest London papers, and the second, which has been shipping still larger quantities to the eastern United States, has recently been bought by an English company. During the last four years newsprint has displaced codfish as the chief export.

Minerals. The deeply-eroded, lava-formed igneous rocks, which cover most of the island, are as promising to the miner as are similar rocks of north Canada, which are giving up their riches in increasing amounts. Two productive mines are already adding much wealth to this poor island. *Bell Island,* a small mass of rock in Conception Bay, has the largest deposits of *iron* ore in the British Empire (#161, #169), and a procession of steamers carries its products to the steel works at Sydney, N.S. Recently at *Buchans* (#161, #169), on Red Indian Lake, a valuable mine has been supplying increasing quantities of zinc, lead, and copper concentrates to Belgium and British smelters. *Coal* beds, near the surface in the south-west part of the island (#161), have been known for many years, and now that the railway across the island enters this district, they should be available.

Farming. For centuries, merchants, governments, and fishermen, not only neglected but opposed farming. The bleak, rocky coasts with long cold springs, short summers, and chilling winters were not conducive to the tilling of the land. In spite of these drawbacks the

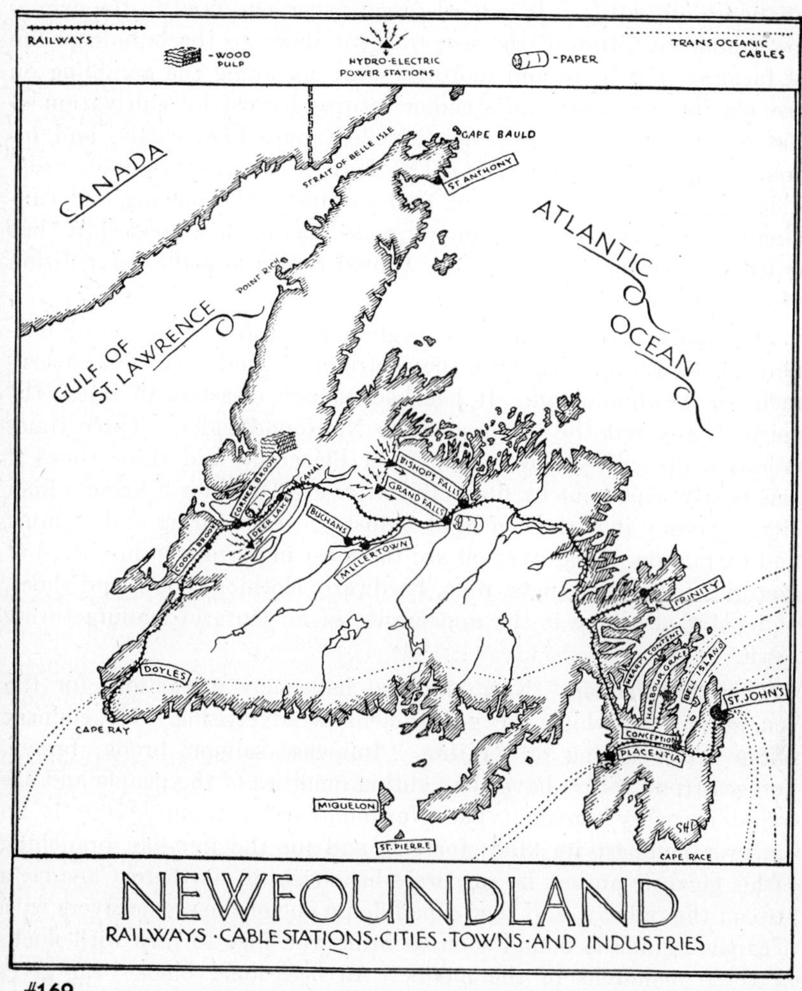

#169

poverty-stricken farmers, driven by the pinch of hunger, learned that on any little pockets of soil among the rocks, potatoes, turnips, cabbages, and sturdy roots generally produced good crops, that cattle could find pasturage in the river valleys, and sheep could crop fodder among the rocks. A more careful study, made possible by the building of railways, has proved that there is also much valuable farm land at the heads of inlets, along valleys of the largest rivers in the interior, and especially along river valleys in the south-west corner of the island (#161). It is hard to get fishermen, used to the excitement and adventure of the sea, to settle down to the humdrum life of farming, but more and more they are forsaking the gambling on the sea for the surer and steadier returns earned by cultivation of the soil. There are now some excellent pure bred cattle, and increasing numbers of sheep (#161). Hay is the most valuable crop, followed by potatoes, cabbage, and turnips. The picking and canning of *blueberries*, growing among rocks, has made a good start, but it has to face high duties in the United States and the competition of New Brunswick.

St. John's. Almost from the beginning, *St. John's* on a snug harbour facing Europe, has been the centre of life, industry, and government in Newfoundland. It has had enough disasters to break the spirit of any but the unconquerable Newfoundlander. Three times it was captured by the French from Placentia, and three times it was nearly wiped out by fire. But after each disaster it arose a finer city. To-day it is the centre of industry. Fish curing and canning and extraction of cod-liver oil are the most important industries, but factories for making nets, rope, hardware, clothing, boots and shoes, and paint also give it the appearance of an Ontario manufacturing town.

Tourist attraction. Newfoundland has many attractions for the tourist (#162), which the government is advertising. No country offers better fishing or hunting. Immense salmon, brook, brown, and sea trout (#162) have the fighting qualities of the people and are found in every stream. The Newfoundland caribou (#162) is an emperor amongst its kind; for size and for the tree-like branching of his gigantic antlers he surpasses his cousin in northern Canada; a trip on the railway, and then a paddle up one of the scenic rivers with a capable guide is bound soon to bring you face to face with flocks of these monarchs of the moors. An occasional black bear gives spice to the hunt. He who likes bird hunting will find grouse everywhere (#162), and ducks, snipe, curlew, and other shore-birds in abundance.

CHAPTER XVIII

THE UNITED STATES

Good neighbours. The United States is of greater interest to Canada than any other country except England. She is our big next-door neighbour, running parallel with our southern border from the Atlantic to the Pacific. Her people speak the same language and most of them are from the same British stock. She is also of great interest to our French-Canadian people, as probably a million of them are at work in the lumber camps and factories of the New England States. Almost all of these came originally from Quebec. With nothing but surveyor's posts to mark the boundary between the countries, the two peoples are blending more and more; many Americans live in Canada, and still more Canadians in the United States. However, in recent years there has been far less freedom in exchange of residence. Nevertheless every summer tourists by the million swarm across the borders, and motor cars from any State may be seen side by side with those from any Province.

The land of big products. There are other reasons why Canadians should know much about the great country to the south. Its production of important commodities, such as wheat, oats, barley, cotton, corn, tobacco, from the soil; copper, lead, zinc, coal, iron, and petroleum, from the mine; lumber from the forest; and steel, automobiles, moving-pictures, rayon, silk textiles, and many other manufactured products, far surpasses that of all other countries. It is by far the wealthiest of countries, most of its people earn higher wages, have better homes, and spend more than those of any other; consequently, Canada reaps a rich harvest of trade by being so near to such a lavish customer.

Mountain and plain. Because the two countries run side by side all across the continent, their structure is very similar. The massive Rockies to the west swell out to broader width and rise to greater heights than in Canada. The rich Canadian Shield alone, abundant in its rich minerals, almost avoids the United States. A small tongue of it, however, extends around the west end of Lake Superior, but what a tongue it is! (#170). Though so small, it contains the most valuable iron and copper mines ever discovered. Another branch of the Shield extends across the St. Lawrence River at the

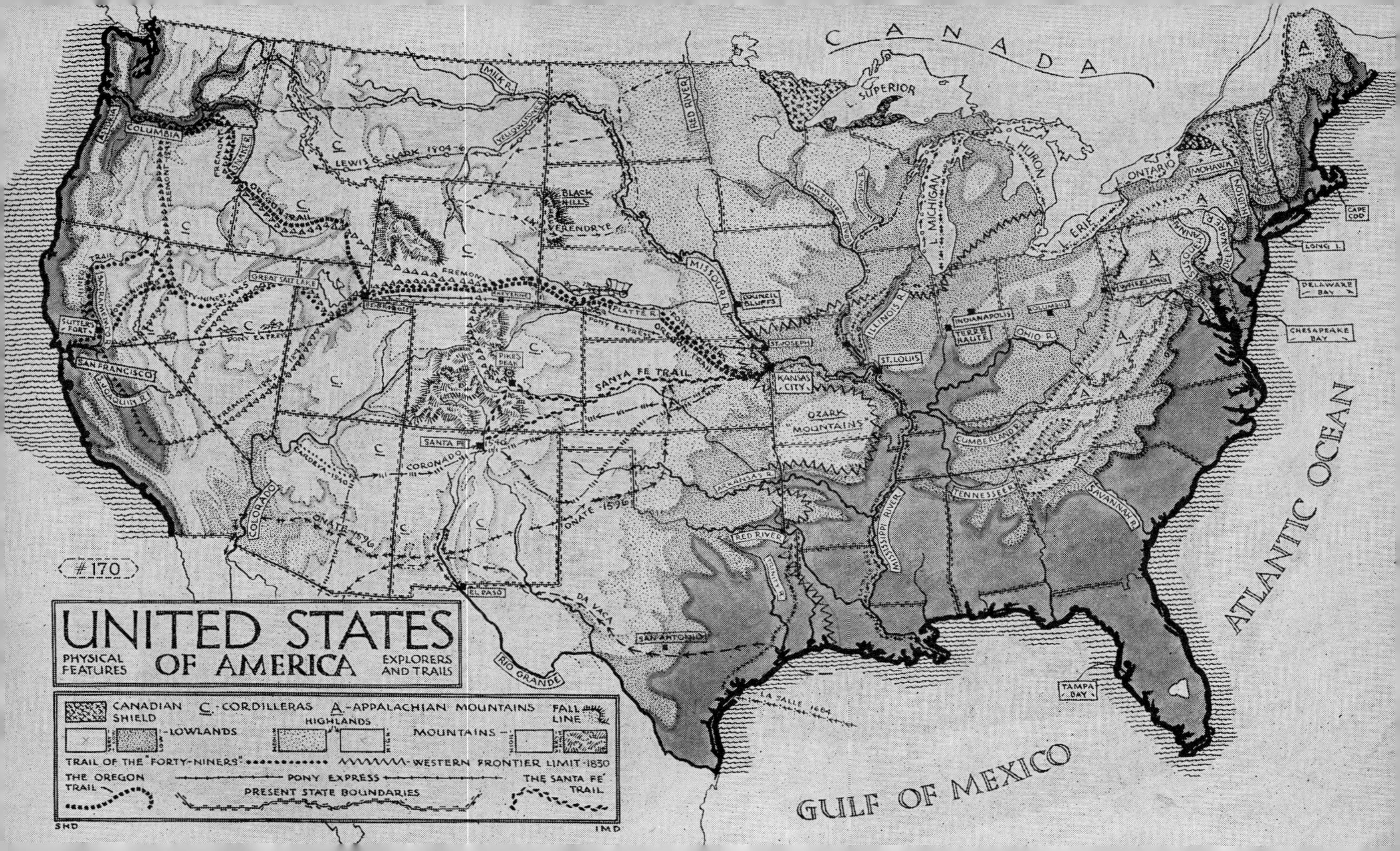
UNITED STATES OF AMERICA
PHYSICAL FEATURES
EXPLORERS AND TRAILS
#170
CANADIAN SHIELD
C-CORDILLERAS
A-APPALACHIAN MOUNTAINS
FALL LINE
LOWLANDS
HIGHLANDS
MOUNTAINS
TRAIL OF THE "FORTY-NINERS"
WESTERN FRONTIER LIMIT-1830
THE OREGON TRAIL
PONY EXPRESS
THE SANTA FE TRAIL
PRESENT STATE BOUNDARIES
SHD
IMD
CANADA
ATLANTIC OCEAN
GULF OF MEXICO
L. SUPERIOR
L. MICHIGAN
L. HURON
L. ERIE
L. ONTARIO
MOHAWK
HUDSON
CONNECTICUT
DELAWARE
SUSQUEHANNA
CAPE COD
LONG I.
DELAWARE BAY
CHESAPEAKE BAY
TAMPA BAY
WHEELING
INDIANAPOLIS
TERRE HAUTE
OHIO R.
ST. LOUIS
ILLINOIS R.
CUMBERLAND R.
TENNESSEE R.
SAVANNAH R.
MISSISSIPPI RIVER
OZARK MOUNTAINS
KANSAS CITY
ST. JOSEPH
COUNCIL BLUFFS
MISSOURI R.
RED RIVER
ARKANSAS R.
TRINITY R.
SAN ANTONIO
LA SALLE 1684
DA VACA
RIO GRANDE
EL PASO
SANTA FE
SANTA FE TRAIL
ONATE 1596
CORONADO
CORONADO 1540
PIKES PEAK
PLATTE R.
OREGON TRAIL
PONY EXPRESS
FREMONT
BLACK HILLS
VERENDRYE
MILK R.
YELLOWSTONE
LEWIS & CLARK 1804-6
SNAKE
COLUMBIA
GREAT SALT LAKE
FREMONT 1843
FREMONT 1845
COLORADO
SAN FRANCISCO
SUTTER'S FORT
SACRAMENTO R.
SAN JOAQUIN R.

Thousand Islands to expand into the beautifully rugged *Adirondack Mountains*, one of the great playgrounds of the United States. Canada also has the lion's share of the fertile St. Lawrence Valley. The Appalachian Plateau, however, which seams southern Quebec, New Brunswick, Nova Scotia, and Newfoundland with low, parallel, rocky ridges, swells out in the United States to become a wide, rough highland, furrowed with valleys cut out of the softer layers of rock.

The Atlantic Coastal Plain

A plain slopes under the Atlantic. On the eastern side of the Appalachian Plateau is a steep edge, called the *Fall Line* (#170), standing out as an escarpment, from the foot of which a very flat level plain slopes gradually away to the Atlantic Ocean and beyond under the water. This coastal plain is very narrow as far south as New York but then rapidly widens until in the Carolinas it is two hundred miles across. In the northern or New England States, the Appalachians are close to the sea, and the plain forms only a fringe around the many excellent harbours. Farther south, where the mouths of the rivers expand to bays, the largest of which are *Chesapeake* and *Delaware*, the plain is much wider and offers better opportunities for farming.

Life history of the Appalachian Mountains. Behind this plain rise the forbidding ridges of the broad Appalachian Plateau, which is the most ancient part of the United States. Millions of years ago a mighty thrust from the Atlantic Ocean began to press in towards the immovable Canadian Shield and gradually crumpled the rocky layers into a series of parallel folds as high as the Rocky Mountains. That was so long ago that the rain, rivers, winds, and weather have had time to wear these massive mountains down to their very roots; then these relics, dying of old age, were again revived and raised to a respectable height and again eroded to their present state. The ridges are made of the more resistant rocks and the valleys have been formed by ice, water, and weather wearing away the softer rocks between.

River gaps. From the first the Appalachians formed a repelling barrier against the advance of the early settlers into the interior. It is true that in the north the rivers had worn gaps through these mountains. Of these *river-gaps* (#172), as they are called, the most inviting is the north-south trench of the *Hudson River* and *Lake Champlain*, and the east-west trough of the *Mohawk Valley* from the Hudson River to Lake Ontario. Other gaps are those of the *Delaware, Susquehanna*, and *Potomac* rivers. But all of these are rough, narrow, and not easy to traverse. Nevertheless it was along these gaps, as we have seen (p. 166), that the French and English marched to attack each other. They bristled with many a fort and stockade the ruins of which can still be traced along the route; farmers, while ploughing their peaceful fields, still turn out stone arrowheads, ancient guns, and gruesome skulls.

Settlement

The coast explored. This coastal strip was the birthplace of the American nation. *John Cabot,* as early as 1498, had sailed along the Atlantic seaboard of the United States as far south as Virginia (#72), *Verrazano*, an Italian, sent out by the French, skirted the whole coast from Florida northward in 1524 and was probably the first white man to enter New York Harbour (#72). Spaniards, about the same time, explored the southern half of the coast (#179). But the lure of the fisheries in the Atlantic, and the fur trade of the St. Lawrence, and the glitter of gold and silver of Mexico and Peru caused this choicer coast to be almost forgotten.

Sir Humphrey Gilbert. This English explorer, as brave a sailor as ever crossed the ocean, made the first attempt to plant settlers in what is now the United States. In the spring of 1583, with the blessing of the great Queen Elizabeth, in high hope he set sail with five ships and two hundred and sixty men for the new world. The motley settlers landed first in Newfoundland, where trouble began. The men were a rough and worthless lot, more interested in robbing Spanish treasure ships than in founding new homes. Many became sick and others so unruly that Gilbert had to despatch a shipload of the most troublesome home. His ships, now reduced to three, sailed towards Sable Island to get supplies, for there were wild cattle and pigs on this sandy but treacherous shore; the largest ship was ripped to pieces on a reef and most of the sailors drowned. Delays had so used up the food, and the season had so far advanced into the fierce storms of autumn, that the two ships, the *Golden Hind*, and the *Squirrel,* no bigger than a sail boat, had to turn east across the stormy Atlantic. Gilbert insisted on sailing in the smaller boat; it was tossed like an eggshell in the pitiless autumn gales, and one stormy midnight the sailors on the *Golden Hind* saw the lights of the frail little craft suddenly go out in the darkness. The brave-hearted Gilbert went down nobly with his men, the first, but not the last martyr to settlement of the United States.

Courtesy wins a queen. *Sir Walter Raleigh* (#171) had the same heroic blood flowing through his veins as Sir Humphrey Gilbert, for

#171

SIR WALTER RALEIGH

they were half-brothers, sons of the same aristocratic mother. Raleigh was tall, handsome, stately, a capable soldier, and a mannerly courtier. The story is probably true that as he saw Queen Elizabeth, dressed in her dainty robes, pause before a mud puddle in the ill-kept London streets, he unclasped his brilliant scarlet cloak, and quickly spread it over the mud to form a carpet for the queen. It was not long before he was her great favourite and showered with wealth. Raleigh determined to follow Gilbert's footseps and plant an English settlement on the Atlantic coast of America.

Virginia. In 1585, at his own expense, this brilliant man fitted out seven ships, which sailed from Plymouth with one hundred men, and landed on *Roanoke Island* (#179) off the coast of North Carolina. They called the whole region *Virginia* in honour of the virgin queen, Elizabeth. Here was planted the first English settlement in the United States. But it was soon in trouble, as the laggard settlers were more eager to quarrel with Indians and burn their villages than to clear and cultivate the land. By the following spring scarcely any clearings had been made, food was scarce, and many were dead from disease and Indian arrows. It is no wonder that when Sir Francis Drake visited them, they insisted on returning to England in his ship.

Virginia Dare. But the wise, ambitious Sir Walter was more determined than ever to establish a new British Empire across the Atlantic, and in 1587 he sent out a new settlement of one hundred and fifty, including some women, to the same spot on Roanoke Island. In August, 1587, a few months after they landed, was born to *Eleanor Dare* the first child of white parents in the United States. She was named Virginia, and *Virginia Dare*, though she was to be massacred by Indians in a few months, is in this respect the most notable of all American babies. From the first there was quarrelling among the settlers, which foreboded no good. The ships sailed out of the harbour towards England, and next year, when they returned, the whole settlement had vanished as completely as the winter's snow, including both Virginia Dare and her mother. No certain record of this ill-fated settlement of Sir Walter Raleigh's was ever found, but undoubtedly the colonists had been cut to pieces by the fury of Indians, whom they had aroused by unjust treatment.

England seeking an Empire. But England was ripe for making settlements in America. Because wool brought better prices than wheat, the cultivated fields of England had been turned into sheep pastures, and millions of farm labourers were thrown on the roads and wandered about as beggars seeking work and finding none. English merchants were eager for new markets. As pirates shut them out from the trade of the Mediterranean Sea, the Portuguese almost completely controlled the rich traffic of spices and silks with India and the east, and the German *Hansa* ports monopolized trade in the

Baltic Sea, they looked to North America as a new field for commerce, provided only there were settlers to make it profitable. Moreover, the golden age of Elizabeth had made Englishmen proud of their country and determined to extend its bounds into a great empire, as Spain and Portugal had already done. Consequently, groups of wealthy merchants were willing to risk large sums to settle these new lands, in the hope that discoveries of gold and silver and produce of the soil, the forest, and the sea, would return to them all and more than they had invested.

Seeking new homes. But most promising of all was the fact that on account of religious persecution in Great Britain, France, and the German countries, noble bands of deeply religious people were eager to seek new homes where they could worship God freely. As has been already pointed out, these people were not coming like the Spaniards and the French to make a fortune and return, but to settle down and found new homes better and freer than those they had left.

A two-months' settlement. After Sir Walter Raleigh had wasted his treasure on the fateful tragedy of Roanoke Island, and the whole settlement had vanished from sight, no new movement was made until 1602, when thirty-two settlers were landed on a delightful little island south of Massachusetts. A fort was built, and seeds were planted, which in two weeks had become seedlings nine inches high. But food became scarce and all returned to England the same autumn.

Virginia Settlement

Jamestown founded. The year 1606 is the greatest of all years in the settlement of the United States. Two influential companies, formed in England to bring out colonists and composed of the most successful and wealthy merchants and powerful noblemen in the country, were so filled with enthusiasm that they were prepared to spend large sums of money and to laugh at losses. One, called the *Southern or London Company*, by December, 1606, had got together in three ships a motley group of one hundred and five men, only twenty of whom were real workers; the remainder were largely black sheep from good families, packed off to get them out of the way. Even on the voyage out, the clergyman had all he could do to keep them from each other's throats. On April 26, 1607, this unpromising cargo of settlers entered the beautiful waters of *Chesapeake Bay* (#170), and they named the two capes guarding its entrance *Charles* and

Henry, after the sons of James I. who was then king of England. They selected a site for a town near the mouth of a river. The town they called *Jamestown*, and the stream, James River, after the king. These were the vanguard of that vast stream of millions of settlers that poured across the Atlantic for three hundred years.

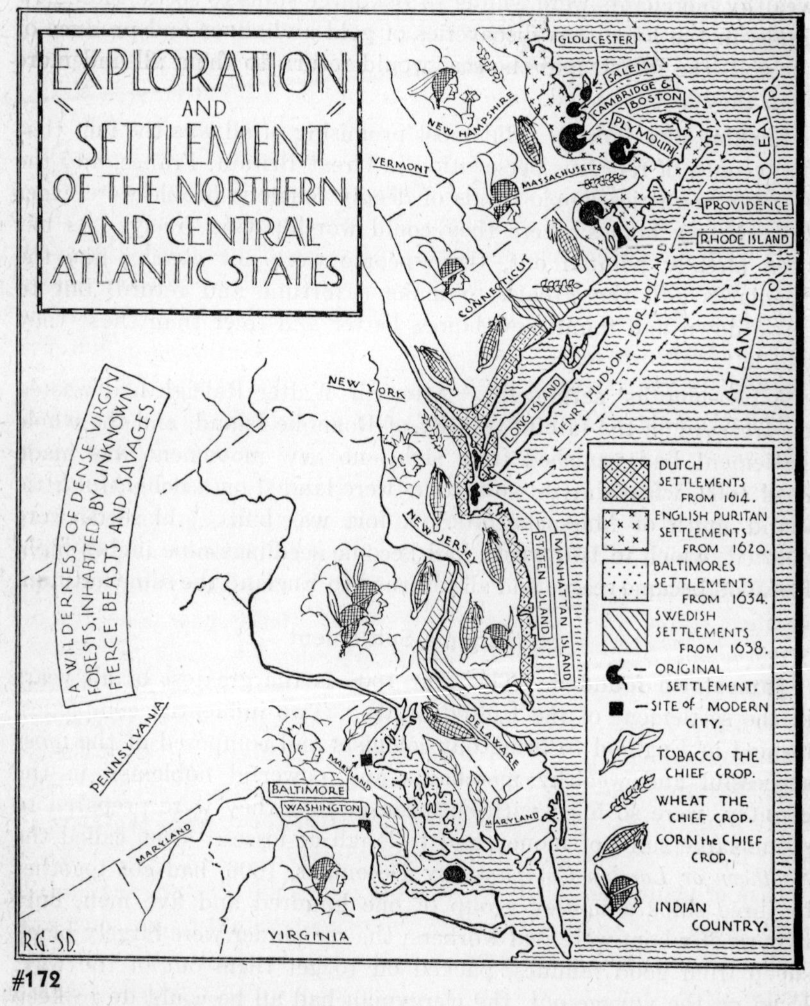

#172

Death stalks through the land. As all the land was to be held in common, and all produce was to be put into a common store, there was no great inducement to work. Fortunately they were well received by the Indians, a party that explored the James River receiving presents of venison, turkeys, corn, strawberries, mulberries, raspberries, pumpkins, nuts, and tobacco. But by September disease and starvation had mounded the little grave-yard with fifty graves, half the total number. There was one resourceful man among them, *John Smith*, who has left us his story of these grim days, and his skill and tact lose nothing in the telling. He had been in many parts of the world and knew how to manage Indians. By trading trinkets for food along the rivers, he secured enough to last until another boat load of settlers and food arrived the following year. These settlers were as worthless as the first band and died off like flies; in spite of Smith's efforts to make them work, scarcely any grain was grown. Although five hundred more arrived in 1609, they were no better than the former group, and it seemed that every merchant and aristocrat in England shipped all his misfits, idlers, and black sheep to be swallowed up by the forests of Virginia. In six more months, hunger, idleness, and disease had caused the five hundred to melt away to sixty.

Tobacco rules. But before 1609 had passed, the tide had turned. The generous and helpful London Company at home spent good money to replace bad, and sent the living to replace the dead. No longer were land and produce held in common, but each was given his farm, the production of which was to be his own. This gave them a new stimulus to work; the colony prospered, and, as one testified, three then did more work than thirty had done before. When *John Rolfe* harvested successfully the first stalk of tobacco (#172), the course of farming in Virginia was set, from which she has never turned back. The long, warm summer and the fine, rich sandy loam were the exact needs in the growth of tobacco. So profitable was the crop that farmers could be induced only by special laws containing threats to spare enough land for wheat and vegetables to keep them from starving to death in winter.

Rise of plantations. In order to obtain enough labour the planters were given fifty acres of land for every man they brought out, and each labourer was compelled to work for the planter a number of years, after which he was given a farm of his own. As the tobacco brought a high price in England, farmers brought out workmen by dozens or by hundreds, and as every new man meant an addition of fifty acres to his farm, it was not long before farms became plantations of thousands of acres, and modest farmers became aristocratic planters living in big country houses, and surrounded with servants.

The lovely Pocahontas. At first the Indians were troublesome, but soon a treaty of peace was signed with a great chief, *Powhatan*, who called himself the emperor. His beautiful little daughter, *Pocahontas*, soon became the most notable person in Virginia. She was kidnapped by the settlers and held until her father returned some of them, whom he had taken prisoners. The winsome Indian maiden was baptized as Lady Rebecca, and soon that clever man, John Rolfe, who had started the tobacco plant on its prosperous

#173

AUCTIONING SLAVES IN SOUTHERN STATES

journey in Virginia, became the fortunate wooer of the Indian maiden. They were married in the church at Jamestown, and shortly after travelled to England, where her graceful manners and beautiful face and form made her the sensation of London. She was presented to the king. But, alas for poor Pocahontas! just as she was about to return, this bride of a year sickened and died. She left one son about whom nothing is known; nevertheless for two hundred years some of the best families in Virginia living in the beautiful country houses on the plantations, boasted to be descendants of this swarthy Indian maid.

A loathsome traffic. In 1619, a Dutch ship landed an ugly cargo at Jamestown. It was composed of twenty fearful, quivering negroes, whom the crew had snatched from their homes and friends in West Africa and offered for sale as slaves (#173). The greedy planters quickly paid the price, marking the beginning of that loathsome traffic which after two hundred and fifty years, was going to drench the United States in blood by causing the fearful war between the North and the South. But it was a long time before slaves became numerous.

Prosperity and treachery. By 1619, Virginia was on such a wave of prosperity that the first parliament on the continent was called together at Jamestown. About the same time, settlers came out by the thousands, including two cargoes of young ladies, who were married almost as quickly as they could be landed, each groom paying one hundred and fifty pounds of tobacco for his bride. In 1622, the faithless Indians, who had up to this time mingled freely and peacefully with the settlers, suddenly one morning turned on the outlying villages along the James River and by noon had hacked to pieces more than three hundred and fifty men, women, and children (#174). After this the settlers made war against them as if they were vermin and almost put an end to the whole tribe.

Planters in control. From that date Virginia had no reverses, and soon great tobacco plantations stretched back from the rivers. Each plantation had its wharf, from which tobacco was loaded to sailing vessels, which exchanged it in England for manufactured goods required by the planters. There was no need for towns, a group of wealthy planters, surrounded by their labourers and slaves, composing a rural self-governing community.

Pilgrim Fathers

Exiles from home. The quiet little village of Scrooby, not far from Nottingham in England, is the acorn from which has grown that massive oak, the people of the New England States. In 1603 King James of Scotland began to rule in England. He was a stubborn man, determined to make all Protestants conform to the Anglican church or "harry them out of the land." But England was being honeycombed by devoutly pious people who objected to the English Church prayer-book and services as being too formal. They were very serious people, who guided their whole conduct by the Bible, and considered dancing around a may-pole to be heathenish, and eating mince pies at Christmas a serious sin. These people were called *Puritans.* Since the king refused to let them have their own

#174

A MASSACRE OF SETTLERS

churches, they had to meet in secret, often in private houses or dark forests. At Scrooby many of the farmers, weavers, carpenters, and other workmen were earnest Puritans, and for this reason were put in jail, fined, and threatened, until life became almost unbearable. But the last thing these brave-hearted people thought of giving up was their religion. Indeed, so devoted were they to it, that they decided rather to give up their homes, their beloved England, and everything they cherished, and to seek new homes in a foreign country, where nobody was persecuted for being a Puritan. They went to Holland, where the Dutch had been fighting the battles of the Protestant religion for half a century.

In a strange land. In this strange land, where everybody spoke a language they could not understand, they settled down in 1606 to earn a living and rear their children. But their hearts were still in their native isle, which they loved in spite of persecution, and they were afraid their children might grow up with no love in their hearts for England and its king, and perhaps speak Dutch instead of English.

Looking westward. Word reached these exiles of the success of the settlement in Virginia, and also that of the Dutch in New York, and they longed to set out to this new free land where from morning to night a foreign language would not be forever in their ears, and where strange houses, strange streets, and ungodly people would not be before their eyes. Above all they longed for this new land in order to worship God freely, and to be away from all temptations to do evil.

A New England Company. The New England Company had just been formed in England with a charter allowing them to settle a large area north of Virginia. The exiled Puritans in Holland asked to be sent out. The Company favoured such a group, unified by the bond of piety and by their common brotherhood ever since their flight from England.

The Mayflower. The company saw that such a group of serious men, suffering under a common persecution, devoted to a common piety, and bound together for twelve years in a foreign land by the silken cord of a common brotherhood, should be capable of founding a successful settlement in the new land. After three long years of bargaining, on September 6, 1620, a small sailing vessel, the *Mayflower*, left Portsmouth with its precious cargo of one hundred and seven fearless, determined pilgrims, men, women, and children, turned its prow westward, and sailed into the stormy unknown. After sixty-three dangerous days, on November 9, they saw the stern features of *Cape Cod* rise dully in the western gloom. They could not have found a worse place at which to land along the whole coast, nor could they have come at a more unfriendly season.

Home, sweet home. After a search along the coast, a suitable place for settlement was found, and on Monday, December 25,

#175

PILGRIM FATHERS' FIRST WINTER

1621, the first of the sturdy band landed at their future home, naming it *Plymouth* (#172), after the last port of old England that they had left. They began to build log houses on the stormy shore, covered with snow and swept by the raw wintery winds of the Atlantic. It was March before the last passenger had been transferred from the crowded cabin of the Mayflower to the cheerless log houses on this desolate coast, where the moan of the winds through the forest was answered by the dismal roar of the sea.

Death sweeps across the land. Bad food had spread frightful scurvy among them, and the chilling of their famished bodies and the wetting of their thread-bare clothes, while exposed to storms, had caused pneumonia and consumption to strike down even the strongest (#175). By spring, the rounded graves covering the ground were more numerous than the ghostly bodies of the living, and among these there were scarcely enough who were well to tend the sick. Fathers and mothers were swept away from their children, and in a few cases whole families were blotted out by these dreaded diseases.

Beautiful spring. But the hearts of the brave pilgrims never lost hope. Spring brought a wonderful change; the air blew balmy, the sky cleared, the grass became green, and the trees opened their shining leaves. They were able to buy corn from the Indians, the forests furnished wild turkeys and deer, the shallow ponds waterfowl, the sandy shore clams and other shellfish, and the coastal waters abundance of fine fish.

Settling with Indians. They bought land from the Indians, who taught them how to plant corn, pumpkins, and potatoes, all new to them, and the colony was soon on its way to success. During the first summer, 1621, new settlers arrived. In the autumn there was danger of Indian attacks. A powerful chief sent them a bunch of arrows wrapped in the skin of a rattlesnake, which meant war. But when the resolute pilgrims returned the skin stuffed with a mixture of gunpowder and bullets, the Indians understood, and decided that people who had such mixtures as that were better left alone. Miles Standish, the pilgrim's military leader, after one brush with the Indians, brought peace and friendly arrangements for many years.

Success. At first all land, houses, crops and other effects, belonged to the whole community and the New England Company. This resulted in much quarrelling and complaining among even these pious people; real progress was never made until the property was divided in 1623, each family possessing its own land, house, and tools. From that date, crops were always a success, and there was no scarcity of food. In 1624 the pilgrims bought out the rights of the New England Company, and Plymouth became a self-governing settlement.

Boston founded. As the Puritans at home, smarting under the pinch of persecution, heard of the peace and prosperity at Plymouth, they began to join the Pilgrim Fathers in America. The first cattle were introduced in 1624. In 1628 *John Endicott,* a very able man, brought out fifty settlers to *Salem* (#172) at the north of Massachusetts Bay, and in 1629 no fewer than four hundred men and women, with one hundred and forty head of cattle as well as goats and other domestic animals added to the growth of Salem. But 1630 was a red-letter day for New England; seventeen ships with one thousand settlers under the great *John Winthrop* settled around Boston Bay and founded that city, which ever since has been the hub of New England, and the reputed centre of learning and culture for the United States.

New colonies. The New England States are rough and rocky, and the settlers found only a very narrow strip along the coast, and a fringe along the rivers suitable for farming. As there was no broad coastal plain, no plantations were carved out of the scanty farm land, but each family had its own small carefully-tilled farm. In such cultivation crude slave labour is wasteful, and though the Puritans had slaves, they were few in number and were used more for house work. Indeed, it was not long before the ever increasing flood of settlers was forced to seek farms away from the first settlements; some went to *Maine* and *New Hampshire,* some crossed the mountains to the fertile valley of the *Connecticut River* and founded the State bearing that name; others driven out by religious persecution, sailed around Cape Cod, and settled at *Rhode Island* (#172).

Industry grows. There was no great staple crop, but on their small and none too fertile lands they grew wheat, barley, rye, oats, corn, and such vegetables as they needed, and raised cattle and sheep. They soon found the rivers and sea swarming with such choice fish as: cod, haddock, shad, menhaden, and sturgeon, that Boston and Gloucester became centres for profitable fishing. The dense forests yielded both hardwood and softwood logs, which were shipped for masts and cut into lumber. Later they built water-wheels under the cataracts along the Fall Line (#170), and those who had been spinners and weavers of wool in the north of England started woollen mills at these falls, and a whole row of towns soon appeared.

The persecuted becomes persecutor. Soon Plymouth and the Pilgrim Fathers were swallowed up in the much larger towns of Boston, Salem, Gloucester, and Providence, and these notable men, the Pilgrim Fathers, ceased to play the chief part in the life of the New England States.

Though they had suffered so cruelly for their religion, these Massachusetts settlers soon proved just as cruel and narrow-minded as the leaders of the Church of England who had driven them out.

#176

SUNDAY DESECRATORS IN THE STOCKS

The Puritan clergymen were all powerful with the people; when they formed a parliament in 1632, no man could vote who was not a member of the congregational (Puritan) church. A very famous preacher, *Roger Williams,* landed among the settlers in 1631. He was a devout, fearless, and effective preacher, with liberal views of his own and not afraid to express them. When he advocated freedom for other churches, he was driven out of Massachusetts by the persecuting Puritans, and set sail to a bay farther south. He built a log cabin in the woods with the help of Indians, and named the place, *Providence* (#172). Many who agreed with him settled along the lower course of this river and founded *Rhode Island.* When the Puritans of Massachusetts founded *Harvard College* at *Cambridge,* near Boston, the people of Connecticut founded *Yale College* at *New Haven.* The first Quakers that appeared in Massachusetts were driven out with a warning, and when more of these devout people appeared, four of them were actually put to death.

Dutch Settlement of New York

Fighters of the sea. A very small country with a very big name has played an heroic part in the history of Europe and America. At first the poor country was a marsh fringing the North Sea, within which were quaking swamps of quicksand, and pools and ponds of salt water. The houses were rather in the water than on the land, and the little villages were like nests of sea birds on the edge of the water, in danger of being washed away by every gale. The little farms and well-kept gardens had been won foot after foot by the filling in of swamps, draining marshes, and beating back the very sea itself by throwing up strong sea walls, behind which the inhabitants pumped out the sea-water with big windmills, and from which rain-water is still pumped by the same machines. From time to time fierce gales and cruel waves have beaten down these walls, or dikes, and great areas clothed with flowers, vegetables, and wheat have been again and again swallowed up by the sea, little white villages swept before the flood, and the brave people drowned. That aquatic country was the *Netherlands* or *Holland*, and those wonderful people the *Dutch.*

A fight for life. About the time that Great Britain was facing the huge Spanish Armada, the Dutch were also fighting the Spanish for their lives and religion, for they had become one of the chief defenders of the Protestant faith. Although it was a very tiny state and its enemy the most powerful country in the world, these brave bands of Dutch, who had conquered the sea, did not cringe before this giant, but fought on until they had driven Spain's best soldiers back, and become the free *Dutch Republic.*

Dutch as traders. Though their best men had been killed, their towns ruined, their ships sunk, and their country flooded, they never lost hope, but soon became the most prosperous trading people in Europe. They took to the sea, which they had already conquered, and soon they had a vast fleet of sailing ships, which wrested from Spain and Portugal the immensely rich trade in spices, silk, and precious stones with the east; after they had founded a great empire in the East Indies, this clever nation was not slow to look to America for still further prizes.

Henry Hudson. The route to the East Indies around Africa was long, threatened by storms, and infested by pirates. The Dutch, like the British, believed that a shorter route to these spice countries could be found around the north-east of Europe; and this route, called the north-east passage, they determined to conquer. For this purpose a great trading company in the Netherlands sent *Henry Hudson* to sail this dangerous but attractive course. Hudson was an Englishman who had already battled with ice in the Arctic Ocean, and if such a route was possible, he was the gallant sailor to worm his way through the vast frozen fields. With his little sailing ship, the *Half-Moon*, he pushed as far east as any man dare, but endless seas of thick ice were too much even for his stout heart, and when his men threatened to rebel, he turned back, but not for the Netherlands or England.

Search for North-West Passage. The famous John Smith of Virginia, had told him of a large river which he had found on the Atlantic coast of America and which, for all he knew, might cut through the continent and open the way to the Indies. Hudson and his men decided that if they could not find the north-east passage, perhaps they could find the north-west one and still gain a short path to the treasures of silk, gems, and spices in the Indies.

Hudson enters New York Harbour. By July, Hudson with his sixteen sailors was off the coast of Maine, and on August 31 they entered *New York Harbour* where sycamore, walnut, butternut, and giant oak forests clothed the land, now covered by the greatest array of giant sky-scrapers in the world. The sweet fragrance of trees and flowers came across the quiet waters, which soon were alive with canoes of friendly Indians who brought corn and tobacco to trade.

Exploring Hudson River. As a wide stream opened to the north, and Hudson was seeking a north-west passage, he sailed up the river, but alas! instead of widening and deepening, it became narrow and shallow. He went just as far north — past Albany — as a small boat could be rowed. Now as he was sure the north-west passage did not open this way, and as the Indians showed signs of unfriendliness, he hastened back between the high rocky banks of this noble river, later named after himself, and was soon again in England.

New Netherlands. When he reported to the Dutch Company of the splendid harbour and majestic river he had discovered, with the rich forests, fair, fertile lands, and natives eager to trade, it decided to send out traders. Beginning in 1610, year after year, it sent out ships, which brought back rich cargoes of choice furs. By 1614 it had trading posts at the mouth of the river and at the head of navigation. The one on what is now Manhattan, the core of New York, was called *New Amsterdam*, and the other, near modern Albany, the capital of the state, was called *Fort Orange*. The whole region from the Connecticut River to Delaware Bay was called *New Netherlands* and claimed as a colony of Holland.

First Dutch settlement. It was not until 1623, when a new group, the West Indies Company, gained control, that the first real settlers, as opposed to traders, were placed in this ideal spot (#172). A ship brought out thirty families of Protestants, who had been driven out of what is now Belgium, and settled them partly at New Amsterdam (New York), and partly at Fort Orange (Albany), far up the River Hudson. The latter made a treaty of peace with the neighbouring Iroquois at Fort Orange, which was never broken. Soon cattle, horses, sheep, and pigs were added to the farms, and the colony was well started; corn as high as themselves was grown the very first year.

New York bought for twenty-five dollars. In 1626, the governor, *Peter Minuit*, bought Manhattan Island for some knives, beads, ribbons, and red cloth (#177); to-day the same land is worth billions of dollars. By 1628, the progressive village at the foot of this island had two hundred and seventy people, including slaves; houses with straw roofs and wooden chimneys, set off by Dutch windmills, were scattered along the Hudson River. Their sailing vessels were combing every bay and inlet along the coast from Delaware to Cape Cod for the rich beaver and otter skins brought down by the Indians. In New Amsterdam they made bricks and lime, but were not allowed by the home government to spin thread or weave cloth. Though the fertile soil, sufficient rains, and warmth of summer were very suitable for growing crops, the glamour of trading furs caused farming to lag.

The Patroons. But 1629 saw a great change in the structure of the colony. The West Indian Company offered vast tracts of land to aristocratic Dutch gentlemen who would settle fifty families on their plantations. Soon both banks of the Hudson River were lined with these vast estates, the owners of which were called *patroons*. Some of them had a front of sixteen miles on the river. One of the most notable of them, *Van Rensselaer*, had a farm twenty-four miles front by forty-eight miles deep on both sides of this river. Another Patroon bought the whole of *Staten Island*. These country gentle-

#177

BUYING MANHATTAN FROM INDIANS

men brought out settlers and bought slaves to work their estates. They built fine country houses, erected forts to defend them from Indians, and besides farming, always had a keen eye turned towards profitable trade in furs with the Indians.

Dictators. Such a method of settlement had its disadvantages. With such large plantations, houses were widely scattered, villages were few and small, and defence against unfriendly Indians was not easy. Every patroon ruled his own tenants like a dictator, and there was no opportunity given for a people's council. The Company's director was the real governor, but his authority did not spread much beyond New Amsterdam, and the patroons resented his meddling with their affairs.

New Sweden. The more liberal Dutch, unlike the strict Puritans of Massachusetts, did not drive people out who differed from them in religion, and it was not long before many persecuted Christians took shelter along the Hudson. Indeed, the settlements spread out until they began to overlap the English on the Connecticut River (#172) and Long Island, and also flowed along the shores of Delaware Bay. This choice region was soon (1638) occupied by thrifty Swedish and Finnish settlers, who called it *New Sweden.*

The rising storm. But the small English farmers of Massachusetts, Connecticut, and Rhode Island waxed fat in wealth and numbers, compared with the sluggish patroons with their broad plantations. They began to look on the Dutch as intruders with no right along that coast, for had not John Cabot discovered it for England more than one hundred years before? The Dutch looked in the same way on the Swedes in New Sweden and in 1655 they made war against them and took possession of their settlements.

Bloody massacres. About 1638 the Dutch began to get in trouble with the surrounding Indians, whom they had armed with guns and often driven to madness with liquor. At last the Algonquins broke into fury, and farm houses and villages went up in flames, a number of people were cruelly tortured, and many cattle slaughtered (#174). For two years these sudden raids from the forest went on until there was danger lest the whole settlement should be wiped out in bloody massacres.

A welcome director. Then in 1646, *Peter Stuyvesant*, a new director, was sent out and became the guide of the people until they were finally conquered by the British. Stuyvesant's previous record was none too good, but he was brave, fearless, and as fair as a furious temper would allow. He was always prompt and energetic in action, though often rash and unwise in judgment. He had lost a leg in a fight in the West Indies, and this bluff, honest old czar with his wooden leg trimmed with silver bands was the fountain of progress in the settlement. During his reign of nearly forty years, persecuted Protestants from every harassed European country settled along the Hudson. Huguenots from France, Moravians from Bohemia, and many sects from Germany, as well as Quakers from England and even liberal-minded Puritans from New England, were welcomed to this haven where religion and trade were both free.

Pride goes before a fall. The woodsmen produced lumber for France, the fishermen searched for whales on the high seas, the prosperous farmers grew the vine and mulberries, as well as corn and wheat, and reared successfully flocks of sheep and herds of cattle. The population by 1664 was about ten thousand, and New Amsterdam rivalled Boston in size and importance.

But the old director wanted no assistance in government and threatened to make any disturber "a foot shorter." The people were restless under this well-intentioned tyrant, and many longed for the self government found in all the adjoining English settlements. Consequently, when in 1664, after war between England and Holland had been declared, British war ships under the command of *Colonel Nicholls* appeared in the harbour of New Amsterdam and demanded the surrender of New Netherlands, Peter Stuyvesant was furious. He tried to stir the Dutch to fight, but as the fort was little stronger than a picket fence and there were not more than one hundred soldiers able to fight, the people begged Peter to surrender, and the old veteran, almost with tears in his eyes, was forced to give up the strong healthy colony that he had nourished from a weakling. Not a shot was fired.

Walking in the evening. The whole Atlantic coast of the United States north of Florida, was now British. Colonel Nicholls was made governor, the name of New Netherlands and of New Amsterdam was changed to New York, and Fort Orange to Albany, all after James, Duke of York and Albany, the brother of Charles II. Poor Peter Stuyvesant was shipped off to the Netherlands, but his heart was across the Atlantic in the land he had cherished. When the war was over, and enmity between British and Dutch had died down, the old man again crossed the ocean and settled as a patroon on his estate a few miles out of New York city. He called it the Bowery, and the name is continued to-day, although it is now one of the most vicious parts of the great city. It was not an unusual sight to see Colonel Nicholls, the new British Governor, and Peter Stuyvesant, the old Dutch director, walking arm and arm in the evening discussing methods of increasing the prosperity of the new land, which they both loved (#178). If old Peter could to-day look out and see the great throbbing city of New York with its fabulous wealth, its myriad of giant skyscrapers, and its seven million people, how his heart would swell with pride at the growth of that lusty infant, which he started on the road nearly three hundred years ago.

Settlement of Maryland

A present from a king. Sir George Calvert, Lord Baltimore, obtained permission from the king of England to start a colony in Newfoundland and soon had a small group of settlers on the southeast peninsula of that bleak island, now named Avalon, but called *Ferryland* by him. After a visit to Newfoundland in 1627, he decided that the winter weather was too severe for successful settlement,

#178

A WALK IN THE EVENING

and he sailed south to seek a more pleasant climate. As he had recently turned Roman Catholic, he was received coldly by the ardent Anglicans of Virginia. However, he was determined, and his skill in managing men, especially rulers, soon secured him a charter for a settlement on the sunny shores of Chesapeake Bay. King Charles I gave him complete ownership of an immense region between Virginia and New England, for which he had to pay yearly two arrows and half the gold discovered.

A colony for Roman Catholics. He desired to found the colony partly for profit in trading and farming, but also as a peaceful home for persecuted Roman Catholics. Just before his plans were completed he died, but his son had two little vessels, the *Ark* and the *Dove* heading westward from the Isle of Wight in November, 1633.

They contained twenty English gentlemen, all Catholic, and three hundred labourers, mostly Protestant, as well as Jesuit priests. In March, 1634, in balmy air they were sailing up the placid waters of Chesapeake Bay, and at night the shores blazed with the camp fires of the Indians. They soon found a beautiful place for settlement on a peninsula and called it *St. Mary's*. As the trees had been cleared away by the Indians for their crops, it was ready for planting (#172).

Success from the start. They bought the land from the Indians and soon had crops of corn looking green and shiny in this delightful climate. They erected their log houses and had such luxuries as chimneys built of bricks, which they had brought with them. The Jesuits soon had a little church smiling down on the settlement from a hill. So successful were their crops, that in the first autumn they had one thousand bushels of corn to sell to New Amsterdam (New York).

A proprietor's colony. This colony differed in several respects from those of New England, New York, and Virginia. It was owned and governed by one man, Lord Baltimore, who was the proprietor. Strange to say, he never set foot in the colony, but was represented by his brother. In this colony there were no religious persecutions. Protestants and Catholics mingled together from the first, like brothers. Lord Baltimore distributed the land in large blocks to English gentlemen, especially to the twenty who came out at the forming of the settlement. Each was given one thousand acres on condition that he should bring out settlers. These, after working for several years, received land for themselves. The colony soon became, like Virginia, rows of tobacco plantations facing the water, the planters, mostly Catholics, living in large, sprawling, wooden manor-houses, which were considered magnificent, compared with the clusters of log houses of the labourers, who did the work on the plantations. As in Virginia, tobacco was king (#172), and so profitable was the crop at first that it was hard to get the people to grow enough wheat, corn, potatoes, and other produce to feed themselves. As the tobacco was loaded directly on to the sailing vessels from the shore of the plantation, there were no commercial towns, and the centre of life for each group of people was the manor-house.

Wasteful farming. As tobacco soon eats the goodness out of land, the crop rapidly decreases unless fertilizer is used. But none of the owners thought of spending money on fertilizers, since with their numerous workers, including slaves, they could more easily clear new land and allow the exhausted fields to run wild and go back to forest. In these waste lands and through the forests herds of cattle and droves of pigs wandered at will as they do to-day, the latter eating the nuts falling from the trees (#179). As the forests

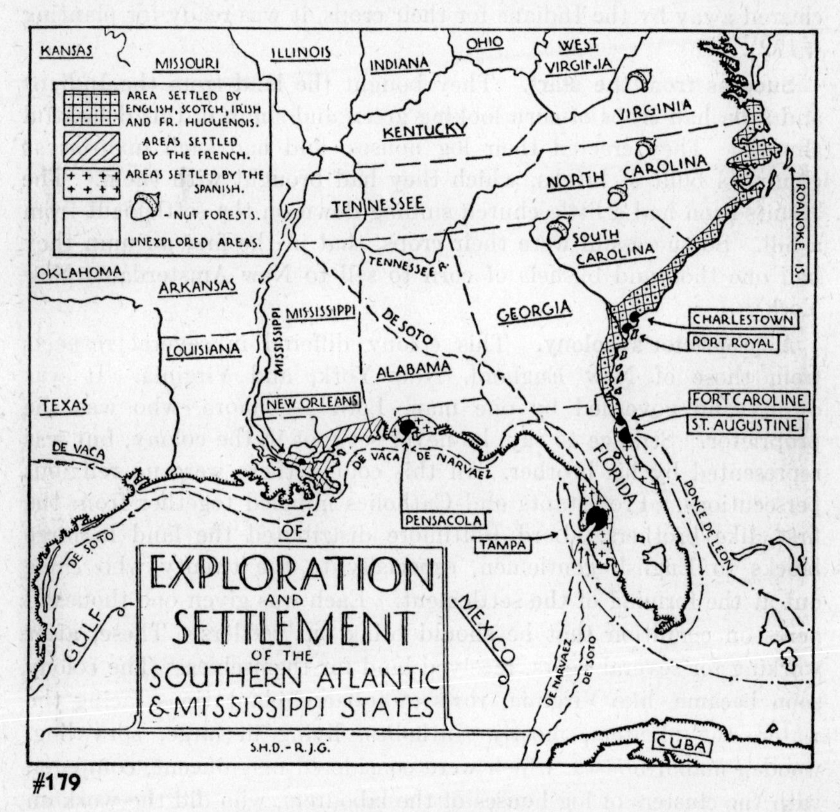

#179

were swarming with game, deer, turkeys, opossums, and wild fowl, the people had plentiful supplies of a great variety of meat.

Maryland, which was named after Queen Henrietta Maria, wife of Charles I, had the most peaceful growth of any of the colonies. As the struggle in England between Protestant and Catholic, between Cavalier and Roundhead, swayed back and forth, there were similar struggles in Maryland, although of a much milder character. The happy colony went on its way, growing quietly and steadily, and the people, year by year, gaining more and more power to govern themselves in the absence of their mild proprietor. By the time Queen Anne died, Maryland had thirty thousand people.

Settlement of Florida

Columbus. Columbus not only discovered America for Spain but also made four notable trips in which he spread out on the map most of the West Indies and much of the coast of Mexico, Central America, and Venezuela, lands bounding the Gulf of Mexico and the Caribbean Sea. He thought, until the day of his death, that he

#180

PONCE DE LEON AND THE FOUNTAIN OF YOUTH

was on the edge of the rich East Indies, and to secure it for Spain, he brought out numerous settlers to Cuba and Hispaniola or Hayti (#72).

The fountain of youth. One of the earliest settlers was an old veteran, *Ponce de Leon* (#180), who came out in 1493. He heard of a strange land to the west which was rich in gold, and, best of all for man of his age, it had a fountain of youth. Any man who drank the clear water gushing out from this spring would at once feel the stiffness of his joints vanish, the pallid wrinkles of old age smooth into the pink of youth, and the whole body become as lithe as that of a young man. He set sail to discover that delightful land (#180). Imagine his thrill when a new shore rose above the horizon. As he approached nearer, the trees looked glorious in the bright sunlight of spring, and the perfumes of scented woods and sweet flowers were wafted over the water. This new land was discovered on Easter Sunday, which in Spanish is *Pasqua Florida.* Now as Florida is the Spanish for flower, and as the name seemed especially suitable on account of the dazzlingly bright flowers which carpeted the shore and decorated the trees, he named the land Florida. He was near *St. Augustine,* on the east coast, which was soon to be famous in American history. He sailed around the tip of the peninsula, struggled with the swift currents of the Gulf Stream, and followed the west coast; when he landed, he found the Indians hostile, and he returned to Porto Rico without discovering the fountain of youth.

Death of de Leon. Though he was delayed from returning, the old man's heart was still with the beautiful land of Florida, and in 1521 he set out both to explore it and to found a colony. With his two ships, two hundred men, fifty horses, and tools of all kinds, he landed at *Tampa Bay,* only to face the fury of bands of savage man-eating Indians. He was fatally wounded, his men were driven back, and the colony never took root. Poor old de Leon had only reached Cuba when he died. On his tombstone is written this suitable verse:

Beneath this stone repose the bones
Of the valiant lion whose deeds
Surpass the greatness of his name.

(The Spanish word Leon, means lion.)

The swamps of Florida. *Pamphilo de Narvaez,* a rich soldier, greedy for gold, set out in 1528 with three hundred Spaniards and forty horses to search Florida for mines, which, they had been told, would rival those of Mexico and Peru. He landed at Tampa Bay, and plunged into the forests to reach a rich city farther north, where the ships were ordered to meet him. He and his men gazed in admiration on that wonderful land with its tall dark pines, spreading cypresses, sweet red gum trees, graceful palmettos, humble palms, majestic magnolias, and live oaks; but when these first white men who tried to traverse Florida endeavoured to advance, their admiration gave way to anxiety, for they had to wade up to their armpits through quaking swamps, made terrible by slimy fallen logs, and through a hostile country where deadly arrows from hidden foes tormented them at every turn. Numerous rivers, swollen with water, plunged across their dangerous paths and had to be crossed on rafts. Their food was playing out, and yet the rich city they had heard about, the high hill, the fountain of youth, and the piles of gold, seemed ever to retreat farther away.

Ships hewn from the forest. When they reached the bay, where their ships were to rescue them, the harbour was empty and the ships never arrived. But nothing daunted these iron-hearted Spanish soldiers. They must have ships, yet they had no tools and no materials. They beat their metal stirrups, spurs, and gun barrels into saws, axes, and nails, cut down trees, sawed them into planks, and in sixteen days they had built five boats, each thirty feet long. These they covered with horsehide, rigged with ropes made of twisted horsehair and fibres from bark; the seams were stopped with tar from the pines, the sails were made from shirts, anchors from stones, and water-bottles from horsehide.

Shipwreck. Two hundred and fifty men weighted down these crude ships almost to the water's edge, as they pushed out into the stormy waters of the Gulf of Mexico in September. The boats, with water gushing in at every seam, were soon wrecked; de Narvaez, the leader, was never again heard of, and only fifteen were left of that body of proud men, and these were hopelessly stranded on the shore of a gulf never trod by white man's feet, without food, and without clothes. But they were not without hope, for among their numbers was *Cabeza de Vaca*, one of the most remarkable explorers that ever risked his life in the dangerous places of America.

The miracle men. The whole party was soon captured by Indians, and when de Vaca was fortunate enough to cure some of their sick, these poor savages looked on him as a medicine man who could perform miracles. They brought their sick in order that he might breathe on them, and so noted did these healers become throughout a vast region of Texas, that they were sometimes followed by thousands of admiring Indians. For four years they worked westward with this procession of eager Indians in their train. They traversed the whole coastal region of Texas, along the Rio Grande River and nearly to the Pacific Ocean in western Mexico; and in July, 1536, they were back among their friends in the city of Mexico (#170).

A proud band of treasure hunters. *Hernando de Soto* had fought bravely in the conquest of Peru and came back to Spain a hero and a very rich man. When he announced in the city of *Seville* that he wanted men to seek treasure in the north which would be equal to the fabulous rooms full of gold found in Peru, the proudest young aristocrats of that proud city sold their estates to equip themselves to follow this daring explorer. With six hundred of the noblest blood of old Spain, he landed his fleet at Tampa. Never had such a highly equipped band set foot on the American shore. The men were mounted on prancing steeds, their armour was complete, and the food which was varied and abundant, included a drove of pigs. How thoroughly this expedition had been prepared can be seen when it is noted that the equipment listed such articles as forges and metals; even chains were carried to bind their captives and bloodhounds were taken along to hunt out dangerous Indians.

Fearful cruelty. As they wandered northward, allured by gold, the treasure seemed to retreat before them. Through the swamps, among fallen timber in the forest, and across swollen streams the horsemen met the greatest hardships, and the drove of pigs wandered through the woods and were as difficult to manage as a flock of startled birds. Indians fought them at every vantage point, and they in turn treated these savages with the greatest cruelty; one who said there was no gold ahead was burnt alive; some were tossed to starving bloodhounds, others pitched alive into fires, and hands and ears were hacked off as punishment for trivial offences.

The father of waters. This proud army wandered on month after month in the jungles and forests north of Florida, looking for a new Peru of gold, but finding nothing but dense forests, wild beasts, and still wilder savages. At one of their winter-quarters they were so suddenly attacked by Indians that their houses, pigs, and clothes, were all burned and they were driven out as naked as the savages whom they fought. By May, 1541, more than two hundred and fifty of these sons of proud Spain were dead. Soon after, they came upon a splendid river of great depth and swift current, along whose muddy waters were swept whole trees and timber. It was the *Mississippi*. A month was spent building rafts of logs to ferry the men and horses across its wide and turbulent current, which they passed near Memphis.

A watery grave. During 1541-42, at one of their winter-quarters near the middle Mississippi, these sons of sunny Spain suffered greatly from cold and deep snow, and their great leader de Soto, not only grew weak in body but was humbled in pride to think that his great expedition was dwindling to nothing; his heart was sick, but his pride would not allow him to return empty handed. On May 21, 1542, this marvellous explorer died; his body was wrapped in mantles heavily weighted with sand and sunk in the Mississippi River, a suitable tomb for this brave Spaniard.

His men sailed down the Mississippi and coasted along the shores of Mexico to *Panuco*, a great Spanish centre; but only 311 ragged worn-out men returned of the 620 proud cavalry officers that pranced into the awful American wilderness.

Spanish Settlement

First in the field. The population of Spain had been terribly thinned by long and bloody wars with the Moors, yet immediately after Columbus had discovered America, she set out to settle the new continent. So zealous and successful was her work, that before France, England, or Holland had a single settler in America, Spain had founded two hundred Spanish cities and towns, brought out 160,000 settlers, and set up nine thousand Indian villages, containing a population of five million. She had also christianized a great many of the natives; and native temples, smoking with human sacrifices and dripping with blood, were replaced by beautiful Christian churches, which spread their goodness to every town and village. The city of Mexico contained fifteen thousand Spaniards and possessed a University, a high school for boys and girls, and four hospitals.

A settlement that failed. This swift colonial activity preceded the founding of Acadia by the French and Jamestown by the English, but only a very few of the settlements were founded north of Mexico. The first attempt on the Atlantic coast was made in 1526, when *Lucas Vasquez de Ayllon* with six hundred people, including negro slaves and three friars, set foot at *St. Miguel de Guaddape* in South Carolina. In less than a year the hot swampy land and disease had brushed them off like flies, and cold weather drove back the fragment still alive — one hundred and fifty — to the West Indies.

The River of May. In 1562 *Jean Ribault*, with a band of French Huguenots, sought a new home where they could worship God without persecution. They landed at one of the most enticing spots on the Atlantic coast of South Carolina, which they called *Port Royal*. The sweet odours of trees and flowers and the beauty of scenery gave unspeakable pleasure; the little river boiled and roared to its glittering bay under a canopy of spreading cypress and graceful palm trees. The streams swarmed with trout, and the adjoining sea with food fish; the forest was alive with deer, wild pigs, bears, lynx, turkeys, partridge, woodcock, herons, bitterns, curlew, and swans, and the Indians were mild mannered and kindly. They settled down along this *River of May*, as they called it, and were more interested in building a fort than in planting seeds. As the season wore on, the game fled, the fish became scarce and shy, and ugly starvation settled down on this shiftless settlement. As their ships had departed, they built a little boat, made sails of sheets, and in this egg-shell made a mad attempt to sail back to France. Storms opened the seams of their badly-built boat, food and fresh water were soon gone, and the living were driven to eat the dead. The remnant of this ship's crew were snatched from a living death by a British ship and returned to France.

Fort St. Augustine. At last, in 1564, began the first attempt at a permanent settlement on the coast of the United States, and no other settlement was ever established in such a welter of blood. Again a band of French Protestants crossed the Atlantic, but this year they landed farther south on the coast of Florida, a little north of St. Augustine. They did not profit much from the experience of the former settlement at Port Royal, as again they built a fort (Caroline), but planted no seeds. Just as they were on the edge of starvation, *Sir John Hawkins* sailed along and brought relief. A little later seven ships arrived under Jean Ribault. But in the meantime a Spanish settlement had been made a few miles away, and a strong fort built, at *St. Augustine* (#179). Ribault, instead of landing his men, went to attack the Spaniards. But in the meantime, a terrible disaster had happened. The Spaniards from St. Augustine, in the dead of night and in a heavy rain, crept up on the sleeping French in Fort Caroline, slaughtered the men who fought, and stabbed to death all that surrendered. When Ribault's men landed near St. Augustine, they also were trapped; those who surrendered were led out by tens with hands tied behind their backs, and stabbed on the sands among the unburied bodies of their friends, who had been murdered a few days before.

First settlement. The Spaniards now had possession of both forts, but not for long. After three years, in 1568, *Dominique de Gourgues*, with three ships, attacked the settlement, cut to pieces four hundred Spaniards, but failed to take the strong fort, St. Augustine, which remained in Spanish hands as the first settlement along the Atlantic Ocean. Fort St. Augustine is as revered to-day by the United States as Fort Royal, N.S. and Quebec, are by Canadians. But the fort had little peace for many years. Sir Francis Drake in 1586 burned the settlement and destroyed the fort; however, it was again almost immediately occupied by the Spaniards.

Spanish missions. Further settlement was not pushed at this time by the Spaniards. As their proud Armada had been destroyed by the British, and they were becoming so much engaged in wars and troubles in Europe, such distant objects as the extension of the Florida settlement were neglected. Their chief work for nearly one hundred years was the establishment of missions, where the Indians were collected in villages, converted to Christianity, and trained by Jesuit and later by Franciscan missionaries to cultivate the soil and learn trades. These missions were strewn along the whole coast of Florida and among the islands fringing the coast of Georgia. But real settlement was very slow, as in 1647 there were just three hundred families in St. Augustine, the only Spanish village in Florida.

Pensacola. When the French began to settle at the mouth of the Mississippi, the Spaniards became alarmed and built a fort at *Pensacola* on the extreme western corner of the gulf coast of Florida. In 1719 it was captured by the French, and in 1763, after the British had won Canada, Florida became British also.

Industries of the Atlantic coastal plain

Settlement moves south. The southern part of the Atlantic coastal plain was soon settled also by English, Scottish, Irish, Germans, and French Huguenots. *North Carolina*, the poorest and most backward, was occupied by a rough and lawless gang who tended herds of cattle and hogs running nearly wild through the rich nut forests. In *South Carolina* it was not long until rice and indigo took the place of tobacco. Georgia was the last of the States to receive settlers, many of whom were poor and had been transported from England for debts. It formed a buffer between the British of the more northern settlements and the Spanish of Florida. Great changes have taken place along this coastal plain since the days of these early efforts at colonization.

Industries of New England. In the New England States, where only a few valleys are fertile, the settlers have had from the beginning to look to the sea for foods, and to the waterfalls for power to manufacture. The cities of *Gloucester* and *Boston* as well as the smaller towns soon became centres of fishing and continue to this day to send out fleets of schooners to catch mackerel, herring, cod, haddock, and halibut, from as far away as the Banks of Newfoundland. But these New England States, as well as New York and Pennsylvania, with their settlements of shrewd, pious Puritan and Quaker workmen, have made this north-eastern corner of the United States the busiest hive of factories on the continent. The weavers from the north of England built waterwheels at the Fall Line (#170), where the water falls from the Piedmont to the coastal plain, to run woollen mills; later, when cotton became cheap, these were partly replaced by cotton mills. To-day such New England cities as *Lowell*, *Manchester*, *New Bedford*, and *Fall River* are centres of manufactures of finest cotton textiles.

Lumber and pulp. This region was also the centre of the lumbering industry of the United States for nearly a century, and every stream still floats down its spruce, cedar, pine, and hemlock to be cut into lumber or ground into pulp at big mills along the lower courses.

Cotton is king. Farther south, where the coastal plain was wider and flatter, the summers warmer and wetter, and the rocks replaced by clayey, sandy, and gravelly soils, there has grown up a great variety of farm industries. Tobacco is still the chief crop in Maryland and Virginia (#181); the rice and indigo crops of the Carolinas and Georgia have been completely, and tobacco partly driven out by cotton, which after the discovery of a cheap process of pulling the fibre from the seed, has rapidly replaced wool and linen in clothing the world. In the more fertile soils of North and South Carolina, Georgia, and Florida, cotton is king (#188). Since it has to be picked by hand, negro slaves were brought into all these states in great numbers to handle this crop, and since they received their freedom they have still been the chief source of labour in the cotton fields. But many other crops are now grown in all these States. In some parts early cabbages and cauliflower are extensively grown for shipment to more northern cities, and a little later in the season appear sweet potatoes, immense water-melons, and great crops of peanuts.

The southern forests. The whole area was covered with forest when settlers first arrived, and one of their greatest difficulties was to clear the most fertile lands. The more barren and swampy parts were left covered with forests, southern pine occupying the gravelly soils, and cypress and redwood the swamps. At present, these furnish valuable lumber, the swamps often being drained first before the lumbermen are put in to cut down the cypress and redwood. The pines are tapped as maple trees are in Canada, and turpentine, which is mixed in paints, is obtained. The old stumps and branches are heated in ovens and tar and resin obtained.

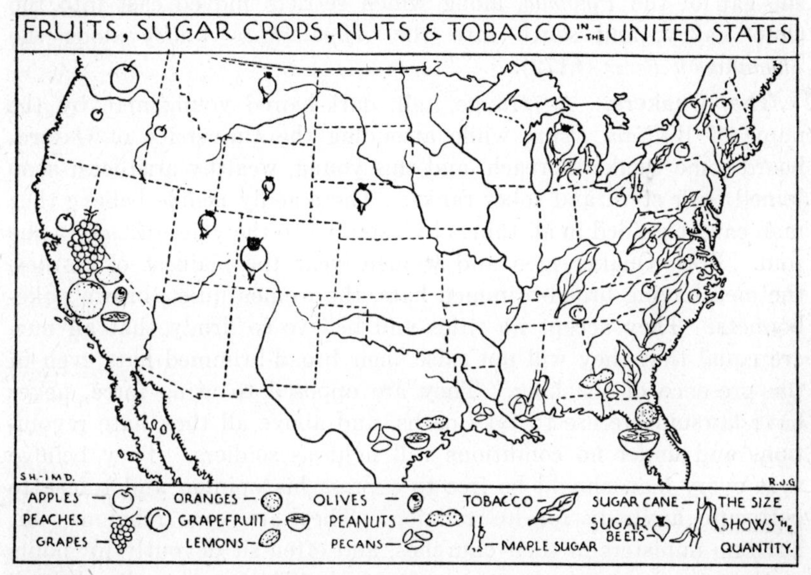

#181

Winter resorts. On account of the delightful winter and spring weather along this south Atlantic coast, many Americans and Canadians flee from the harsh winter and spring weather of the north to bathe in the tepid water and bask on the sunny beaches of these more congenial regions. *Atlantic City* in New Jersey, and *Jacksonville, St. Petersburg*, *Tampa*, and *Miami* in Florida, swarm at these seasons with wealthy northerners who are able to spend their money freely.

The Appalachian Highlands

The settlers along the coastal plain, as they looked westward, saw rising in the distance a continuous line of steep mountains, mantled to the top with forest. Not a single inviting pass appeared to the eye. But as the restless settlers multiplied, they began to be pushed by land hunger towards these forbidding highlands. It has already been stated that there were several gaps along the rivers by which the mountains could be crossed. Four of them are of great importance, the *Hudson*, through which it has already been shown that first the Dutch and later the British penetrated almost to Lake Ontario, the gaps of the *Delaware* and *Susquehanna* through which the Quakers and Germans reached the plateau of Pennsylvania, and the gap of the *Potomac*, along which settlers moved east into the Ohio Valley, and south into the *Appalachian Valley* along the *Shenandoah River* (#172).

The Quakers. A strange, tall, dark-haired young man by the name of *William Penn*, while attending the University of Oxford, heard some Quakers preach, and this young, wealthy aristocrat soon joined their stern and sober ranks. These godly people believe that man can be guided in all things by listening to the voice of God in the soul. Both Quaker men and women wear the plainest of clothes, the men having broad-brimmed hats, the women quiet, brown poke-bonnets. They accept no titles and believe so firmly that all men are equal that they will not raise their broad-brimmed hats even in the presence of the king. They are opposed to using force, never have lawsuits, refuse to take oaths, and above all they hate revolutions and under no conditions will fight as soldiers. They believe that every man should be free to express his opinion and refuse to persecute anybody for his religion. They are very religious, but have no ministers in their churches, and often sit devoutly for hours at their services, nobody uttering a word. They are usually shrewd, hard workers, and many become successful business men. But men who would not take off their hats to kings and lords, who refused to pay tithes to the English Church, and who insisted on preaching their strange beliefs, soon fell foul of the authorities in all parts of Europe.

Pennsylvania bought for two thousand pounds. Penn was thrust in jail more than once, but he always seemed to have influence with England's rulers. As King Charles II owed Penn ten thousand pounds, which he showed no intention of paying, a bargain was struck that the Quaker should receive a great area of land on the Atlantic coast between New York and Virginia. Penn decided to settle it with the persecuted and down-trodden of all the countries of Europe. In 1682 he landed at the mouth of the *Delaware River* with about one hundred English Quakers, and the next year in a peaceful fertile spot in the forest he surveyed lots for a model city, which he called

Philadelphia. He gathered the Indian chiefs together, asked that there might be peace between the white and the red man, and although he had already purchased the whole of Pennsylvania from the king, he again bought every square foot that the white men occupied from the Indians. This good beginning led to a long peace between the two peoples.

A refuge for exiles. When, throughout Europe, the news spread abroad that William Penn, the Quaker, had opened an asylum for the oppressed of every nation, and after he had travelled through the chief European countries gathering the children of misfortune and bringing them to the new free land, Pennsylvania grew by leaps and bounds. The hard-working godly pioneers pushed up the Delaware River and spread over into the States of New Jersey and Delaware. Philadelphia, the port at the head of navigation, also grew almost like a mushroom, and gained more population in three years than New York did in half a century. Though it was carved out of the forest only in 1682, by 1684 it had a population of eight thousand, and in six years it had grown to twelve thousand. The growth of Pennsylvania is of particular interest to Canadians, as many of the German Mennonites and Quakers of Pennsylvania later came to Ontario as has been already described (p. 323).

Scottish, Irish, and German pioneers. In the midst of the Appalachian Plateau is a wide fertile valley running almost from end to end of this highland; the *Shenandoah River* runs along this trench and empties into the Potomac River. Large numbers of Scottish and Irish settlers entered the United States by the *Delaware* and *Potomac* rivers and migrated southward into this valley, and these, together with German settlers, were the real pioneers that fought their way across this savage barrier between the Atlantic plain and the vast central flat land in the interior of the country.

Forests of the mountains. The whole Appalachian Plateau receives sufficient rain, and because of its height, the southern part is cooler than along the coastal plain. From Pennsylvania south the surface is covered with chestnut and oak, which are valuable ornamental woods used for furniture, making when finished a beautifully coloured and handsomely grained floor, door, or house trim.

Dawn of a better day in the valleys. The descendants of the Scottish and Irish, who early occupied these valleys in the southern Appalachians are still there, and as these valleys are deep and the

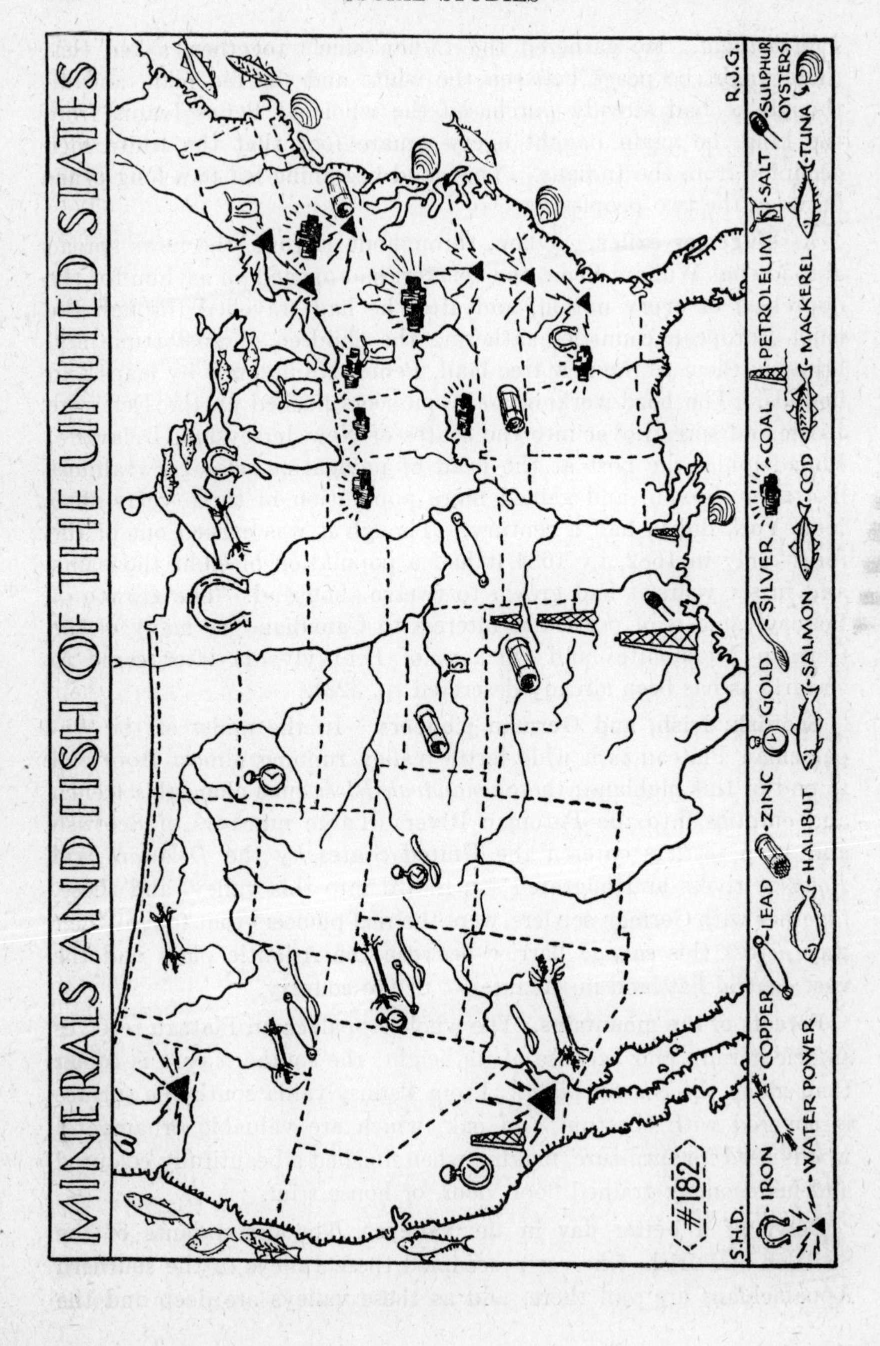
MINERALS AND FISH OF THE UNITED STATES
#182
S.H.D.
R.J.G.
IRON
COPPER
LEAD
ZINC
GOLD
SILVER
COD
COAL
PETROLEUM
SALT
SULPHUR
OYSTERS
WATER POWER
HALIBUT
SALMON
MACKEREL
TUNA

ridges between them steep, they have been shut away from the great changes of the last century as much as though they were living in another world. They are ignorant, poor, and eke out a miserable life, growing a little corn and vegetables, hunting opossums and other wild game. Today the motor car and the railway are doing much for them; modern highways are now carrying the life-blood of progress into the backward valleys, and the blight of ignorance is being overcome by contact with the rest of the country. A great hydro-electric development on the Tennessee River is working wonders. Their houses are now lighted with electricity, cotton mills are starting to give them work, schools are changing the outlook of the children, and the dawn of a new and brighter day is beginning for this darkest spot in the United States.

Coal seams make cities. On the long western slope of the Appalachians are the most wonderful beds of coal in the world (#182). They lie so near the surface that thick seams stick out along the banks of the rivers. Here has sprung up during the last hundred years the largest coal mining industry in the world. As iron ore was found near the coal, the smelting of this metal was started early. To-day in Pennsylvania and Ohio in the north, and in Alabama in the south, are found great blast furnaces by the thousand, belching forth their flames night and day, and pouring out searing streams of molten steel, which are rolled and moulded into rails, locomotives, machines, steel bridges, ships, and a thousand useful implements for farmers, miners, and all other industrial workers. *Pittsburg* and *Youngstown* are the steel centres in the north and *Birmingham* (#190) in the south. However, the coal of Pennsylvania also supplies the cities on the Great Lakes with fuel for blast furnaces. Iron ore brought down the lakes from Duluth, on Lake Superior, to the coal of Pennsylvania and Ohio has converted *Cleveland, Toledo, Buffalo, Detroit, Chicago,* and *Milwaukee* (#190) into thriving manufacturing and commercial cities.

Great Central Plain

Sediments from many sources. Extending from the Rocky Mountains to the Appalachian Plateau is a broad flat plain sloping gradually from the mountains towards the Mississippi River (#170). Much of the southern half of it is formed of sediment washed down by the streams to the mouth of the Mississippi, which at one time was near the present mouth of the Ohio; all the ancient sea between that point and the present mouth in the Gulf of Mexico has been filled with a fine rich sandy soil. Much of the western part of the plain is overlaid with sediment washed down from the mountains by streams and settled out, when the current slackened, on the plain; large areas in the north are covered by the deposit, which we have already learned oozed down to the bottom of ancient Lake Agassiz (p. 344.)

Mountains sticking through the plain. The whole surface of this vast area, covered by such fertile soils, is underlaid by nearly horizontal layers of rock. Only in two places do mountains break the uninteresting flatness of the plain. *The Black Hills* of Dakota are the flat-topped granite remnants of much higher, wider, and larger mountains that covered the district. The other, the *Ozark Mountains* of Missouri, are broader, but lower and more like a highland than a mountain.

Metals from the mountains. It is in these worn-down mountains that the chief metals of the Great Plains are found. The Black Hills of Dakota, fifty years ago, had become the centres of great *gold* and *silver* discoveries. But the mines are nearly exhausted. In the Ozark Mountains the more modest metals, *lead* and *zinc*, are mined in great quantities; the town of *Galena* is the centre of the lead mines and *Joplin* of the zinc (#182).

Petroleum and sulphur. The horizontal rocks underlying the level part of the great plain are far more fruitful in such necessary minerals as coal and oil. #182 shows the location of these products. A steady stream of black crude petroleum from *Texas* and *Oklahoma* is pumped for thousands of miles through huge pipes to supply the great cities and dense farm population of the States bordering on the Great Lakes. In some mysterious way thick layers of pure, yellow *sulphur* have collected under layers of quicksand in Texas. It is impossible to dig a shaft through such treacherous sands, as the poisonous gases would kill the workers and the quicksands would flow in on the shaft as soon as formed. However, the sulphur is obtained by forcing down through a steel tube water heated sufficiently to melt the sulphur and then by compressed air forcing the liquid mixture of water and sulphur up through a second tube surrounding the first. Every sulphide pulp-mill in Canada has a pile of Texas sulphur in its yard.

Settlement of the Great Plain

The call of the Ohio Valley. It has already been described how Marquette and Joliet, (#116), and most of all La Salle, (pp. 308, 309), explored the Mississippi River to its mouth, and how Radisson, (p. 305), La Vérendrye, (p. 354), and Henry opened the region of the Missouri River, and De Soto (p. 451) and Coronado, (p. 469), the south-west. The fight for the Ohio Valley has also been described, (p. 175). After the British had conquered Canada, settlement in the Great Plains was forbidden, as it was left as a hunting ground for the Indians. But one might just as well try to stop the advance of a great tidal wave as to dam back eager, aggressive, land-hungry Scottish and Irish pioneers from the rich and fertile prairies, simply waiting to be ploughed, and teeming with herds of buffalo that shook the earth as they passed, as well as deer, beaver, otter, mink, and wild turkeys to give them furs and meat. However, the real rush of settlers across the mountains did not begin until the United States won its independence in 1783.

Daniel Boone, the Indian fighter. The first important settlement took place in the beautiful blue grass region near the Kentucky River, a southern branch of the Ohio (#183). The man who explored the country, found the path through the mountains, led the settlers in, and defended them against the Indians, was the famous Indian fighter, *Daniel Boone.* He was brought up in the back woods of North Carolina, with a gun in his hand. Big and brawny of body, clear-headed and fearless of mind, and as quick and silent in his movements through the forest as a deer, he was more than a match for the fierce savages just beyond the mountains. This restless young man in a black buckskin coat, with a gun over his shoulder, loved to wander through the forests for months or even years at a time. He slept on the ground, depended on the game which he caught for food, and on the skins and furs of the deer he shot and trapped to support his family.

A two-year hunting trip. In 1770, during one of his trips, he passed through the Appalachian barrier by the historic pass called *Cumberland Gap.* This soon became the bottle neck through which surged the thousands of settlers, who were in a few years to cover *Kentucky* and *Tennessee* with early colonists. He was pleased beyond measure at the delightful land, covered with lush blue grass and swarming with big game of all kinds. There for the first time he saw herds of buffalo. He remained two years, and then turned back towards Cumberland Gap, loaded with furs and deer skins to pay off his debts; but he was pounced upon by Indians, who stripped him of everything.

#183

HOUSES IN BOONESBORO

Soon he had the urge to return to those choice new lands; he organized a band of settlers and started westward, only to find that the region was on fire with a raging Indian war; the party had to turn back. His own son, seventeen years old, was scalped by the Indians.

Settlers move west. By 1775 a great tract of land had been purchased by Judge Richard Henderson, one of the many land speculators who were making fortunes buying government land at low prices. He induced Boone to assist him by leading a band of settlers into this promised land, which was called the "*Dark and Bloody Ground*", because it was bordered north and south by fierce savages who met there to settle their fights. Boone, with his own family, hacked out a rough road for the settlers' carts through the Cumberland Gap and over the rough ridges out to the new land, which was called *Transylvania.* There, where a creek emptied into the Kentucky River, a fortress was built, and a settlement formed, called *Boonesboro* (#183). The cottages were built side by side to form the walls of a fortress, with no windows facing outward excepting loop-holes. Here the settlement was established, but only by the most desperate fighting. Once Boone was captured and held as prisoner by the Indians for about five months. His long absence from the settlement had suggested to his wife and friends that he was dead; but all that time he was really directing the Indians in such a way as to protect the fortress until he escaped.

Boone, the pathfinder. But as the settlement grew, game became scarce, and danger disappeared; Boone became restless, felt crowded in this pioneer settlement, and decided to move westward into a new wilderness. Before he died at the age of eighty-five, he had finally settled in his old age on the Missouri River in Kansas, where he was honoured by the whole settlement, and made judge and governor of the community. He was truly a great *pathfinder* of the west.

The Ohio beckons westward. There was another entrance to Kentucky. When Daniel Boone arrived with his settlers at Boonesboro, he found that another pioneer hunter, *Harrod,* had come down the Ohio and already formed a settlement. This stream, formed by the junction of the *Allegheny* and *Monongahela* rivers, after sending out threads of water into the back settlements of Pennsylvania and Virginia, beckoned the people westward; then the strong current of the Ohio would sweep them onward for nine hundred miles, past smiling meadows interlocked with choice forests of maple, oak, hickory, chestnut, and elm.

Drifting westward on barges. The pioneers built awkward, box-like, flat-bottomed boats (#184), with a deck to shelter the passengers. On these they loaded their horses and cows, their beds, tables, and chairs; they pushed out from the shore, and the swift current of the Ohio did the rest. A great paddle sticking out behind was used to steer the barge. At the end of the journey downward, the scow was broken up. During 1788 more than one thousand of these strange ships drifted towards the Mississippi River with tens of thousands of settlers. Over 170,000 people had taken up land in the Ohio Valley by 1790.

#184

DOWN THE OHIO ON BARGES

Flat boats on the Mississippi. Soon, on these farms in the forest, the hardy pioneers, by hard work, were producing more butter, cheese, hides, wool, linen, cotton, potash, and whiskey than they required and were looking for a market. No boat could go up the Ohio against the swift current, and it was too far and the roads were too rough and muddy for pack horses to carry a large quantity of goods to the settlers on the Atlantic coast. But these shrewd Americans were not to be defeated. They built rough boats, loaded on their freight, drifted down the Mississippi, and sold their goods in New Orleans. Their ships were broken up at this port, and they rode back on horseback along the streams. The dangerous journey took six months.

Robert Fulton creates a revolution. In 1803 a great event took place; all land west of the Mississippi had belonged first to Spain and then to France. On that date it was purchased from Napoleon and a new vast field for settlement opened up. But in order to fill up these wide flat plains with farmers who could dispose of their goods, better means of transport were necessary. In 1807 a decisive event

#185

FULTON'S FIRST STEAMBOAT

occurred. A shrewd engineer, *Robert Fulton*, put a steam-engine on a small boat on the Hudson River. This engine turned two immense side-wheels, which propelled the boat (#185). As it moved upstream, with its sidewheels splashing the water, the people on the shore viewed it with open-mouthed wonder. In 1811 the first steamboat went down the Ohio and Mississippi. It was not long until hundreds of broad flat-bottomed steamers were carrying the farmers' cotton, tobacco, grain, hides, and other produce down this great river and returning with all kinds of European goods. Now ships could also go up the Ohio.

The stage-coach. Roads were built very slowly. A company would construct a highway, erect toll-gates, and charge drivers a few cents every few miles along its course. The rough *Cumberland road* was the most notable of these. It ran from Cumberland in the midst of the Appalachian Mountains almost to the Mississippi River and passed through such important towns as *Wheeling*, *Columbus*, *Indianapolis*, *Terre Haute*, and *St. Louis*. Stage-coaches, crowded with passengers, were soon lurching over half a dozen such rough roads between the Appalachian Mountains and the Mississippi River.

Early canals. The steamboat brought the canal. In 1825, the *Erie Canal* from *Buffalo* to the Hudson River and New York City opened the first waterway from the west to the Atlantic coast, and at once began to deflect freight from the Mississippi and New Orleans to New York City. Canals were soon cut from the Ohio River to Lake Erie, from the Mississippi through the Illinois River to Lake Michigan, and by 1850 a complex network of waterways was filled with steamships.

King of the road. But just as the steamboat and the canal supplanted the stage-coach, an event was taking place at Baltimore that was destined almost to empty many rivers of their steamboats. In 1828 the first sod was turned of a railway line to run from Baltimore over the Appalachian Mountains to Ohio and the west. Although on its first trip the Tom Thumb (#186), as the little puffing iron horse was called, was beaten by a one-horse stage-coach, it was not long until the locomotive with its train of cars began to crowd the stage-coach off the road, and with its superior speed it left the river and the canal boats so far behind that it was soon king of the highways.

A steel network. A wild craze to build railways took possession of the United States, and railway after railway first connected the Great Lakes and the Ohio with the seaboard at New York, Boston, Philadelphia, and Baltimore; then, the water routes proving too slow, the railways spread westward to the Mississippi, and the new west was opened up.

Settlement of Texas

Cities of gold. While De Soto and his band of mail-clad explorers on horseback were rapidly melting away before treacherous Indian attacks, gaunt hunger and dread diseases; news was leaking through to the mining settlements in northern Mexico of a fairyland to the north called *Cibola,* which contained seven glorious cities glistening with gold. The doors of the natives' houses were studded with beautiful turquoise, and whole streets were lined with goldsmiths' shops, so enormous was the quantity of gold.

#186

ONE OF THE FIRST AMERICAN LOCOMOTIVES

Procession of wealth and colour. An expedition, rivalling that of De Soto's was fitted out to seek this El Dorado (#179). *Francisco de Coronado,* clad in shining golden armour, was placed at its head, and three hundred gentlemen served under him. It was, indeed, a procession of wealth and colour that in 1540 started out from a small Spanish settlement in Mexico to find the fabulous cities of gold. First came a band of mounted men in polished silver armour, each with a helmet and a shield of bullhide. The horses were decked with gaily coloured blankets almost trailing the ground. Footmen armed with crossbows and rifles followed. Behind them were hideous bands of naked savages, their bodies splashed with black, yellow, and crimson, and their cruel faces encircled with a hood of yellow and green parrot feathers. Pack mules, cannon, one thousand horses, droves of cattle, sheep, goats, and swine brought up the rear.

A dirty village. It was a long, dry, and difficult journey across the rugged mountains, but at last the *Rio Grande River* was reached, and the weary, dusty procession followed up its welcome stream, every eye straining for the first sight of the magical cities. The first view was like a dash of cold water in the face; when the seven great cities groaning with gold shrank to a single dirty, little crowded Indian village, its huts so crowded together that it seemed to be crumpled up. They took possession of the fort in less than an hour. It was very close to the present thriving city of *Santa Fé*, which from that day became the centre of Spanish settlement in what is now the state of New Mexico.

Discovery of Grand Canyon. Though the brave heart of Coronado was sad, he was not discouraged. He soon heard of new wonders both east and west. The Indians told of a famous river to the west, and one of his assistants, *Cardenas*, was sent to explore it. After many adventures he had the great joy of being the first white man to gaze upon one of the wonders of the world, the *Grand Canyon of the Colorado River*. The two steep cliffs, striped with many colours, were miles apart, and the river so far down that it looked like a narrow ribbon of water only a few feet wide, but the Indians stated that from shore to shore it was more than a mile. His men tried to descend the steep cliff but failed.

A new fairyland. The news from the east was still more startling and alluring. Coronado was told of a land, *Quivira*, far to the east, which contained a wonderful river in which fish as large as horses amused themselves by leaping into the air. The great chief of Quivira each afternoon took his siesta under a wide spreading tree on whose branches grew little golden bells instead of flowers, which tinkled in the wind and lulled the great chief to sleep. So abundant was gold that dishes of the common people were made of that metal.

Across the prairie. Coronado decided to seek this new golden country. Before he did so he had to stamp out a rebellion of Indians, which he did in true Spanish fashion. He stuck two hundred wooden crosses into the ground, fastened an Indian captive to each, and burnt them all alive. After thus teaching these savages a lesson, he moved east with a group of his most trusted men; gradually the dreary desert changed to greenness, and at last he was on the broad prairies of Texas with lush grass up to the waist. Soon he met for the first time great herds of woolly-headed buffalo. He now had abundance of good food, and the flat prairie made progress swift, though he was somewhat alarmed, for the grass in their wake rose up and left no track by which they could retrace their steps.

A miserable retreat. At last he reached Quivira, the land of tinkling bells, giant fish, and dishes of gold. Alas! they were but the inventions of Turko, their Indian guide, who wished to draw them away from his village and into the wilderness to starve. Instead, he found unfriendly Indians and a scarcity of food, which compelled him to retrace his steps. But now the grass was dried, the buffalo were gone, and it was a sorry band of half-starved, almost naked, discouraged men that dragged themselves to the Rio Grande, and back to Mexico. Nevertheless, a great area from the Colorado almost to the Mississippi River had been explored for the first time, and Texas had been prepared for settlement.

First settlement at Santa Fé. Though De Vaca crossed the south of Texas in 1533, and Coronado the north in 1541, there were no real settlements formed until seventy-five years later. *Juan de Onate*, a noble, with a band of Spanish dandies, including one hundred settler soldiers with their families, left the city of Mexico in the spring of 1596, to struggle along the harsh Mexican Plateau towards a lake of gold, supposed to be burnishing under the desert sun of New Mexico. It was not until April, 1598, that the dusty procession of soldiers, negroes, Indians, two thousand head of stock, and eighty-three wagons loaded with baggage reached *El Pazo*. A few months later they had a church built near *Santa Fé*, where the first settlement was placed. He, like Coronado, went down the Canadian River toward Quivira, the fabled land of gold. He reached almost as far as Omaha in Kansas, but the Indians drove him out. On his return to Santa Fé, he robbed the Indians of their food, and slaughtered whole tribes in order to make the settlement safe.

Rivals in the field. From Santa Fé as a centre, missions spread out until twenty-five were strung along the Rio Grande and sixty thousand Indians were converted. Though these settlements, located along the valley of this river running through the hot semi-desert of New Mexico, struggled against drought, heat, and Indian raids, no advance was made into the much more attractive grassy prairies of Texas until the alarming news arrived that the French were settling along the coast to the east and west of the Mississippi, which La Salle had discovered in 1684. Then with feverish haste, missions and military stations were formed in eastern Texas, only to be abandoned again, when the neighbouring French became threatening.

Islands of peace in a grassy sea. Finally, in 1718, attracted by a bubbling spring of clear, cool water, the first permanent settlement of Texas was formed at *San Antonio*, soon to be followed by a necklace of missions and settlements stretching from El Paso nearly to the Mississippi River; by 1720, these missions were permanently established to do their civilizing work for the Indians, and the number of settlements slowly but steadily increased.

The missions and presidios. A mission, together with its presidio, (Spanish for a garrisoned post), formed the centre of a Spanish settlement. A friar or priest had complete control of the whole mission. He had five or six soldiers to protect his people from the savage tribes. The Indians, without a moment's warning, were always ready to make a sudden pounce on the cattle or sheep, and then melt away in the tall grass in the prairie like a passing cloud. The missionary gathered in Indians either by coaxing, threats, or force, and they became little better than slaves as far as their freedom was concerned. Soon the priest had built a cluster of stone buildings, the centre of which was a beautiful church, around which were gathered the school, workshop, and the community houses of the Indians. The men planted grain, vegetables, and cotton, and tended the flocks, and the women spun and wove the cotton and wool into clothes and shawls. Altogether they lived a peaceful, sleepy life, became civilized and Christian, and were devoted to the church and the friar or priest. In time white settlers became attracted by the peace and tranquillity of the scene.

Presidios. Besides the missions there were *Presidios*, or military settlements, in which rough soldiers offered a great contrast to the self-sacrificing priests of the missions. They ruled the timid Indians with a rod of iron and compelled them to cultivate the farms and supply the soldiers with all that they required.

Americans crowd them out. When the bold pioneers of the United States had crossed the Mississippi River, the broad prairie cattle lands of Texas allured them farther west, and by 1821 they had begun to settle in the Mexican province of Texas, which seemed large enough for all. *F. Austin* took up the first land, and before long this hustling American had a ranch containing many thousand acres and tens of thousands of cattle. By 1830, more than thirty thousand alert American ranchers had taken so much of the best land that the Mexicans became alarmed. They issued an order that no more Americans were to enter the province and began to crowd the country with Mexicans. The Americans stirred up a rebellion, and under the able leadership of *General Sam Houston* dealt the Mexicans a smashing blow, from which they never recovered. Texas became an independent republic in 1837, and a few years later joined the United States. The highest monument in the United States has just been erected in Texas in memory of Houston, who is the State hero.

Settlement of the Dry Plain

The Long Trail. Settlers reached the Mississippi in large numbers between 1800 and 1820, and there they paused for nearly fifty years. The broad, dry, boundless plains with tall, coarse grass, roved over by millions of buffalo, were considered little better than a vast, drought-blasted desert. Even after the railways had crossed it on their way to the California mines, its wealth of pasture was ignored. At last, about 1860, the stimulus came from the south, where settlements from Mexico had spilled over into Arizona, New Mexico, and Texas. Men began to drive their herds of long-horned cattle northward, and the forward march swelled to that of a vast succession of armies along definite routes called the *Long Trails*, which were beaten deep by the hoofs of countless cattle. First they moved towards *Iowa* and *Illinois*, then to the States of *Kansas*, *Missouri*, and *Nebraska*, and then began ascending the dry, rough lands in *Nebraska*, *Colorado*, and *Wyoming* (#190). The whole plain from the Mississippi River to the Rocky Mountains became a boundless, unfenced pasture field for cattle, and the *cowboy* was king (#187).

The round-up. The cattle of all owners were mingled together, feeding at will on the grassy plains. Every calf was branded with a distinguishing mark by a red-hot iron that neither rain, nor sun, nor rubbing, nor time, could ever wear off. The strong, sunburned cowboys, on spirited ponies, watched over the herds; twice during the year there was a "round-up" of all the cattle. These were the busy and exciting days for these expert riders. The first took place in spring to brand the calves. Into the wild, chaotic swaying mass of arched horns the cowboys ploughed their way to search for every cow marked with their brand. The calf following this cow was gradually forced out of the stormy sea of cattle, lassoed with a rope, dragged up to a fire, and branded. Dozens of cowboys tugging their opposing calves, made a tangle of ropes around the fires. In July and August, during a second round-up, the young, fat steers were sorted out and driven to the government stores to be used for food by the soldiers and on the Indian reserves, or to the railways to be shipped to slaughter houses. It was during these times that the great meat-packing houses contributed to swell the size of such cities as *Chicago*, *Kansas City*, *St. Louis*, and *Omaha.*

#187

THE COWBOYS ARE IN TOWN

The farmer drives out the cowboy. The immense herds of buffalo were soon destroyed by the lasso and the more fatal bullet of the white man, and the small settler came in and built a barbed-wire fence around his several hundred acres. As the farmers learned better how to grow grain on this dry land, millions from every country in Europe flocked to these free lands and settled on farms, until the whole plain became a fine-meshed crazy quilt of Irish, Scots, Finns, Swedes, Germans, Norwegians, Danes, and Poles.

Soon the open ranch land was greatly cut up by these farms and by railway fences, and then the large ranchers themselves began to fence in their immense pasture lands. With this change, the picturesque cowboy, or cow-puncher as he was sometimes called, dis-

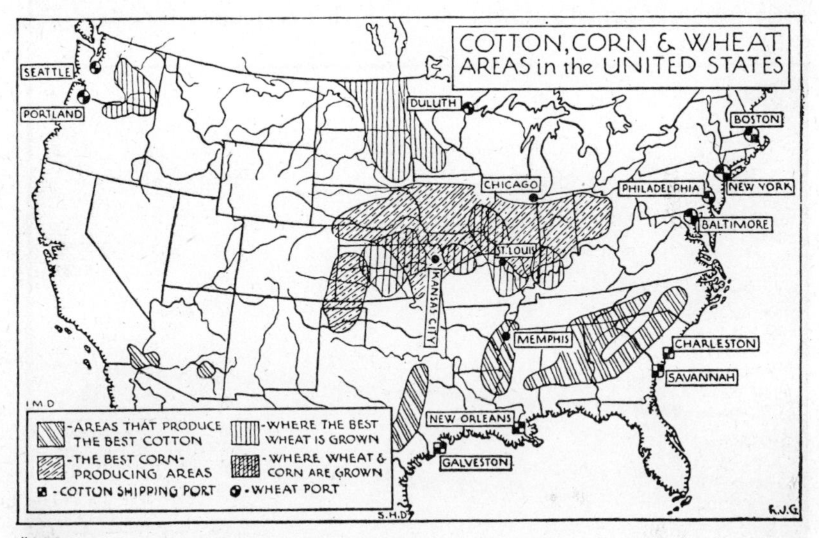

#188

appeared, and the wide open lands from Texas to Canada have been changed into large fenced wheat and corn farms and still larger fenced ranch lands, well stocked with cattle and sheep.

Wheat belts, corn belt, cotton belt. Thus to-day this wide level plain has become the most extensive and productive farm centre in the world. In the south is a *cotton belt* (#188), a sea of white, stretching from Texas to North and South Carolina. In the hot, humid lowlands of Louisiana are the chief *sugar-cane* and *rice* lands of North America. Farther north is a *tobacco belt* stretching from Virginia to Kentucky. From Ohio to Nebraska fields of *corn*, often ten feet high, spread out in every direction like a swaying, green sea. Here also millions of fat *pigs* and choice beef *cattle*, as well as immense numbers of *horses*, *sheep*, and *poultry*, are made ready for market by being corn fed. Two immense wheat belts lie in this plain. Between the corn belt in Iowa and the cotton belt in the south is an area from Kentucky and Tennessee on the east to Kansas and Nebraska on the west that has wheat planted in autumn and harvested in June or early July; just to the north of the corn belt from Minnesota to Montana is another belt in which wheat is planted in April and harvested in August.

Between these grain belts and the Rocky Mountains, where the cattle formerly ranged at will on the high dry plains, there are now

#189
A DUST STORM ON SOUTHERN GREAT PLAINS

smaller closed ranches, where cattle, sheep, horses, and goats in the south, are reared in great numbers.

Dust storms. A new danger is threatening this high plain east of the Mississippi River. The rainfall comes in warm, moist air masses that drift up from the Gulf of Mexico. The farther from the source, the less the rain. In the old days, before the plain was overrun by cattle and ploughed to raise crops, it was provided with a thick covering of grass, which prevented moisture from being evaporated by the sun, and a dense tangle of roots which held the soil together. But the cattle and the sheep have cropped the grass and shrubs so

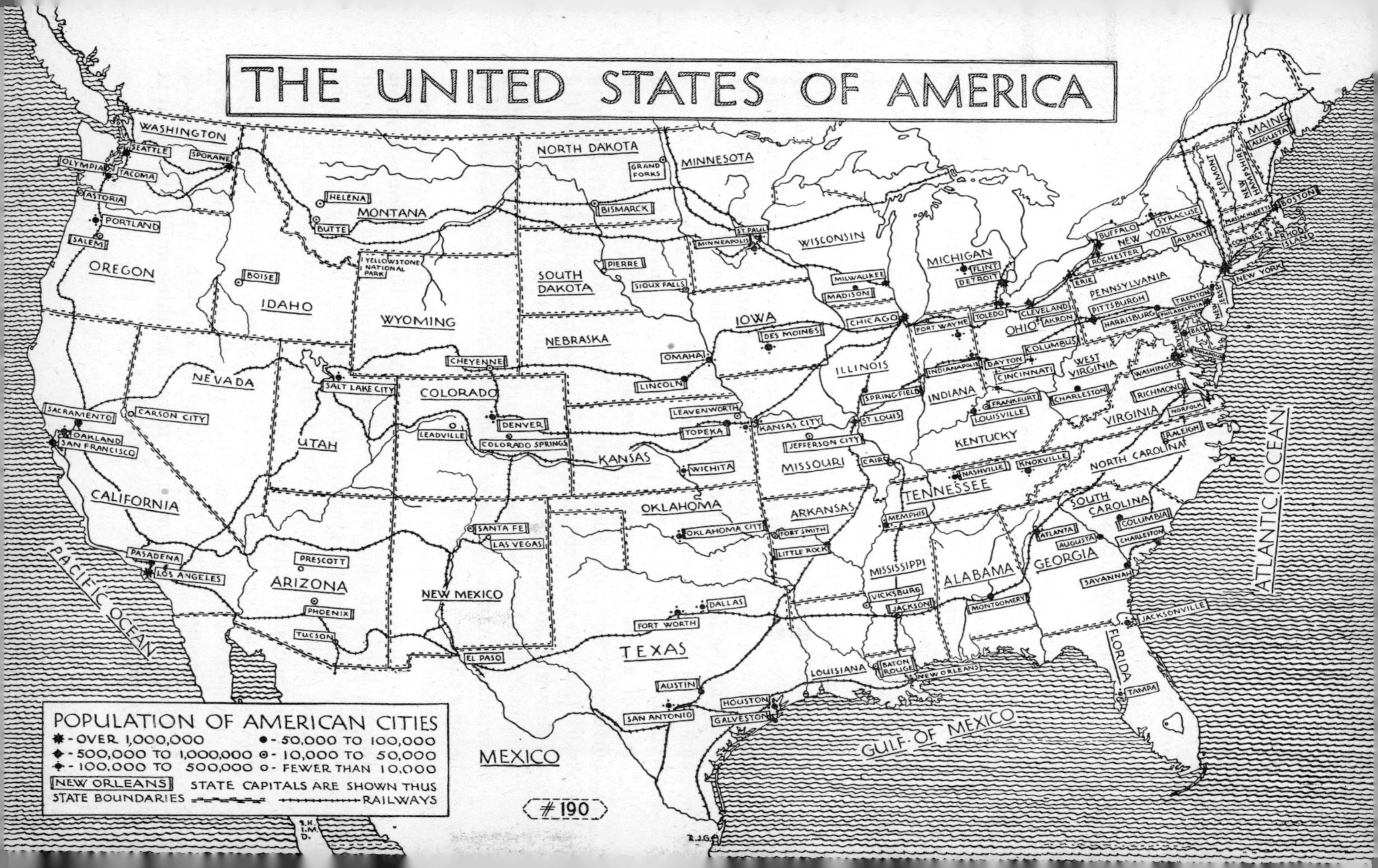
THE UNITED STATES OF AMERICA
WASHINGTON
SEATTLE
SPOKANE
OLYMPIA
TACOMA
ASTORIA
PORTLAND
SALEM
OREGON
HELENA
BUTTE
MONTANA
BOISE
IDAHO
YELLOWSTONE NATIONAL PARK
WYOMING
NORTH DAKOTA
GRAND FORKS
BISMARCK
MINNESOTA
SOUTH DAKOTA
PIERRE
SIOUX FALLS
ST. PAUL
MINNEAPOLIS
WISCONSIN
MILWAUKEE
MADISON
MICHIGAN
FLINT
DETROIT
NEBRASKA
OMAHA
LINCOLN
IOWA
DES MOINES
CHICAGO
ILLINOIS
SPRINGFIELD
ST. LOUIS
NEVADA
CARSON CITY
SACRAMENTO
OAKLAND
SAN FRANCISCO
CALIFORNIA
PASADENA
LOS ANGELES
SALT LAKE CITY
UTAH
CHEYENNE
COLORADO
DENVER
LEADVILLE
COLORADO SPRINGS
KANSAS
LEAVENWORTH
TOPEKA
KANSAS CITY
WICHITA
JEFFERSON CITY
MISSOURI
CAIRO
ARIZONA
PRESCOTT
PHOENIX
TUCSON
NEW MEXICO
SANTA FE
LAS VEGAS
EL PASO
OKLAHOMA
OKLAHOMA CITY
FORT SMITH
LITTLE ROCK
ARKANSAS
TEXAS
DALLAS
FORT WORTH
AUSTIN
SAN ANTONIO
HOUSTON
GALVESTON
LOUISIANA
BATON ROUGE
NEW ORLEANS
MISSISSIPPI
VICKSBURG
JACKSON
MEMPHIS
TENNESSEE
NASHVILLE
KNOXVILLE
ALABAMA
MONTGOMERY
GEORGIA
ATLANTA
AUGUSTA
SAVANNAH
FLORIDA
JACKSONVILLE
TAMPA
SOUTH CAROLINA
COLUMBIA
CHARLESTON
NORTH CAROLINA
RALEIGH
VIRGINIA
NORFOLK
RICHMOND
WASHINGTON
WEST VIRGINIA
CHARLESTON
KENTUCKY
LOUISVILLE
FRANKFURT
INDIANA
INDIANAPOLIS
FORT WAYNE
OHIO
TOLEDO
CLEVELAND
AKRON
COLUMBUS
DAYTON
CINCINNATI
PENNSYLVANIA
ERIE
PITTSBURGH
HARRISBURG
PHILADELPHIA
TRENTON
NEW JERSEY
NEW YORK
BUFFALO
ROCHESTER
SYRACUSE
ALBANY
VERMONT
NEW HAMPSHIRE
MAINE
AUGUSTA
MASSACHUSETTS
BOSTON
CONNECTICUT
RHODE ISLAND
ATLANTIC OCEAN
PACIFIC OCEAN
GULF OF MEXICO
MEXICO
POPULATION OF AMERICAN CITIES
- OVER 1,000,000
- 500,000 TO 1,000,000
- 100,000 TO 500,000
- 50,000 TO 100,000
- 10,000 TO 50,000
- FEWER THAN 10,000
NEW ORLEANS STATE CAPITALS ARE SHOWN THUS
STATE BOUNDARIES
RAILWAYS
#190
S.H.
I.M.
D.
R.J.G.

close that the hot sun in spring turns the soil into dust, and as there is not enough tangle of roots to hold it together, the strong winds blow away the soil in clouds so dense that it is almost as dark as night (#189). Even the seed that has been sown is carried away. As a result a great deal of the land has been turned almost into a desert.

Cities of the Plain

In this immense chequer-board of farms, spreading from the Appalachian highlands to the Rocky Mountains, cities first arose at key points along the waterways, and later, when the steamboat gave way to the locomotive, these cities had become large enough to draw the railways towards them (#190).

Chicago. In 1812 the present site of Chicago was occupied by *Fort Dearborn*, a tumble-down stockade, from which the small group of American soldiers were driven and massacred by Tecumseh's Indians, who cut out and ate the Captain's heart to make them brave. It was at the farthest inland tip of navigation on the Great Lakes, and when the westward migration began, the flood of immigrants from Europe caused it to spring into importance. It became the pivot of a fan, from which the veins of traffic spread outward to north-west, west, and south-west. When, in the seventies, the long trails from Texas were gorged with moving cattle to supply the rancher with stock, Chicago became the focus for fat cattle, and its immense slaughter houses were started and have been expanding ever since. As the forests of Wisconsin and Illinois were cleared, it became the great mouth through which lumber was fed to a regular procession of steamboats and barges, which distributed their cargoes to eastern ports. A little later the farmers of Iowa, Minnesota, and Dakota began to ship wheat, barley, oats, corn and flax seed, and Chicago became the greatest grain port in the world. But the farmers of the west required farm implements, furniture, metal goods, and a thousand other things, which Chicago, with timber in the surrounding states, coal mines just to the south, and an endless supply of iron ore conveniently located at Duluth on Lake Superior, began to manufacture. Thus has Chicago grown with the settlement of the plains, until to-day, with its more than three million people, it is the second largest city in the United States.

Minneapolis and St. Paul

Minneapolis and St. Paul. At the head of navigation on the Mississippi River are these twin cities, Minneapolis and St. Paul. Already ships drawing not more than six feet of water can sail from them to New Orleans, and soon a nine foot channel will be completed. In the early days of water traffic, it was natural that at the head of navigation of such an important river a settlement should be formed, especially when farther progress was blocked by a waterfall, which could also be used for running grist and sawmills.

Father Hennepin. La Salle's companion, Father Hennepin, was captured by the Sioux Indians and, in 1680, was brought to the waterfall, which he christened by its present name, *St. Anthony.* He was, as far as is known, the first white man to see this fall and to discover the site of the twin cities.

Farmers' towns. The United States established a fort here in 1819 to protect the fur traders from unruly Indians. A lumber mill was erected in 1822, and a flour mill the next year, but there was really no settlement until about 1850. When the wheat farmers began to crowd into the flat prairie, formed from the rich sediment that had settled to the bottom of Lake Agassiz (p. 344), these cities were their natural markets. Soon the transcontinental railways, four of them, connected these commercial centres with all important points east and west.

Flour and linseed. The water front to-day is lined with mills (#191) that can grind nearly forty-two thousand barrels of flour daily, and with immense elevators, one of which may be as long as several city blocks; the capacity of all is nearly one hundred million bushels, and the mills never run short. Its linseed mills, which crush the seeds of flax, express linseed oil, and compress the residue into cakes for cattle food, are also vast. In addition, this farmers' city stands first in the manufacture of butter and powdered milk.

The City of Omaha. Omaha, on the Missouri River, has watched most of the exciting scenes in the westward sweep of American migration. In 1541, Coronado, (p. 469), reached Nebraska and may have come close to the site of this city. Lewis and Clark, (p. 484), in the first expedition to the Pacific Ocean, stopped near Omaha, and as it was on the route of the fur trade from the west to the east, the United States as early as 1819 built a fort near the present position. The surge of hundreds of thousands of settlers

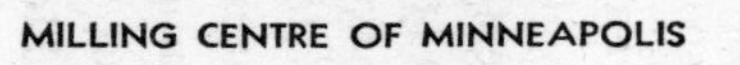
#191 MILLING CENTRE OF MINNEAPOLIS

over the Oregon Trail, (p. 494), passed through this city, which became an important frontier trading post. The Mormon migration from Ohio, (p. 483), to Salt Lake City settled down for the winter near Omaha, but its adherents were persecuted so cruelly that with the opening of spring they were forced to move on. When the cattle rush from Texas to the north, (p. 473), had reached Nebraska and Dakota, Omaha became a centre of the meat-packing industry, and when the same regions changed into wheat and stock farms, Omaha became a centre of flour milling and butter making. To-day both Omaha and Minneapolis claim to be the largest butter producers in the world. Omaha had its greatest stimulus when the first transcontinental railway, the *Union Pacific*, passed through its limits. As this road lies directly on the course between Chicago and the Pacific, Omaha has always been one of the most important railway cities in the west.

The Cordillera

Seas of mountains. The western highland of the United States is called the Cordillera. It extends more than one third of the distance across the country and consists of many ranges of mountains, most of which run north and south, making the passage from east to west very difficult. The highest ranges are the *Rockies* on the east, and the *Sierra Nevada* and the *Cascade* on the west (#170). Between these is a high, dry, dreary plateau, networked with mountains running in every direction.

A valley of fruits. Two fertile lowlands are at the west. One in the south between the Sierra Nevada and the Coast Ranges is the wonderful *Valley of California* (#170). It is almost like a Garden of Eden. Its level valley is carpeted with orchards of oranges and grapefruit, groves of gnarled olive trees, vast vineyards of grapes for making currants, raisins, and wines, plum orchards, whose fruits, when spread out and dried, make prunes, and many other fruits and vegetables (#181). Many of these crops, including wheat, grow during the winter when the rains occur and are sent to supply the northern United States and Canada with fruits and vegetables when prices are highest. Oranges, grapes, figs, plums, and other summer fruits are watered from reservoirs up in the mountains.

Fertile lava beds. The other lowland is in the lower valley of the Columbia River in Washington and Oregon. It is the home of the finest apple orchards on the continent (#181). Farther east in the same States, on a volcanic plateau where the rock has crumbled into fertile soil, is one of the most valuable wheat belts (#188) on the continent.

The interior, almost desert in its dryness except on the highest parts, grows a thin scattered covering of coarse tufts of grass, which supports vast flocks of sheep.

Utah, more particularly that part of it surrounding Salt Lake City, is a good example of what an industrious, zealous group of people can do to make the desert blossom like the rose.

Salt Lake City and the Mormons

The golden book. One of the strangest, sternest romances of United States history is the rise, western migration, and rapid growth of the *Mormons*. *Joseph Smith*, a poorly educated, none too honest dreamer, believed that he received a message from heaven telling him that if he went back to his home district in New York State he would find hidden in the earth a box, containing a book of thin golden plates, on which was inscribed a message from God. He professed to have found the plates, and by means of a pair of spectacles containing two strange stones, which he called *Urim* and *Thummin*, he was able to translate them into a volume called the *Book of Mormon*, which was the guide to a new religion.

The leader murdered. This strange religion spread rapidly. Three missionaries soon had a group of converts in Ohio and built the first Mormon church. But these odd people, with their still odder religion, were greatly disliked, and often mobbed and persecuted by their rough pioneer neighbours. Consequently, the first colony in Ohio was driven to Missouri, where after a few years they became so unpopular that they and the Gentiles had a pitched battle, in which many were killed on both sides. Hence, in 1838, they moved north and founded a new city in Illinois on the Mississippi River. There were now twenty thousand of these fanatical people, who wished only to work hard and to be left alone. Smith here in 1843 received another doubtful revelation, which stated that men should have a number of wives. The leader at once began to put the message into practice. This led to his undoing, for he soon found himself and some other leaders in jail, with a furious mob without thirsting for their blood. Before their opponents' anger had cooled, they broke into the jail and killed Smith and his brother.

Out into the wilderness. *Brigham Young*, who had been a missionary in Great Britain, now came to the front as the brains of the Mormon church. He saw by 1845 that, with their strange beliefs and practices, they could never have peace in an American community, and he decided to move the whole settlement into the western wilderness. During the winter of this year they crossed the Mississippi River on the ice, (#170) dragged wagons through the snow, and moved beyond the pioneer border out into the dry, lonely, Indian-infested high plains.

A dreary home. They remained at Council Bluffs, near Omaha, during the first winter, and while there they were persecuted by the people of this frontier town. In April, 1847, Brigham Young, with a chosen band of nearly one hundred and fifty men, three women, and two children, started westward towards their new homes. They travelled in covered wagons, drawn by oxen and mules, carrying food, seed, ploughs, and other tools. The savage Indians were startled by their hymn singing and their musical instruments, and these fighters of the plain were greatly amazed that these travellers did not wish to fight and injure them. In the surprisingly short time of one hundred and six days this zealous band had overcome the fierce mountains and yawning canyons of Nebraska and Wyoming and had arrived at their new home. It looked so lonely and uninviting that they could never think that anything but wild beasts would ever molest them in such an unwelcome spot. Great grey mountains, their tops often among the clouds, hemmed in dry, alkaline valleys, dull with scattered, sickly sage-brush, but almost empty of other life. Silence and loneliness hovered over the scene, and away to the west lay that freak of nature, *Great Salt Lake*, the "Dead Sea" of Utah, with its sullen shores of salt.

Farming the desert. On the day of their arrival they offered up a prayer to God, and then began to plough the bare, dry sand and to plant wheat and potatoes. Like many desert soils, this sand was really very fertile, and all that it needed was the magic touch of water. The Mormons built a dam, deflected the water over the fields, and the first and greatest battle had been won. Though they were plagued by crickets and grasshoppers that ate their crops, and spells of great drought, they never lacked faith, and they became the first great irrigators in the United States. Even the birds helped them. There is a monument erected in Salt Lake City to the seagulls, (#192) which saved their crops by devouring swarms of crickets.

Desert blossoms as the rose. Wave after wave of Mormons moved westward into Utah, irrigated the land, and toiled on the blistering sands under the scorching desert sun. Though they had no forests to clear, they had to bring water down from the mountain streams, and before they could build homes, they had to mould and sun-bake the clay into bricks. To-day the valleys smile with luscious crops of sugar-beet, grain, potatoes, alfalfa, grapes, apples, pears, peaches and berries, and in the lower and warmer valleys, cotton, almonds, figs, and pomegranates. The farms are small and thoroughly cultivated, the houses are close together, and the families are large.

The show city of the desert. *Salt Lake City* (#192) is the hub of Utah. The Mormons have spread like bees from this mother hive and now form a large part of the population of all the adjoining states, even reaching into southern Alberta. This desert capital of Utah is one of the show cities of the world. It was surveyed into ten-acre blocks, separated by streets one hundred and thirty-seven feet wide. As each block was to have only eight houses, every lot would contain more than an acre. However, the pressure of rapid growth has crowded the houses more closely together, but still the city shows an openness and a fine array of ornamental trees. Each north-south street has two open brooks of clear water flowing along its sides.

Within the Mormon church enclosure there is a round domed tabernacle containing a wonderful organ. It is in this building that the members of the church hold their meetings. In the same enclosure is the beautiful Mormon Temple which took forty years to build and into which only members of the church are allowed to enter.

A varied boundary. Its surroundings are more varied. Great Salt Lake is a little to the west, great peaks of the *Wasatch Mountains* thrust up their snowy heads to the east and north, and the valley of the *Jordan River*, rioting in greenness and plenty, stretches away many miles to the south and west.

No commercial or manufacturing city to compare with Salt Lake City is found for many hundred miles around. Transcontinental railways connect it with San Francisco and Los Angeles, and mineral ores from a wide area are brought for reduction to some of the largest smelters in the world.

The Oregon Trail

Lewis and Clark. When the wide tract west of the Mississippi River, called Louisiana, was sold by Napoleon to the United States,

#192 ENTRANCE OF SALT LAKE CITY

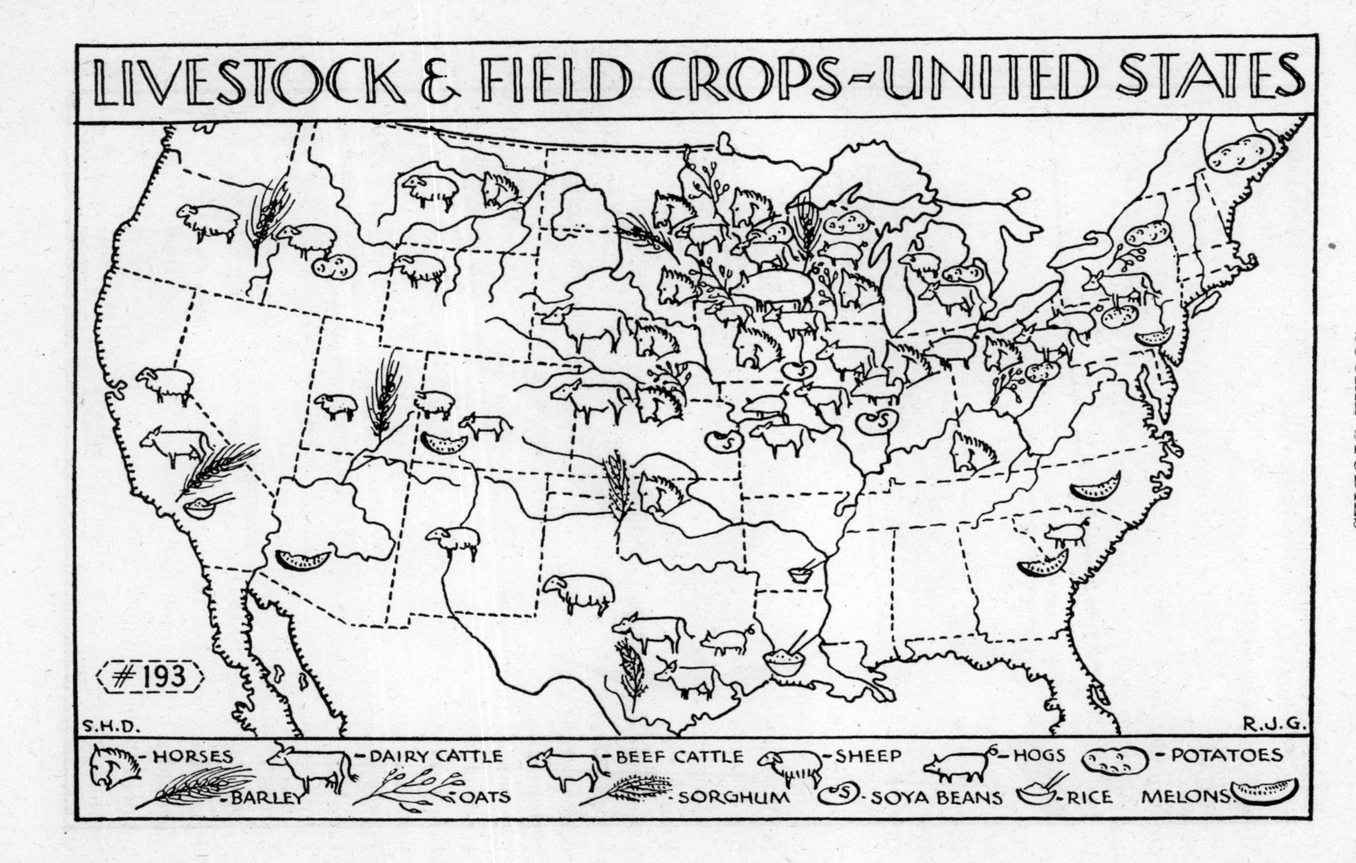
LIVESTOCK & FIELD CROPS - UNITED STATES
#193
S.H.D.
R.J.G.
- HORSES
- BARLEY
- DAIRY CATTLE
- OATS
- BEEF CATTLE
- SORGHUM
- SHEEP
- SOYA BEANS
- HOGS
- RICE
- POTATOES
MELONS

(p. 467), it was almost as empty of white men as was darkest Africa. A fringe of hunters and trappers had pushed westward from the Mississippi over the plain, and traded with the Indians and often lived amongst them. President Jefferson believed it was going to be of the greatest importance, and in 1804 he formed an exploring party under *Meriweather Lewis*, a skilful hunter and woodsman with a good education, and *William Clark*, a practical man, who was a tireless hunter, explorer, and Indian fighter (#170).

Islands of buffalo in a sea of green. In May of that year a well equipped expedition of forty-five men in three boats started up against the current of the muddy and turbulent waters of the *Missouri River*, (#170). They were soon passing through green prairies almost as level as the sea and spotted with great living black islands of grazing buffalo (#198). Herds of deer, elk, and antelopes, with wolves and bears feeding on them, became a daily sight. After travelling all summer against a strong current, the two explorers covered sixteen hundred miles and settled down among the *Mandan Indians* for the winter, as death would be certain if they traversed the snow fields of the stormy Rockies at this season.

A winning race. Next spring *Sacajawea*, the little Indian wife of a French trapper, undertook to lead these easterners through the high passes of the Rocky Mountains towards the head-waters of the Columbia River. Without her help they would hardly have succeeded in traversing the terrible peaks, zigzagging across the uncertain passes, crossing rushing torrents in deep canyons, and unravelling their course through the braided network of confusing streams at the sources of the rivers. But at last they reached the broad stream of the majestic *Columbia River*, and after a year and a half reached the sea. As we have seen, Simon Fraser, (p. 383), and David Thompson, (p. 385), in British Columbia were racing with these explorers to trace the Columbia to its mouth. Lewis and Clark won, and thus gained for the United States the rich states of Washington and Oregon.

Pike discovers Pike's Peak. A few years later, in 1807, *Zebulum Pike* marched west from the Missouri up the steep slope of the Cordillera in Colorado and discovered that giant dome, *Pike's Peak*, named in his honour. He turned south into what was then the Spanish colony of New Mexico, and was taken prisoner to *Santa Fé*, which we have already learned was a very old Spanish town.

Santa Fé Trail. Shrewd United States traders heard of this Spanish settlement and the surrounding missions, whose chief markets

#194

ON SANTA FÉ TRAIL

had been over a long, tedious trail to Mexico. They decided to capture the extensive trade of this lonely group of missions from the distant Spaniards. The traders followed the famous *Santa Fé Trail*, over the dry and dusty prairie, (#194). Every spring long lines of white-topped covered wagons, often four or eight abreast, moved out of *Kansas City* and slowly creaked along the dusty, deeply furrowed trail, for eight hundred miles. Their wagons were well loaded with calico, gingham, glassware, dishes, needles, hardware, and tools to be traded for buffalo robes, blankets, and silver. After weeks of sand storms and burning heat, it was a delight when, nearly

#195

KIT CARSON

exhausted, they mounted the last crest to see the cluster of flat-topped mud houses of Santa Fé glistening in the shimmering sun. The Mexican men and women, dressed in their gayest clothes, received them with a cheer, the goods were unpacked, and for weeks the trading went on. For twenty years these annual marches along the Santa Fé trail continued, but at last the prosaic railway put an end to these romantic but nevertheless hard and trying journeys.

A great traveller. *John Charles Fremont* explored, surveyed, and described a broader area of the United States Cordillera than any other man (#170). In 1842 he was on a boat on the Missouri preparing to explore and map a route to the Columbia River. In describing his plans to fellow passengers he expressed his need of a suitable guide. The voice of a wiry young fellow spoke up and said that he was Fremont's man.

Kit Carson. It was the famous *Kit Carson* (#195). He had started learning the trade of saddle maker in a small town near Kansas City, the terminus of the Santa Fé Trail. As he heard the teamsters of the trails and the cowboys tell their wild tales while he mended their saddles, he was thrilled by their adventures. One day in 1826, the local paper carried the following advertisement:

"Notice is hereby given to all persons that Christopher Carson, a boy about 16 years old, small of his age, but thick set, light hair, ran away from the subscriber, living in Franklin, Missouri, to whom he had been bound to learn the saddler's trade, on or about the first of September last. He is supposed to have made his way towards the upper part of the State. All persons are notified not to harbour, support, or assist said boy under the penalty of the law. One cent reward will be given to any person who will return him.

David Workman."

But Kit never went back to the saddler's shop. He became a hunter, trapper, teamster, Indian fighter, and guide, and soon had travelled over much of the highlands between the Rocky Mountains and the Pacific coast. In his wanderings he had many hair-breadth escapes and furious fights with Indians, and had laughed at hardships of hunger, thirst, heat, cold, and fatigue, which would have crushed an ordinary man.

With Carson as his guide, it is no wonder that during 1842 and 1843 Fremont surveyed and mapped the famous Oregon Trail to the Columbia River (#170), struck south from the Columbia through the desperately dry and hot deserts of Nevada, crossed the Sierra Nevadas, heaped with snow in the middle of winter, and arrived at Sacramento in California. Less than a year later, he was back at Great Salt Lake, and in 1845 he was again in California in time to take an important part in winning that state from Mexico. He later became a noted general and ran unsuccessfully for president of the United States.

The great question. After Lewis and Clark had blazed a trail to the Columbia River, (p. 484), hunters, trappers, and fur traders gradually reached the *Oregon country*, as it was called, and at *Astoria* a fur trading centre was established by *John Jacob Astor*. But the Hudson's Bay Company had also developed a vigorous trade in this region, and the great question arose, was it to become British or American?

A cry from Macedonia. A delegation of Flathead Indians, strangely enough, had much to do with the decision. They so longed for the book from heaven, as they called the Bible, and for missionaries to

#196

A SENTINEL BY NIGHT ON OREGON TRAIL

teach them Christianity that in 1832 a small band made the trying and dangerous journey of three thousand miles to St. Louis to ask for missionaries. When the Macedonian cry reached the churches of New England, missionaries were sent. *Dr. Marcus Whitman* and his beautiful, courageous young bride started on the long four thousand mile trail, and after heart-breaking hardships, began preaching the gospel in the rich valleys near *Walla Walla*.

Oregon country attractions. A few years later, after Fremont had surveyed the Oregon Trail, it was necessary for Whitman to return east to Washington, and everywhere he told of the delightful climate, the vigorous rains, the fertile soil, and boundless free lands of the Oregon country and urged the people that if they would move west he would be their guide.

On the trail. What was his surprise on his return to St. Louis to find one thousand eager men, women, and children ready for the great adventure westward. These were the vanguard of three hundred thousand settlers that staggered their way over this long weary trail during the next ten years (#196).

#197

COVERED WAGON

Prairie schooners. They used great boat-shaped wagons, called *prairie schooners* (#197). Each was covered with a broad hooped top to make a roof for a very crowded house, in which the family lived for the six months necessary for the journey. Every piece of the wagons was made of wood, any part could be repaired or rebuilt on the journey, and as they forded streams these serviceable homes floated like flat boats, occasionally upsetting as they lurched through the water.

A stampede. These animals would sometimes stampede across the prairies and on such occasions created an awe inspiring spectacle with their mad onrushing charge.

The travellers in their long rows of white, lurching, creaking wagons would first hear a low rumbling sound like thunder in the distance. Then like an approaching storm, the rumble would grow to a roar ever louder and louder. The ground itself would seem to tremble as a dark cloud began to sweep closer over the prairie. Quickly the travellers would rush their wagons away from the path of this ominous cloud, knowing only too well that to stay in the

#198 INDIANS HUNTING BUFFALO

path of a crashing buffalo stampede would spell disaster and perhaps death for all.

A weary march. As they moved farther west, they found the smiling prairie gradually change into a frowning desert. Trees and grass cannot exist in the dry dusty ground without a drop of water, and only grey tufts of poor sagebrush are scattered over the hot and hungry soil. The tired oxen, with bowed heads, straining necks, bulging eyes, wide nostrils, gaping mouths, and dry tongues hanging out, tug the shrieking wheels through sand six inches deep. Dense clouds of choking dust bury the whole caravan in a black cloud from morning to night.

A fortified camp. But on they go. Every night, when they reach the camping ground, the two rows of wagons diverge in a curve and then come together again to form a circle; the chains from the oxen bind the wheels together to form a solid fortress against Indian attacks. The camp fires soon dot the circle, and supper is cooked; then the old folks gather to chat; some one plays a fiddle, and the young people have a lively dance. Soon weariness subdues the noise, the sentinels stand at guard, the teamsters snore under the vehicles, and families are asleep in the covered wagons.

Across the mountains. Day after day they creep upward towards the west, scanning the skyline for their greatest enemy, which is still ahead. At last a faint blue line running north and south along the western horizon puzzles them as to whether it is real. Yes, the next day the stern barrier of the Rocky Mountains lies across their path. The nights are getting colder, the land is getting rougher, and the rivers are now down in deep ravines. The poor patient oxen strain at their chains against the steep grades. At one place a sheer drop of one hundred feet lies right across their path; the oxen are unhitched, the chains must be joined together, and the wagons lowered separately. All goods have to be carried down a steep and narrow path on the backs of men, women, and children, and the oxen are led down the same trail.

Rafts on the Columbia River. The last mountain is passed, the slopes far below are covered with trees, and the Columbia River, like a silver ribbon, flows at the bottom of a deep gorge. As they creep down cautiously, there is gladness in every face; soon they are floating down the Columbia River on rafts, and then pushing ashore and unloading their goods as they come to the land which they select for their farms. Such was the manner in which the Oregon country was won for the United States.

California

Pushing northward by sea. Shortly after Magellan's expedition had crossed the Pacific Ocean to the Indies, the Spaniards discovered the *Philippine Islands* and began bringing back shiploads of their valuable spices. But after the long journey across the broad Pacific, supplies were sure to be nearly exhausted, half the crews were often dead or dying of scurvy, and the ships were usually badly battered by Pacific storms. It was necessary to have safe harbours on the Pacific coast, where the sick could be nursed back to health, and the ships repaired and supplied with food and water, *Juan Cabrillo* was sent north in 1542 to explore the coast for suitable harbours. He discovered beautiful *San Diego Bay,* but died during the expedition; his assistant, however, went as far north as Oregon, but missed the two safest harbours, *San Francisco,* and *Monterey* bays.

Confusion of the California coast. Nothing further was done until the Spaniards' claims to the coast were threatened by the British. In 1577, that most daring sailor, *Francis Drake,* started with four ships on his wonderful trip around the world. By the time he had weathered the stormy straits of *Magellan,* only one ship was left, but nothing daunted this fearless Englishman. With his little sailing vessel, the *Golden Hind,* he spread terror along the whole Pacific coast of South America by plundering Spanish towns and treasure ships. In 1579 he appeared on the California coast, much to the vexation of the Spaniards. After obtaining a cargo of spices in the East Indies, he returned safely to England, the first Englishman to sail around the world. He divided the rich plunder with the great queen; and Elizabeth in return held above his head a sword, as he knelt before her, and said, "Rise up, Sir Francis Drake." The plain sea captain rose a distinguished knight. A few years later, *Cavendish,* another English buccaneer, plundered the chief Spanish treasure ship returning from the East Indies.

Exploring the coast. To protect the Spanish treasure ships returning from the Philippines from such enemies, the Spaniards sent *Sebastian Vizcaino* in 1602 to explore the whole coast to find a safe harbour for treasure ships from the Indies. He discovered *Monterey Bay,* examined minutely most parts of the California coast, but went no farther north than did *Ferrelo* sixty years earlier. There was not a single settlement yet in California, nor was there to be for more than one hundred years.

A new threat from the east. Under the stirring government of *Peter the Great,* Russian trappers pushed eastward across Siberia, soon reached the Pacific coast, and in 1725, *Vitus Bering* was exploring the shores of Alaska, searching for a north-east passage to the Atlantic Ocean. He established Russians on the coast and adjoining islands, who slaughtered seals and sea otters and shipped their costly furs to the Chinese markets.

A great state founded. This grim Russian threat from the north at last aroused the sluggish Spaniards of Mexico to the action that finally led to the settlement of California. In the person of *Jose de Galvez*, King Carlos chose the right man to perform this difficult task. Between Mexico and California lay several of the hottest, harshest, and barest deserts on the continent; across the pathway the snow-clad Sierra Nevadas threw their threatening ridges, and much of the route bristled with fierce, blood-thirsty savages ready to jump, like beasts of prey, on whoever strayed from the path or lingered behind for a moment. In 1769 an expedition of settlers, under *General Gaspas de Portola*, started in two divisions, one on three ships, and a second along the hostile sands and rocks of Lower California. The land party, one hundred and twenty souls, were clad in long six-ply buckskin coats, and carried two-ply bull hide shields to repel the arrows of savages. The ships arrived in *San Diego* harbour with most of the crew dead, or sick with scurvy; later they were joined by the overland party, and the first mission and presidio was formed in California. Mass was celebrated by *Sirra*, the missionary, a salute was fired, and California was founded.

First glimpse of San Francisco Bay. Portola marched north in an exploring expedition along the coast of California and travelled the route now followed by the railway. It was a trying journey over mountains in the intense heat and dust of the middle of summer. The party were struck with awe as they marched through the forests of giant redwoods; and as some of them ascended the mountains to hunt game, they looked to the north, and there for the first time the white man gazed on that magnificent body of water, *San Francisco Harbour*, with its splendid rocky pillars standing guard at the narrow entrance, the *Golden Gate*.

Stripping the sick. When he returned to the mission at San Diego, he found it in a bad state; food was scarce, and the Indians so troublesome that they had even stripped the clothes from the sick. When they were in blackest despair, a ship was seen cleaving the clear blue twilight bringing relief, and the first mission was saved.

Fruitful missions. Next year, 1770, a mission was formed on *Monterey Bay* (#199), six years later, 1776, a mission and presidio at *San Francisco*, in 1781 one at *Los Angeles*, and in 1782 one at *Santa Barbara*. It was not long until a new overland route was surveyed from Mexico, and before the end of the century a whole line of prosperous missions was strung along the coast. Each was composed of a group of substantial flat-roofed stone and adobe buildings clustered around the dignified graceful church with stately columns, rounded domes, and tall spires. The absolute ruler in each was the Franciscan priest; the Indians performed all the labour in the olive and orange groves, vineyards, wheat and barley fields, and on the ranches for cattle, sheep, goats, and horses. Irrigation ditches were built to bring down water from the mountains in summer, and to nourish the crops, which grew to perfection under the strong sun from a clear sky.

#199

SPANISH MISSION CHURCH IN CALIFORNIA

A stately procession. *Junipero Sirra*, the pioneer Franciscan priest, was the good genius of all the missions, and his spirit of charity and piety was a sweet savour, which made the Indians faithful workers and pious worshippers; there was no more peaceful and prosperous community than his own mission at *San Carlos* on Monterey Bay. He lived a simple, austere life in a single small cell near the church. It contained only a table, a chair, a bed of bare boards, with one blanket to keep out the cold air from the Pacific breezes, as he slept at night. But the church was as magnificent as his cell was plain, and the procession to mass on special days revealed the true nature of the mission. Gay Spanish gentlemen and soldiers in slashed and gilt-laced trousers, with a flashy shawl over the shoulders led the van; then came demure-looking Spanish beauties with jet black hair peeping out from the black lace shawls over the head and shoulders, and with keen glances of their dark eyes cast in the direction of the knightly courtiers. Next came the Spanish peasant settlers under huge brimmed hats, humble looking friars with long gray gowns and sandaled feet, Indian cowboys, mule drivers, and a long line of Indian labourers and their families clad in coarse, dull cotton smocks. The scarlet, gold, bright blue, black, white, and dull brown and gray procession marched along the shore of the sea, touched with sapphire and amethyst by the sun, and bordered on the opposite side by the gold of hemp fields and the rich green slopes of other crops, or through the columns of giant redwoods towards the red-tiled church, whose musical bells pealed far out over the silent water and the lazy hills.

Peace gives way to prosperity. For nearly seventy-five years this long line of whitewashed missions lived their peaceful, prosperous lives with hardly an Indian raid to break in on the sleepy existence of these hermit communities. But in 1848 California was joined to the United States, and in less than two years this forgotten land became a golden Mecca, into which men from all lands poured in wild confusion; its peaceful tranquillity was destroyed forever, and some of these drowsy missions have become great hives of industry and commerce.

The "forty-niners"

A great discovery. *John Sutter* had been the uncrowned king of what is now the city of *Sacramento.* It was then *Sutter's Fort.* This rancher owned all the land for many miles around and had immense herds of cattle, sheep, and horses. To get lumber for a flour-mill, he sent a skilled but dreamy carpenter, *James Marshall,* up to the forests of the Sierra Nevada Mountains to build a sawmill. Marshall selected a suitable spot on the turbulent *American River* and began to clear out a channel for the water-wheel. As he looked at the gravel one day, he was electrified to see the glint of yellow specks. Was it gold? He gathered some lumps as big as grains of wheat, hammered one into a thin yellow sheet, and threw some grains into the corrosive lye of Mrs. Wimmer's boiling soap pot. Next morning in the bottom of the hardened soap were particles of glittering, yellow gold: he put some on the coals in a hot stove, and there next day in the ashes were the glistening golden particles, untarnished by the heat. Every test he knew proved it to be pure gold. The particles of loose gold seemed scattered on the ground all around the mill yard. He rode back with a bag of the little lumps to Sutter's Fort; Mr. Sutter tested it with acid, and nothing changed its colour or brightness.

Marshall had discovered the greatest gold mine of all time. El Dorado, the golden city, which Spanish explorers had searched for in mountains and plains, swamps and deserts, had at last been found. This was in January, 1848.

"Off to the diggings." About seventy miles away, on the beautiful Golden Gate, was the quiet little Mexican village of San Francisco, with eight hundred and twenty people and two hundred tile-roofed white houses. It was not long until settlers around Sacramento had heard of the news, began mining, and soon drifted with their bags of gold into the city. One day *Sam Brannan*, a Mormon shopkeeper of Sutter's Fort, dashed down the sluggish streets of San Francisco like a crazy man, wildly waving his hat in one hand and a flask of gold nuggets in the other and triumphantly shouting: "Gold! Gold! Gold from the American River." The old town, which had slept for fifty years, rocked as though it had felt an earthquake. Excited merchants shut their shops, dull sailors forsook their ships in the harbour, busy carpenters dropped their hammers, unexcitable farmers rushed from the harvest fields, even faithful soldiers deserted their barracks, and in no time the village was emptied, and every man with a pick, shovel, and pan was "off to the diggings." Any man could wash out from twenty-five to fifty dollars' worth in a day.

The "forty-niners." Though there were no roads, no railways, and no telegraphs, the news, in some mysterious way, seemed to leap across the continent in no time. By the next spring, 1849, the whole of eastern Canada and the United States was wild with excitement; settled men by the thousands sold their houses and farms, bought equipment, and set out to make their fortunes at the "diggings." Some went in prairie schooners across the plains and mountains, partly by the Oregon Trail, partly through Utah: the poor Mormons, who had wandered away to Salt Lake City a year before to get away from all mankind and to make their living in peace and loneliness, were directly on the path of this great struggling mass of sometimes ungodly men travelling westward to the diggings. The Mormons sometimes took advantage of the migrating hordes, who often were in a desperate state by the time they reached Utah; these fanatics drove hard bargains and waxed fat off the evil fortunes of these reckless travellers.

Many went in swift and graceful full-rigged schooners around Cape Horn on a six months' journey, and a third group went across the plague ridden marshes of the Isthmus of Panama. From every direction this endless procession of human beings, mostly in rags, focussed on "the diggings" around the American River.

Six hundred million dollars in gold. During the first two or three years there were many great fortunes made, and gold by the million was obtained. In 1848 the output was five million dollars, in 1849, twenty-three millions; in 1853, it swelled to sixty-five million dollars and then it began to dwindle. However, about six hundred million dollars' worth of gold was dug out of the gravel during the first twelve years.

Riches greater than gold dust. With the decline in gold production, the shiftless part of the people began to depart for new excitement. Those who remained soon began to find that the real riches were still there, in the superbly fertile soil of the valleys. The riches of fruit, vegetables, and grain that have poured forth in ever increasing quantities from this garden of California, make the value of gold shrink to almost nothing.

The last frontier. The rush of the farmers on the Oregon Trail to Washington and Oregon, the Mormons to Utah, and the "forty-niners," as they were called, to California led to the opening up of the whole west and made the last frontier of the United States the Pacific Ocean.

A city of confusion. San Francisco became the commercial centre of the gold rush (#200). People poured in so fast that it grew like a mushroom from two thousand to twenty thousand in one year. There was no government, no city council, there were no judges; robbers, thieves, and gamblers took possession; every man carried a pistol, and miners were robbed and shot without fear of punishment. Streets were lined with shacks, and during the winter rains, were impassable with mud.

Half a continent a barrier. The hustling Easterners, eager to share in the wealth of California, were separated from the wild, throbbing mass of humanity at the "diggings" by the wide, dry plains and the wild emptiness of the fierce network of complex mountains, many reaching above the snow-line. The great problem was to throw rapid transportation across this empty gap, for the six months necessary by boat around Cape Horn was far too long for the feverish haste of the gold seekers.

#200 MAIL DELIVERY IN SAN FRANCISCO IN DAYS OF GOLD RUSH

The stage-coach. First the service from the Missouri River to San Francisco was by *stage-coach.* Huge oval four-wheeled carriages, painted green and red, and drawn by four or six swift horses, made the rough journey in twenty days. Nine passengers packed within and two more on the high seat in front with the driver, rushed wildly over desert and plain, through deep mud and deeper sand, up the sides of steep mountains, and wormed their tedious way through high, cold passes, stormed at with sleet and snow. At lonely inns on the route the passengers rested while horses were changed.

Attacks by thieves and Indians. Since many passengers carried bags of gold dust eastward and much valuable money was expressed westward, reckless robbers lurked in many lonely places and rushed out on horseback firing with repeating rifles. Sometimes the expert driver was able to outdistance these highwaymen, or drive them off with his own firearms, but often everybody was robbed, and any that resisted were shot.

The route lay through the land of many untamed Indian tribes, who also attacked these travellers savagely, and many a pitched battle was fought between the passengers in the stage-coach and these enemies on the route. But business could not wait for danger, and in spite of robbery and death the stage-coach connected east and west for many years.

#201

PONY EXPRESS

Pony express. The stage-coach was too slow for valuable letters. A horse-back rider requires no road and can travel more directly and faster than a coach and horses. This led in 1860 to the rise of the famous *pony-express* (#201). The route was in almost a direct line from *St. Joseph* on the Missouri River to San Francisco (#170). Fast, muscular, hardy ponies were kept at stations every twenty or thirty miles along the route. Boys of twenty years, the pick of the frontier, were chosen to ride the ponies. Each was as alert, as muscular, and as tireless as the horse he rode, and he must be cool and fearless, for danger lurked at every turn. The mail from the post-office, not more than twenty pounds in weight, was placed in the saddle-bags, the boy leaped to the saddle, and the pony shot down the street of St. Joseph in a cloud of dust; at the same moment another was leaving San Francisco. Twenty miles distant another horse was saddled and waiting; the rider leaped from one horse to the other, the mail was transferred, and again he was off like the wind. Each rider would cover about one hundred miles a day without stopping for rest, and as riders were moving in both directions day and night, the distance was cut down to ten days.

Dangers on the way. These lonely riders stopped for neither heat nor cold, storm nor darkness, but always pressed forward. Since the postage on every parcel was at least five dollars, the pony express carried only the most valuable mail; Indians and robbers knew this only too well and were more likely to attack these lone riders in their solitary routes than they were the stage-coach. But these skilful riders were full of courage, expert shots, and with horses that could run faster than those of their enemies usually arrived safely, although occasionally they were robbed and even killed.

Buffalo Bill. The most famous of the riders of the pony express was *William Cody*, better known as *Buffalo Bill.* He had been born and brought up on the frontier and was more at home on a horse's back than on a sidewalk. He had the most dangerous route, through bad Indian country, which was infested with dangerous robber bands, but he was so full of tricks that he usually outwitted even the most desperate robber. Once, suspecting trouble, he hid the letters under his saddle-blanket, and filled the saddle-bags with waste paper. At a lonely place two highway men stepped out, covered him with their pistols, and demanded the mail bags. Buffalo Bill tossed them the bags, and, while they were eagerly opening them, put the spurs to his pony, tore down the road with the valuable bank-notes hidden under his blanket, and arrived safely at the next station. At another time he was warned that there were signs of Indians around. Buffalo Bill with his keen eyes watching every bush and rock, flew down the trail; at a lonely spot he thought he saw something move behind a rock. Almost in an instant a whole troop of savage Indians was rushing across the valley towards him, yelling like mad demons. Bill Cody, with teeth set, never thought of giving in, but lying flat on his brave pony, he sped like an arrow down the rough trail. The Indians were gaining on him and arrows were showering on every side when he came in sight of the next station and the Indians withdrew. Though danger lurked everywhere, there was always a supply of brave young men, and the mail always went through.

The new wealth of the west. As new discoveries of gold brought increasing population, and farming in the fertile valleys of California and Oregon became mcre profitable than mining, the romantic stage-coach and the glamorous pony express were quite unable to give good service. Then the railways were pushed through the high passes over the mountains, and one transcontinental line after another joined east to west with iron bonds. These have led to the discovery of one great mining region after another in the Cordillera, so that to-day immense quantities, not only of gold but also of silver, copper, lead, mercury, and petroleum, are found widely scattered over this rough region (#182). Many parts of the high drier parts have also been found suitable for sheep and goats, and to-day the chief wealth is from the forest, the mine, and the ranch, and, as has already been stated, the fruit, vegetables, and wheat of the irrigated valleys of California, Washington, and Oregon.

The export of lumber, wheat, minerals, fruits, and vegetables, and the import of silk, petroleum, rice, spices, sugar, pineapples and many other commodities across the Pacific has led to the growth of active commercial cities at every suitable inlet on the Pacific coast. Besides San Francisco there are *Los Angeles* and *San Diego* in the south and *Seattle* and *Portland* in the north.